AF446762

Current Topics in
Developmental Biology

Volume 70

Current Topics in Developmental Biology

Volume 70

Edited by

Eric T. Ahrens
Department of Biological Sciences
Carnegie Mellon University
Pittsburgh, Pennsylvania 15213

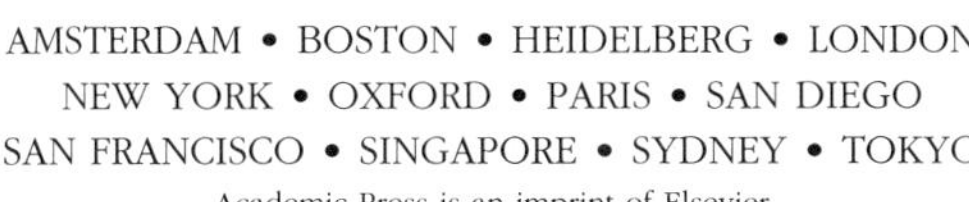

Cover Photo Credit: Courtesy of Eric T. Ahrens, Carnegic Mellon University.

Elsevier Academic Press
525 B Street, Suite 1900, San Diego, California 92101-4495, USA
84 Theobald's Road, London WC1X 8RR, UK

This book is printed on acid-free paper. 

For all information on all Elsevier Academic Press publications visit our Web site at www.books.elsevier.com

ISBN-13: 978-0-12-153170-6
ISBN-10: 0-12-153170-8

PRINTED IN THE UNITED STATES OF AMERICA
05 06 07 08 09 9 8 7 6 5 4 3 2 1

Contents

1

Magnetic Resonance Imaging: Utility as a Molecular Imaging Modality
James P. Basilion, Susan Yeon, and René Botnar

2

Magnetic Resonance Imaging Contrast Agents in the Study of Development
Angelique Louie

3

^{1}H/^{19}F Magnetic Resonance Molecular Imaging with Perfluorocarbon Nanoparticles

Gregory M. Lanza, Patrick M. Winter, Anne M. Neubauer, Shelton D. Caruthers, Franklin D. Hockett, and Samuel A. Wickline

4

Loss of Cell Ion Homeostasis and Cell Viability in the Brain: What Sodium MRI Can Tell Us

Fernando E. Boada, George LaVerde, Charles Jungreis, Edwin Nemoto, Costin Tanase, and Ileana Hancu

5

Quantum Dot Surfaces for Use *In Vivo* and *In Vitro*

Byron Ballou

6

In Vivo Cell Biology of Cancer Cells Visualized with Fluorescent Proteins
Robert M. Hoffman

7

Modulation of Tracer Accumulation in Malignant Tumors: Gene Expression, Gene Transfer, and Phage Display
Uwe Haberkorn

8

Amyloid Imaging: From Benchtop to Bedside

Chunying Wu, Victor W. Pike, and Yanming Wang

9

In Vivo Imaging of Autoimmune Disease in Model Systems

Eric T. Ahrens and Penelope A. Morel

Contributors

Numbers in parentheses indicate the pages on which the authors' contributions begin.

Eric T. Ahrens (215), Department of Biological Sciences and Pittsburgh NMR Center for Biomedical Research, Carnegie Mellon University, Pittsburgh, Pennsylvania 15213

Byron Ballou (103), Molecular Biosensor and Imaging Center and Department of Biological Sciences, Carnegie Mellon University, Pittsburgh, Pennsylvania 15213

James P. Basilion (1), Center for Molecular Imaging Research, National Foundation for Cancer Research (NFCR) Center for Molecular Analysis and Imaging, Department of Radiology, Massachusetts General Hospital and Harvard Medical School, Boston, Massachusetts 02115

Fernando E. Boada (77), Magnetic Resonance Research Center, Department of Radiology, University of Pittsburgh Medical Center, Pittsburgh, Pennsylvania 15213

René Botnar (1), Department of Medicine, Cardiovascular Division, Beth Israel Deaconess Medical Center and Harvard Medical School, Boston, Massachusetts 02115; Department of Nuclear Medicine, Technical University of Munich, 81675 Munich, Germany

Shelton D. Caruthers (57), Division of Cardiology, Washington University Medical School, St. Louis, Missouri 63110

Uwe Haberkorn (145), Department of Nuclear Medicine, University of Heidelberg, Clinical Cooperation Unit Nuclear Medicine, German Cancer Research Center, D-69120 Heidelberg, Germany

Ileana Hancu (77), Magnetic Resonance Research Center, Department of Radiology, University of Pittsburgh Medical Center, Pittsburgh, Pennsylvania 15213

Franklin D. Hockett (57), Division of Cardiology, Washington University Medical School, St. Louis, Missouri 63110

Robert M. Hoffman (121), AntiCancer, Inc., San Diego, California 92111

C. Jungreis (77), Magnetic Resonance Research Center, Department of Radiology, University of Pittsburgh Medical Center, Pittsburgh, Pennsylvania 15213

Gregory M. Lanza (57), Division of Cardiology, Washington University Medical School, St. Louis, Missouri 63110

George LaVerde (77), Magnetic Resonance Research Center, Department of Radiology, University of Pittsburgh Medical Center, Pittsburgh, Pennsylvania 15213

Angelique Louie (33), Department of Biomedical Engineering, University of California Davis, Davis, California 95161

Penelope A. Morel (215), Department of Immunology, University of Pittsburgh School of Medicine, Pittsburgh, Pennsylvania 15213

Edwin Nemoto (77), Magnetic Resonance Research Center, Department of Radiology, University of Pittsburgh Medical Center, Pittsburgh, Pennsylvania 15213

Anne M. Neubauer (57), Divisions of Cardiology and Bioengineering, Washington University Medical School, St. Louis, Missouri 63110

Victor W. Pike (171), Molecular Imaging Branch, National Institute of Mental Health, National Institutes of Health, Bethesda, Maryland 20892

Costin Tanase (77), Magnetic Resonance Research Center, Department of Radiology, University of Pittsburgh Medical Center, Pittsburgh, Pennsylvania 15213

Yanming Wang (171), Department of Medicinal Chemistry, College of Pharmacy, University of Illinois at Chicago, Chicago, Illinois 60612

Samuel A. Wickline (57), Division of Cardiology, Washington University Medical School, St. Louis, Missouri 63110; Philips Medical Systems, Cleveland, Ohio

Patrick M. Winter (57), Division of Bioengineering, Washington University Medical School, St. Louis, Missouri 63110

Chunying Wu (171), Department of Medicinal Chemistry, College of Pharmacy, University of Illinois at Chicago, Chicago, Illinois 60612; National Laboratory of Nuclear Medicine, Jiangsu Institute of Nuclear Medicine, Jiangsu 214063, China

Susan Yeon (1), Department of Medicine, Cardiovascular Division, Beth Israel Deaconess Medical Center and Harvard Medical School, Boston, Massachusetts 02115

Preface

Fluorescent probes for *in vitro* assays have fueled much of the discovery in cellular and molecular biology over the past two decades. In the future, researchers and clinicians would like the ability to perform the same types of sophisticated biological readouts in the intact organism. With the maturation of non-invasive imaging modalities, along with the development and commercialization of novel imaging probes, *in vivo* cellular and molecular imaging techniques are poised to greatly expand our understanding of biology, and the etiology and treatment of diseases. These new capabilities will be facilitated by the continued development of novel imaging reagents that target specific cell types are chemically responsive to physiology, or are responsive to the presence of specific molecules, such as nucleic acids or enzymes.

In clinical medicine, non-invasive cellular and molecular imaging will redefine our ability to diagnose major human diseases such as cancers, cardiovascular disease, and diabetes. In addition, these imaging capabilities will be used to monitor the delivery of new generations of cellular and molecular therapeutics, such as those based on stem cells or viral vectors. In the future, therapeutic strategies will be closely aligned with *in vivo* cellular-molecular imaging. Pharmaceutical research and development will increasingly rely on cellular and molecular imaging to develop new therapeutic agents due to the economy and rapidity compared to traditional histopathological approaches.

In this special volume we have assembled a diverse collection of articles describing some of the most exciting methods and applications of emerging non-invasive imaging technologies using magnetic resonance imaging (MRI), various optical photon approaches, and positron emission tomography (PET). These methodologies extend eleven orders of magnitude across the electromagnetic energy spectrum. This volume emphasizes recent developments in reagent design that impart unique abilities to these imaging modalities to elucidate biological processes and disease states *in vivo*. It is our hope that this volume will further stimulate research in these exciting areas.

I gratefully acknowledge the contributors for their time and motivation in preparing their chapters, without which this book would not have been possible. In the preparation of this volume I would like to acknowledge valuable assistance from Gerald Schatten, Byron Ballou, Lauren Ernst, Kevin Hitchens, Ulrike DeMarco, and Michelle Waters. Finally, I would like to thank the staff of Elsevier Inc., particularly Cindy Minor and Tracy Grace, for their help.

Eric T. Ahrens
Carnegie Mellon University
Pittsburgh, Pennsylvania 15213

1

Magnetic Resonance Imaging: Utility as a Molecular Imaging Modality

James P. Basilion, Susan Yeon,[†] and René Botnar[†,‡]*
*Center for Molecular Imaging Research, National Foundation for Cancer Research
(NFCR) Center for Molecular Analysis and Imaging, Department of Radiology
Massachusetts General Hospital and Harvard Medical School, Boston
Massachusetts 02115
[†]Department of Medicine, Cardiovascular Division, Beth Israel Deaconess Medical
Center and Harvard Medical School, Boston, Massachusetts 02115
[‡]Department of Nuclear Medicine, Technical University of Munich
81675 Munich, Germany

Current Topics in Developmental Biology, Vol. 70   0070-2153/05 $35.00
     1   DOI: 10.1016/S0070-2153(05)70001-6

Significant scientific effort has gone into the deconvolution and understanding of complex biological systems. These efforts have yielded much information about the molecular changes that are causative or arise as a result of disease. Molecular imaging is a relatively newer field that is attempting to use these molecular data to generate images that report on changes in gene expression. It has been demonstrated that generating images based on molecular differences rather than anatomical differences between tissues has resulted in more sensitive detection of diseased tissues and has allowed imaging of drug efficacy against particular drug targets. This chapter discusses the application of magnetic resonance imaging (MRI) to molecular imaging. It begins with a review of the basis for magnetic resonance image generation and how manipulation of different parameters of the system can be applied to molecular imaging. It then specifically reviews some of the problematic areas for magnetic resonance application to molecular imaging and how these can be resolved by manipulating the magnetic resonance system, altering magnetic resonance probe characteristics, or exploiting the biology to be imaged. It concludes with several examples demonstrating the utility of MRI to generate high-resolution, noninvasive images of molecular events occurring *in vivo*. © 2005, Elsevier Inc.

I. Introduction

A major goal of scientific study over the past several decades has been to understand the underlying workings of biology and alterations of that biology which result in disease. Advances in molecular biology and cellular biology techniques, such as the ability to decode the entire genome of many different organisms (including humans) and improved techniques for defining molecular pathways, have enabled the elucidation of complex biological systems. Many of these are directly related to the pathology of one or more diseases and have assisted in the ongoing search for new potential therapeutic targets. However, relating these discoveries to detection or prediction of disease outcome has lagged behind. Only recently have technologies enabled discovery of molecular markers and molecular profiles required for predicting disease status. Several studies have demonstrated that expression profiles, or groups of genes whose expression changes together, have tremendous utility for detecting and predicting the course or therapeutic response of particular cancers (Alizadeh *et al.*, 2000; Ben-Dor *et al.*, 2000; Butte *et al.*, 2000; Golub, 2001; Golub *et al.*, 1999; Wei *et al.*, 2004; Yeang *et al.*, 2001). Clearly, a molecular approach to management of disease could dramatically alter the way medicine is practiced in the future with the effective use of molecular information becoming vital to everyday medical practice. Molecular imaging holds great promise in helping to bridge the gap between molecular understanding of disease and clinical application of molecular information.

Molecular imaging is a relatively new discipline whose main goal is to study the molecular activities of biological systems noninvasively and in real time in the *in vivo* setting. It is hypothesized that from this perspective the effect of medical intervention on interacting biological systems present within the entire organism can be better understood. This would increase (1) understanding of the *in vivo* interactions of biological systems/pathways that until now have been studied in isolation, (2) understanding of the pathology of disease, and (3) understanding of how drugs or other therapies alter this pathology in a complex biological setting. Molecular imaging, therefore, holds tremendous promise but also faces significant challenges. Some of these promises are, however, already being fulfilled as demonstrated in several examples, including monitoring Gleevec efficacy and disease progression (Gayed *et al.*, 2004; Goerres *et al.*, 2004; Joensuu, 2002; Reddy *et al.*, 2003; Stroobants *et al.*, 2003; Van den Abbeele and Badawi, 2002), imaging progenitor stem cell migration (Anderson *et al.*, 2004; Lewin *et al.*, 2000), imaging enzymatic activity *in vivo* (Laxman *et al.*, 2002; Louie *et al.*, 2000; Mahmood and Weissleder, 2003; Weissleder and Ntziachristos, 2003), subclinical detection of disease (Harisinghani *et al.*, 2003), and monitoring gene transfer in human gene therapy studies in man (Jacobs *et al.*, 2001a,b).

Each of the modalities listed in Table I (positron emission tomography [PET], single photon emission computed tomography [SPECT], optical, magnetic resonance imaging [MRI], and ultrasound) has been demonstrated to have utility for molecular imaging. Each has its own advantages and disadvantages and therefore the choice of imaging modality is dictated by the specific question to be answered. As indicated in Table I, molecular targets that would require highly sensitive imaging in deep tissues might best be detected using nuclear imaging modalities. However, deep targets that require high spatial resolution for detection may be best visualized using MRI. In this chapter, discussion is restricted to application of MRI to molecular imaging.

Table I Some Characteristics of Imaging Modalities

Modality	Resolution	Depth	Cost	Sensitivity
PET	++	++++	++++	++++
SPECT	+	++++	++	+++
Optical	++	+	++	++++
MRI	++++	++++	+++	+
Ultrasound	++	+++	+	++

II. Magnetic Resonance Imaging and Molecular Imaging

The high level of spatial resolution and anatomical definition attainable using MRI makes it a valuable molecular imaging modality (Johnson *et al.*, 1993; Smith *et al.*, 1994). High resolution is particularly important for the many molecular imaging applications developed in small animals. In contrast to other imaging modalities, however, MRI has inherently lower sensitivity for probe detection, complicating its use for imaging of molecular markers present at low concentrations (see Table I). Additionally, MRI methods generally require significant time to collect the data necessary to generate high-resolution images. Often the time required for data acquisition is long enough to encompass several heartbeats or respiratory cycles so motion may dramatically diminish image quality. Therefore, significant work has gone into developing "gated imaging" protocols to eliminate the movement artifacts during data acquisition (see the section on gated imaging below).

A. Principles of Magnetic Resonance

MRI is based on the principle of nuclear magnetic resonance. Nuclei with an odd number of protons and/or neutrons have a magnetic moment. Hydrogen nuclei in water molecules are the most abundant in the body. When placed inside a strong magnetic field, some of the nuclei align, establishing a net longitudinal magnetization. However, the spins rotate around the main magnetic field (B_0) axis with random phase so that no net field is detected in the transverse plane. The nuclei precess about the B_0 axis at a frequency directly proportional to the strength of B_0. To generate an MR signal, a weak magnetic field (B_1) is transiently applied along an axis transverse to the main magnetic field. Coherent spins subsequently rotate around the B_0 axis producing detectable oscillating transverse magnetization. Differences in the recovery of magnetization in the longitudinal direction (T1), the decay of magnetization in the transverse plane (T2), and the combined effect of all dephasing processes including B_0 inhomogeneities (T2*) are the basis of soft tissue contrast in MRI.

B. Magnetic Resonance Signal Intensity

Signal intensity in MRI primarily depends on the local values of the longitudinal (1/T1) and transverse (1/T2) relaxation rate of water protons. Depending on the pulse sequence used, signal generally tends to increase with shorter T1 (higher 1/T1) and decrease with shorter T2 (higher 1/T2) relaxation times. The relaxivities R1 and R2, which are commonly expressed in $mM^{-1} \times s^{-1}$, describe the increase in 1/T1 and 1/T2 per millimolar of the contrast agent as demonstrated in Fig. 1.

$$1/T1 = 1/T1_0 + R1 \text{ [contrast agent]}$$
$$1/T2 = 1/T2_0 + R2 \text{ [contrast agent]}$$

with $T1_0$ and $T2_0$ being the relaxation times of native tissue (before contrast administration).

The environment in which the nuclei are located also helps determine the MR signals created. Therefore, by manipulating the chemical environment around the protons, the signal can be altered. MR contrast agents have been developed as a way to modulate the chemical environment inside an organism. Gadolinium (Gd)-based contrast agents usually increase 1/T1 and 1/T2 in similar amounts (R2/R1 $\cong$ 1–2), whereas iron particle–based contrast agents have a much stronger effect on increasing 1/T2 (R2/R1 $>$ 10). Gadolinium-based contrast agents therefore lead to a positive contrast effect (bright), whereas iron particle–based contrast agents usually cause a negative contrast effect (dark). MR pulse sequences that emphasize differences in T1 and T2 are commonly referred to as T1- and T2-weighted sequences. In addition to their impact on 1/T2, iron particles also increase 1/T2* due to

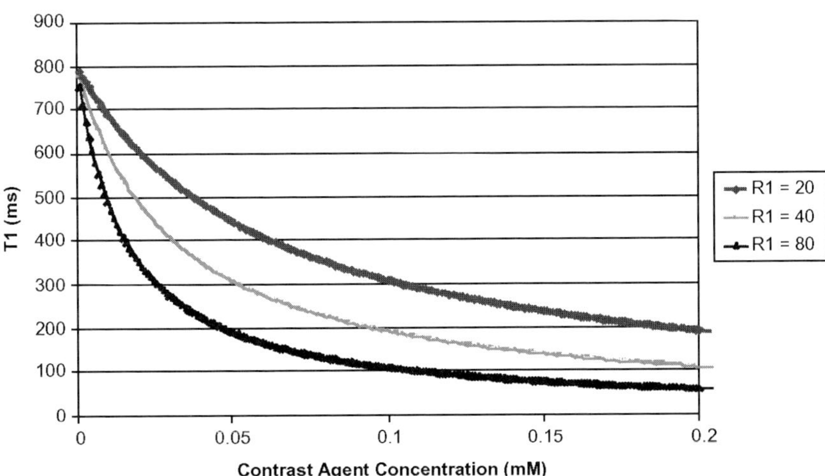

Figure 1 T1 relaxation time for various R1 values plotted for increasing contrast agent concentrations. $1/T1 = 1/T1_0 + R1 \times$ [contrast agent]; $T1_0$ of native tissue (i.e., myocardium) was assumed to be 800 ms. To achieve a T1 of 200 ms that is usually sufficient for contrast agent detection, $<$50 μM Gd is required for R1s $\geq$ 80 $mM^{-1} \times s^{-1}$. For an R1 of $\leq$ 20 $mM^{-1} \times s^{-1}$, this concentration increases almost 4-fold to $\geq$200 μM Gd. (See Color Insert.)

their effect on the local magnetic field B_0.

$$1/T2^* = 1/T2 + \gamma \Delta B_0$$

Iron-based contrast agents are therefore best imaged using T2*-weighted imaging sequences. For signal quantification, T2-weighted multi-echo spin echo (SE) sequences can be used to generate T2 maps. Typical R1 and R2 values of currently approved Gd-based contrast agents are in the range of $R1 = 3$–5 mM $\times$ s^{-1} and $R2 = 5$–6 mM $\times$ s^{-1}. The relaxivities of iron-based contrast agents are significantly higher: $R1 = 20$–25 mM $\times$ s^{-1} and $R2 = 100$–200 mM $\times$ s^{-1}(Caravan *et al.*, 1999).

Due to the generally low concentrations of targets for molecular imaging, high relaxivities are an important means to achieve adequate signal amplification and are further discussed in the subsection "T1 Effects" in "VI. Probe Detection by MRI."

C. Magnetic Resonance Image Acquisition Cardiac and Respiratory Motion Compensation: "Gated Imaging"

Because MRI is an inherently slow imaging technique, acquisition of data for high-resolution images is often performed over multiple cardiac and respiratory cycles (segmented data acquisition), so synchronization with electrocardiography and the position of the diaphragm is required. These issues are especially important for molecular MRI, since molecular targets found in small regions at low concentrations can generally be detected by only high spatial resolution imaging. Since the electrocardiogram (ECG) becomes distorted (with elevated T wave) when recorded from a subject in a high magnetic field, state-of-the-art MR scanners use a ≥ 4 lead R-wave detection algorithm (≥ 2 ECG traces) to differentiate between the R wave and the so-called T-wave artifact (Fischer *et al.*, 1999). The T-wave artifact is caused by the magnetohydrodynamic (MHD) effect. Rapidly moving ions are deflected in the magnetic field, causing an additional voltage that is superimposed on the ECG signal. The MHD artifact is strongest during maximal flow in systole and increases with increasing field strengths. Small-bore animal scanners are commonly equipped with less sophisticated ECG gating hardware and software. In most cases, R-wave detection is performed by simple threshold algorithms. In addition, both clinical and small animal systems are equipped with respiratory sensors (bellows) that allow for respiratory motion gating. However, simple gating mechanisms may cause interruption of MR data acquisition and thus signal variations due to altered MR steady-state conditions. This can lead to artifacts, especially if inversion recovery sequences are used in concert with contrast agents. Gating schemes that acquire data at a near-constant repetition time (TR) and subsequently label data as accepted or rejected based on the respiratory position of the diaphragm help overcome this limitation. A drawback of all gating schemes is that they function at the cost of increased scanning time.

III. Imaging Sequences

A. Spin Echo Sequences

ECG-triggered and nontriggered T1- and T2-weighted SE sequences are part of the standard sequence repertoire of every MR scanner. These sequences are used extensively for neurological, abdominal, or musculoskeletal imaging because they provide excellent image quality and can provide varying amounts of T1 or T2 weighting by adjusting the echo time (TE) and repetition time (TR). T1-weighted SE sequences are characterized by short TEs (5–15 ms) and TRs (300–700 ms), whereas T2-weighted SE sequences are characterized by long TEs (50–150 ms) and TRs ($>$2000 ms). In the presence of a T1-lowering contrast agent, high-resolution images with excellent soft tissue contrast with concomitant T1-weighting can be achieved. For applications requiring morphologic detail or hypointense blood (black blood) appearance along with identification of contrast enhancement, SE approaches are often the method of choice. A disadvantage of fast spin echo (FSE) sequences is that beyond a certain contrast agent concentration, image intensity decreases rather than increases with increasing contrast agent concentration (Fig. 2). In addition, suboptimal contrast between target and background may be observed due to the relatively high signal from surrounding tissue. The maximum MR signal is reached at Gd concentrations of approximately 1 mM for a typical contrast agent with a relaxivity R1 of 4 mM $\times$ s^{-1}. For higher concentrations, the T2 effect begins decreasing the maximal achievable signal due to the finite TE. On the other hand, T2 weighting continues to increase with increasing contrast agent concentrations.

B. Fast T1-Weighted 3D Gradient Echo Sequences

Non–ECG-triggered fast radiofrequency (RF)-spoiled three-dimensional (3D) gradient echo sequences (TE $<$ 5 ms, TR $<$ 10 ms, flip angle = 30–50 degrees) are heavily T1 weighted and exhibit a near-linear relationship between contrast agent concentration and MR signal intensity (Fig. 3). These sequences are therefore especially well suited for higher contrast agent concentrations. Due to their short scan times (5–60 s) and excellent background suppression, these sequences are the workhorse in first-pass contrast-

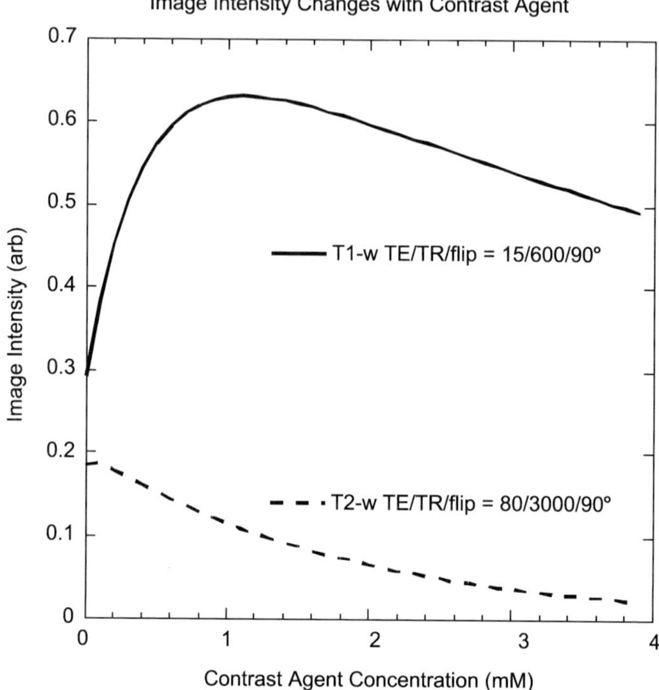

Figure 2 Effect of contrast agent on signal intensity. Effect of contrast agent on T1-weighted (TR = 600 ms) and T2-weighted spin echo images. From Caravan *et al.*, 1999.

enhanced angiography of the large vessels and in molecular imaging of nonmoving tissues and organs. A disadvantage of this approach is the hyperintense (bright blood) appearance of blood, which makes it a suboptimal sequence for molecular imaging of the vessel wall. The use of saturation pulses can help minimize the inflow (blood signal enhancing) effect as demonstrated in a study of molecular MRI of fibrin (Botnar *et al.*, 2004c).

C. T1-Weighted Inversion Recovery 3D Gradient Echo Sequences

T1-weighted inversion recovery sequences are particularly useful if ECG triggering is required for suppression of cardiac or respiratory motion artifacts. Typical scan parameters include TE < 5 ms, TR = 5–10 ms, flip angle = 30–50 degrees, 10–30 RF excitations per heartbeat, and bandwidth = 100–300 Hz/pixel. The choice of the inversion repetition time ((TR_{IR}), e.g., ≥ 1 heartbeat determines the inversion delay (TI):

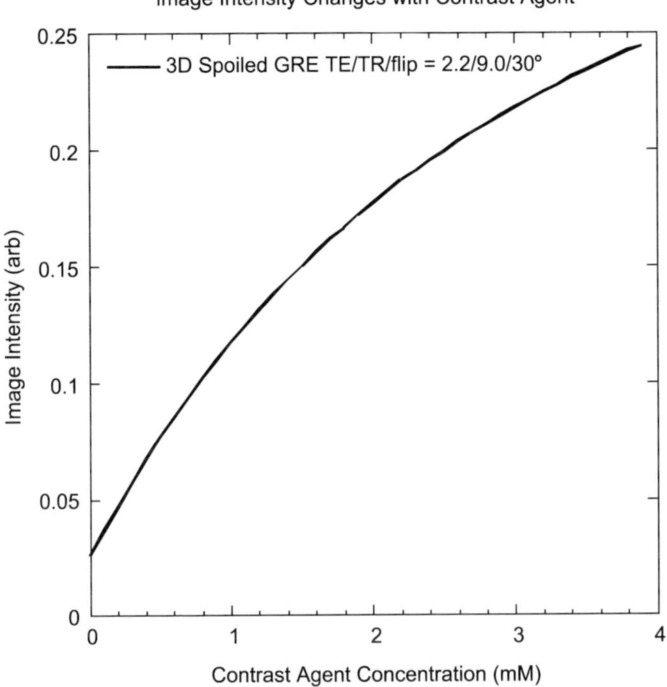

Figure 3 Effect of contrast agent on signal intensity using a short-TR fast spoiled gradient-echo sequence. From Caravan *et al.*, 1999.

$$T1 = \ln2 \times T1 - T1 \times \ln(\exp(-TR_{IR}/T1) + 1)$$

and thus the maximum achievable signal intensity of the administered contrast agent. T1 is the longitudinal relaxation time of the suppressed tissues. Longer TR_{IR}s (>1 heartbeat) allow longer inversion delays (TI) and thus higher signal intensities at the site of contrast uptake (Fig. 4). The drawback of such an approach is an increase in scanning time. The advantages of inversion recovery sequences are excellent background suppression and flow insensitivity. Due to negligible signal contamination from surrounding tissues, these sequences are particularly useful for visualization of small amounts of contrast uptake at a specific target site. A drawback of this approach is the lack of morphologic information provided in the image. This shortcoming can be remedied by obtaining a superimposable companion anatomically detailed image (such as an SE image) using identical spatial coordinates. Unlike nontriggered 3D gradient echo sequences, triggered inversion recovery 3D gradient echo sequences produce a well-suppressed blood signal and can thus enable targeted imaging of the endothelium, vessel wall, and/or thrombus (Fig. 5) (Botnar *et al.*, 2004a).

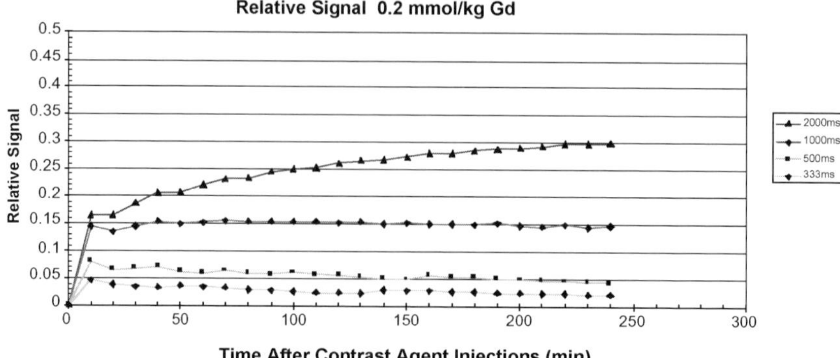

Figure 4 MR signal strength on IR images for different IR repetition times and with respect to time of contrast administration. Simulation of the Bloch equations based on expected Gd concentrations in blood demonstrating the relative signal versus time after contrast agent injection [minutes] for four different IR repetition times (333–2000 ms). For longer IR repetition times, a higher relative signal level can be expected. Furthermore, the simulation demonstrates that the relative signal level remains constant over a relative long period of time after contrast administration. (See Color Insert.)

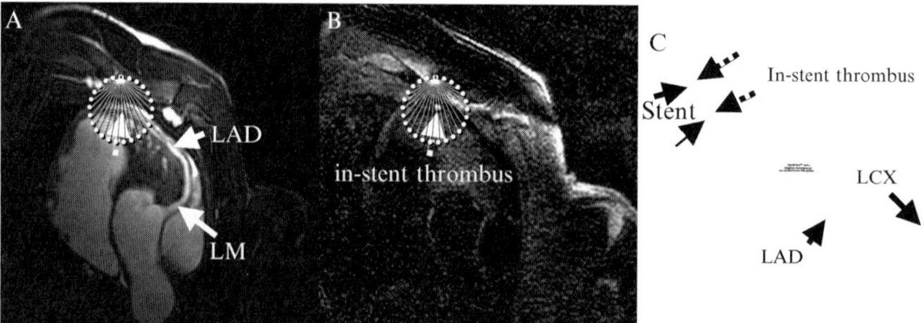

Figure 5 Example of black blood T1-weighted IR sequence for imaging of coronary thrombus. *In-vivo* MR molecular imaging of coronary in-stent thrombosis. Bright blood images of left main (LM)/left anterior descending (LAD) coronary artery after (A) stent placement and injection of a fibrin-avid Gd-labeled contrast agent, EP-2104R. No apparent thrombus and no stent artifacts are visible. (B) Black blood thrombus images after stent placement and injection of EP-2104R. A bright spot (arrow) is visible after intracoronary injection of EP-2104R. Thrombus was subsequently confirmed by x-ray angiography (C). From Botnar *et al.*, 2004a.

D. T2*-Weighted Gradient Echo Sequences

Due to the prominent T2* effect of iron-based contrast agents, imaging to detect iron contrast is predominantly done using T2*-weighted gradient echo sequences. Typical parameters include TE $= 10$–20 ms, TR ≥ 1 heartbeat, and flip angle = 40–50 degrees. Due to the generally low signal obtained in T2*-weighted images, morphologic information is very limited and therefore requires additional scans to allow for co-localization between the site of contrast uptake and the corresponding morphologic image.

IV. Data Analysis

A. T1 Measurements

Measurements of the T1 relaxation time are usually performed using inversion recovery sequences. By changing the inversion delay TI between the nonselective inversion prepulse and data acquisition, signal from tissue A ($T1 = T1_A$) will be nulled if the inversion delay TI fulfills the condition $TI = T1_A \times \ln2$. Most T1 measurement approaches are based on the Look and Locker sequence, which acquires multiple images (6–10) along the T1 relaxation curve after an initial inversion prepulse (Look and Locker, 1970). Several new approaches have been proposed to reduce imaging time (Henderson *et al.*, 1999) and to enable T1 measurements in moving organs such as the heart (Messroghli *et al.*, 2004).

B. T2* Maps

T2* maps are acquired by sampling the signal along the free induction decay (FID) curve using multiple TEs at a constant TR. The most common approaches are based on gradient echo sequences with signal sampling along Cartesian trajectories (Clare *et al.*, 2001; Reeder *et al.*, 1998). The drawback of this approach is the relatively long scanning time it requires. In a recent study, Dahnke *et al.* (2004) proposed a faster approach by taking advantage of the undersampling properties of radial imaging (Dahnke *et al.*, 2004). Undersampled radial subimages with differing TEs were reconstructed from a complete radial data set that was acquired with multiple TEs. An exponential pixel-by-pixel fit of the FID as derived from the undersampled sub-images then allows generation of T2* maps.

V. Target Selection: What to Image?

While there is clearly a significant gap between basic biological research and its applications to medicine, technologies are now available to begin to close that gap. How should one select the target to image? Potential targets may be selected from a variety of molecular markers based on their relevance to various physiological or pathphysiological processes. Target selection may be informed by genomic and proteomic screens that identify markers or groups of markers that are associated with a particular disease, physiology, or biochemical pathway. Often the information derived from genomic analysis of disease does not identify single markers but rather identifies groups of markers whose variable expression as a cohort is diagnostic. Therefore, in the future molecular imaging probes and technologies will have to be designed to survey the expression of several targets simultaneously in order to meaningfully report on the biological state.

Cells surface markers (such as CD4, CD8, Mac1) on immune cells may be targeted by labeled antibodies to image areas of immune response such as murine encephalitis (Pirko *et al.*, 2004). Monoclonal antibodies against tumor-associated antigens (e.g., Ra96, HER-*2/neu* receptor) may serve as probes to identify the presence of tumor cells (Artemov, 2003; Artemov *et al.*, 2003a,b; Gohr-Rosenthal *et al.*, 1993). However, if targeting molecules are not yet available once an appropriate molecular target has been identified, candidate ligands for the target must be identified and selected. High-throughput screening (HTS) methodologies are being developed to screen for small molecule–based MRI agents (Hogemann *et al.*, 2002) and methods for creating and screening libraries of peptides, such as phage display technology, are useful for identification of optimal peptide ligands for various proteins. Using such approaches, ligands with high affinity (low K_d value) for the target can be identified and these can be combined with MR contrast agents to generate MR images of target expression *in vivo*.

VI. Probe Detection by Magnetic Resonance

A. General Considerations

MR contrast agents are not detected directly but by their effect on water protons. The sensitivity of molecular imaging probes designed for MRI depends on the distribution of the probe (intravascular vs. extracellular; extracellular vs. intracellular), the strength of the probe signal (R1 and R2), and the means of detection of the probe (choice of MR imaging sequence). The distribution of a probe is determined by the mode of its administration, its compartmentalization (degree of passage into various intravascular, extracellular, and intracellular spaces), and its mechanisms and speed of clearance. Probe distribution together with probe–target affinity determines the concentration of probe at the target for detection. For magnetic resonance, the strength of the achievable probe signal is determined by the concentration of contrast agent and the relaxivity of the agent.

However, probe signal may not increase proportionately to increases in probe concentration (see below). In addition, the distribution of probe in nontargeted regions is an important consideration for timing of imaging following contrast agent administration and for selection of methods for probe detection that optimize the target-to-background signal ratio. The plasma half-life of an agent must be sufficient to expose target receptors to the agent. On the other hand, it may be necessary to wait for plasma concentrations of agent to decrease sufficiently to distinguish luminal from vessel wall contrast uptake, although methods may be employed to reduce this requirement (Sirol et al., 2004).

B. T1 Effects

Most MR contrast agents are based on either Gd complexes (Caravan et al., 1999; Laniado et al., 1984) or less commonly, iron oxide particles (Weissleder et al., 1990a). Gd(III) is ideally suited for use as an MRI contrast agent because it not only has seven unpaired electrons but the symmetry of its electronic states produces an electron spin relaxation time slow enough to interact significantly with neighboring water protons (Caravan et al., 1999).

Relaxivity is affected by a number of contrast agent properties, including hydration number, distance between the ion and the solvent proton, solvent exchange rate, electronic relaxation time, and rotational correlation time. The hydration number (number of water molecule coordination sites) for Gd chelates is generally >1 (Caravan et al., 1999).

The rotational correlation time (τ_R) is ~ 0.1 ns for approved agents. Since increases in rotation correlation time τ_R enhance relaxivity, various efforts in contrast agent design have focused on increasing this parameter. The τ_R is lengthened by formation of conjugates between the metal ion complex and slowly moving structures such as proteins. Since molecular MRI probes frequently involve attachment of Gd complexes to slowly moving ligands (e.g., peptides and antibodies) that in turn attach to slowly moving targets (e.g., proteins), lengthening of the τ_R is a convenient means of amplifying the detection of contrast agents positioned at molecular targets. Since the unbound fraction of molecular probe retains a lower R1, a good target-to-background signal ratio is achieved.

It may also be possible to improve T1 effects, despite lack of improvement or lessening of relaxivity per Gd atom, by increasing the number of Gd atoms per targeted complex (multivalency of Gd attachment) to sufficiently offset a flat or even diminished relaxivity effect. This effect can be exploited for both Gd chelate-antibody (or peptide) probes as well as for nanoparticle probes, although the magnitude of the effect is much greater for nanoparticles (e.g., Gd/antibody ratio of ~ 20 Gd/antibody molecule vs. 100 Gd/20 nm colloid nanoparticle vs. 300,000 Gd/300 nm nanoparticle). Thus nanoparticles offer the potential for greatly improved sensitivity over conventional chelated complexes because of the higher achievable concentration of Gd or iron oxide. Since contrast agent relaxation theory emphasizes the importance of the interaction of electrons from Gd's outer shell with surrounding water molecules, it might be anticipated that the T1 signal generated for a Gd colloid would be significantly less than the same mole concentration of chelated Gd. However, this is not the case, suggesting other factors that are not currently fully understood must be involved in signal generation.

Estimates for minimal Gd concentration required for detection depend on the relaxivity of the given Gd complex, which varies with the field strength. The maximum relaxivity attainable decreases with increasing field strength. Aime *et al.* (2002) found that for an agent with high relaxivity (R1 $\cong$ 80 mM^{-1} $\times$ s^{-1}), the threshold for detection was $4 + 1 \times 10^7$ complexes/cell or $\cong 15$ μM. In an animal study of coronary thrombosis, we found that Gd concentrations between 100 and 150 μM translated into a signal-to-noise ratio (SNR) of approximately 11 $\pm$ 2, which allowed for target detection (Botnar *et al.*, 2004a). The R1 of the used contrast agent was approximately 30–40 mM^{-1} $\times$ s^{-1}.

As alluded to previously, the effects of MR contrast agent concentration are nonlinear (e.g., Figs. 1 through 3). While contrast agent distribution and contrast agent compartmentalization affects global and local concentrations, binding to biological targets may further increase local concentrations and typically leads to increased contrast agent relaxivities (Caravan *et al.*, 2003), thereby leading to local signal amplification effects. With increasing Gd concentrations, T1 shortens and thus MR signal increases. At very high concentrations, Gd will reduce T2 to the order of the TE, thereby canceling the T1 effect and decreasing MR signal intensity. Therefore, signal intensity on T1-weighted spin-echo images increases and then decreases with increasing Gd concentration. On the other hand, fast T1-weighted gradient echo sequences (especially 3D sequences) typically have a larger scalable range than spin echo sequences (see Figs. 2 and 3).

C. T2* Effects

The synthesis and use of stable, nano-sized iron oxide particles for use as MR contrast agents have been extensively described (Ferrucci and Stark, 1990; Weissleder *et al.*, 1990a). Iron oxide particles have differential effects on 1/T1 and 1/T2 depending on their size. Superparamagnetic iron oxides (SPIO) produce much larger increases in 1/T2 than in 1/T1, so they are best imaged with T2-weighted scans, which reveal signal decrease (Ferrucci and Stark, 1990). SPIO produces a marked disturbance in surrounding magnetic field homogeneity, especially apparent when a nonhomogenous distribution produces a T2* susceptibility effect. On the other hand, ultrasmall superparamagnetic iron oxides (USPIO) have a greater effect on 1/T1 than SPIO, so they can also be used for T1-weighted imaging (Small *et al.*, 1993).

Although iron oxide–based agents have greater relaxivity per metal atom than Gd-based agents, Gd-based agents provide a larger potential scalable range for detection. Furthermore, Gd-based agents provide positive T1 signal enhancement, which is more readily distinguished from artifacts than negative signal effects detected with iron oxide–based probes.

D. Field Strength

For development of T1-lowering contrast agents, the field strength dependency of the longitudinal relaxivity (R1) plays a critical role. This field strength dependency is also referred to as nuclear magnetic resonance dispersion (NMRD). If the product between the Larmor frequency ω (=field strength) and the correlation time (fluctuation of local magnetic field induced by contrast agent) τ_c exceeds 1, R1 begins to decrease (Caravan *et al.*, 1999). Most contrast agents in clinical use today were optimized for use at 1.5 Tesla (1.5T) ($\tau_c \cong 2.5$ ns). When developing contrast agents for 3T or higher field systems, the correlation time must be decreased in order to achieve maximal longitudinal relaxivities R1. In contrast, the transverse relaxivity R2 behaves differently and even can increase at higher field strengths. Thus, at higher field strengths, the ratio R2/R1 usually increases. Iron particle–based contrast agents (=T2* agents) therefore should be well suited for the use in 3T and higher field MR systems.

E. Magnetic Resonance Signal Amplification-Biological Schemes

An alternative to administering high levels of contrast agents is to alter the chemical environment in which the probe is found by exploiting the biological processes of an organism to activate (Allen and Meade, 2004; Louie *et al.*, 2000; Perez *et al.*, 2004b) or specifically accumulate probe molecules (Hogemann-Savellano *et al.*, 2003; Moore *et al.*, 1998). Biology can be exploited in various ways to increase probe concentration and therefore its influence on signal (Fig. 6). First, if the target to be imaged is overexpressed, an accumulation of contrast agent at the target site can be achieved by conjugating contrast agents to molecules with affinity for the target. However, if the targeted gene product is present in finite amounts on the cells and can only interact once with the imaging probe, the degree of contrast observed is directly proportional to the level of target overexpression. This

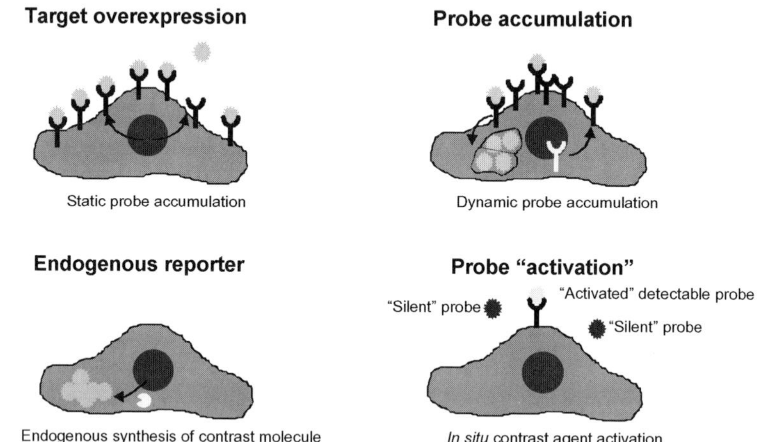

Figure 6 Strategies for MR imaging of gene expression. From Hogemann and Basilion, 2002.

accumulation is termed static (Hogemann and Basilion, 2002). If, however, the targeted gene product is an internalizing cell-surface receptor, it is possible that a single receptor participates multiple times in accumulating targeted probe. The dynamic nature of the accumulation can result in concentration and accumulation of the probe within vesicles in cells (Shibata *et al.*, 1992; Weissleder *et al.*, 1997). Probe accumulation in excess of absolute target number is termed dynamic accumulation. This process has two amplifying results: (1) a numerical increase in the number of probe molecules accumulated by the target (Hogemann-Savellano *et al.*, 2003), and (2) compactation of probe within the vesicles alters the chemical environment enhancing probe detection (Shibata *et al.*, 1992). Thus, dynamic accumulation generally requires lower absolute numbers of target gene product for adequate MR-detectable signals than static accumulation (Hogemann-Savellano *et al.*, 2003; Ichikawa *et al.*, 2002).

A third biologic scheme for signal amplification involves use of endogenous reporters, exploiting natural gene products to enhance MR signal within cells. An example of this application has been demonstrated by (Enochs *et al.*, 1997). In this system, increased tyrosinase expression results in increased melanin, which has a high metal-binding capacity for iron, thus increasing MR signal intensity. A fourth example, probe "activation," describes a strategy in which a contrast probe remains undetectable until activated by a cellular protein (Louie *et al.*, 2000). In the application described by Louie *et al.*, the contrast agent Gd(III) is chemically shielded from the water protons it is designed to influence. Only after release of the shielding molecule by interaction with a cellular enzyme (e.g., beta-galactosidase in the reference cited), can Gd(III) interact with water and thus affect signal intensity. This is a particularly intriguing application because the unacti-vated probe is "silent" until it interacts with target enzyme. Therefore, there is relatively low background signal and interpretation of the signal is more straightforward. The platform for this class of "smart probe" agents is probably amenable to substituting any type of enzyme substrate; further studies to demonstrate the utility of this strategy for mammalian systems are warranted.

Another possibility to increase MR detection of contrast agents is to chemically alter the agent *in situ*, thereby changing its magnetic properties. This avenue has been explored by attempting to generate polymers from individual molecules of contrast agent. The resulting polymers have different magnetic properties, resulting in greater detection sensitivity by MRI.

Two examples for polymerization-based amplification have been re-ported and the principles of the approach are summarized in Fig. 7. First, Bogdanov *et al.* (2002) have demonstrated that modified Gd chelates can be polymerized by enzymatic reduction of chemical side groups engineered into the Gd chelate. The substrate, chelated Gd covalently bound to phenols, serves as an electron donor during enzymatic hydrogen peroxide reduc-tion by peroxidase enzyme-antibody complex. *In situ* polymerization of the

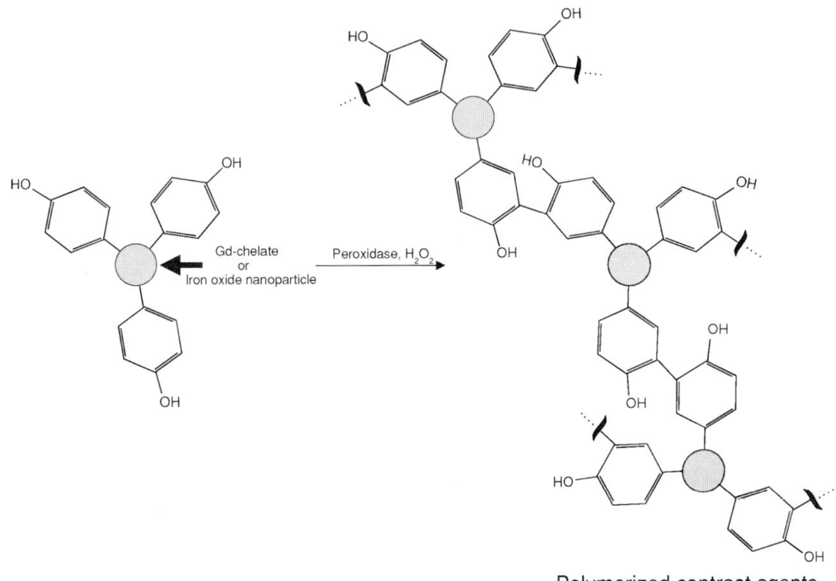

Figure 7　Principle of enzymatic polymerization of contrast agent. Peroxidase in presence of H$_2$O$_2$ results in condensation and polymerization of phenolic contrast agents. Adapted from Perez *et al.*, 2004b. (See Color Insert.)

modified chelate by peroxidase results in a greater than 60% increase in MR signal intensity.

An interesting outgrowth from this initial example was developed by Perez and co-workers, who used superparamagnetic iron oxide nanoparticles for polymerization. In this scheme, iron oxide particles are derivatized with phenolic groups selective for the disease-related peroxidases, Using iron oxide probes derivatized with either serotonin, a substrate for myeloperoxidase, or dopamine, a substrate for horseradish peroxidase, it was demonstrated that peroxidase selectivity could be achieved (Chen *et al.*, 2004; Perez *et al.*, 2004b). Both of these strategies for polymerization of probe are interesting and have potential, but it remains to be seen if adequate levels of the reaction components are formed *in vivo* to polymerize administered probe.

Another important and interesting application of polymerization of MR probes for detection of molecules has been demonstrated by Josephson and co-workers (Perez *et al.*, 2002a). In these studies the researchers have again used iron oxide particles but initiate polymerization via binding reactions between particles and biomolecules rather than enzymatically altering contrast agent. For this approach, iron oxide nanoparticles are derivatized with multiple copies of one of two binding molecules. The two binding moieties bind to different regions of the same target molecule. Upon interacting with the target, the multivalency of the particles combined with two distinct groups of particles carrying different targeting agents results in formation of a lattice structure that significantly alters the T2 characteristics of the probe (Perez *et al.*, 2002a). This approach yields sensitivities for detecting polymerized probe approaching attomole concentrations. This general principle has been applied to demonstrate the utility of this approach to measure RNA and DNA levels (Perez *et al.*, 2002b), to detect single nucleotide polymorphisms and to monitor protein–protein interactions (Perez *et al.*, 2004a), to measure telomerase activity (Grimm *et al.*, 2004), and to measure viral load (Perez *et al.*, 2003). Binding-dependent polymerization schemes, however, have only been demonstrated *in vitro* and significant efforts are underway to develop these probes for *in vivo* study of target molecules.

VII. Applications of Magnetic Resonance in Molecular Imaging

The development of robust signal amplification schemes for application of MRI to molecular imaging continues as an active area of research. However, despite this requirement, investigators have identified several molecular markers whose expression is sufficient for detection by MRI (see Table II and references therein). A variety of targeted MR contrast agents have

Table II Examples of MR Molecular Imaging Probes

Biological Processes	Targets	Ligand	Carrier	Signal-Generating Component	Size/Weight	Relaxivities $(mM \times s)^{-1}$,/ wavelength)	Disease	References
Thrombosis	Fibrin	Anti-fibrin F(ab)′ fragment	Perfluoro-carbon nanoparticle	10,000–50,000 Gd^{3+}	~250 nm	$R1 = 0.18$–0.54 ml × $s^{-1} \times pmol^{-1}$ nanoparticle	CVD	Flacke *et al.*, 2001
	Fibrin	Peptide	Peptide	4 Gd^{3+}	~4000 kDa	$R1 = 21/Gd^3$; $R1 = 84$/molecule	CVD	Botnar *et al.*, 2004c
	Platelets	RGD-peptide	USPIO nanoparticle	USPIO			CVD	Johansson *et al.*, 2001
Angiogenesis	αvβ3	Peptidomimetic vitronectin antagonist	Nanoparticle	~90,000 Gd^{3+}	~270 nm	18/25 (Gd^{3+}) 1.7* $10^6/2.4$* 10^6 (nanoparticle)	CVD, cancer	Winter *et al.*, 2003a,b
	E-selectin	Anti-human E-selectin F(ab′)₂ fragment	CLIO nanoparticle	CLIO	~40 nm	0.3–0.6 mg Fe/ml T2 = 29–40 ms bound T2 = 1500 ms unbound	CVD, cancer	Kang *et al.*, 2002
Apoptosis	Phosphati-dylserine	Annexin-V	CLIO nanoparticle	CLIO	~40 nm	0.3–0.6 mg Fe/ml bound T2 = 1500 ms unbound	CVD, cancer	Schellenberger *et al.*, 2002, 2004

(*Continued*)

Table II *Continued*

Biological Processes	Targets	Ligand	Carrier	Signal-Generating Component	Size/Weight	Relaxivities $(\text{mM} \times \text{s})^{-1}$,/ wavelength)	Disease	References
Vascular inflammation	E-selectin	Anti-human E-selectin $F(ab')^2$ fragment	CLIO nanoparticle	CLIO	$\sim$40 nm	0.3–0.6 mg Fe/ml T2 = 29–40 ms bound T2 $\cong$1500 ms unbound	CVD, cancer	Kang *et al.*, 2002
Neoplasia	Macrophage		USPIO nanoparticle	USPIO	$\sim$20–30 nm	R1 = 7; R2 = 81	CVD, CNS	Dousset *et al.*, 1999; Rogers *et al.*, 1994
	HER-*2/neu* receptor	Biotinylated anti-HER-*2/neu* antibody	Avidin	Gd^{3+}	70,000 mAb		Cancer	Artemov *et al.*, 2003b
	TfR	Transferrin	CLIO nanoparticle	CLIO	$\sim$40–60 nm	0.3–0.6 mg Fe/ml T2 = 29–40 ms bound T2 = 1500 ms unbound	Cancer, gene therapy	Hogemann *et al.*, 2000; Hogemann-Savellano *et al.*, 2003; Weissleder *et al.*, 2000

recently been developed for receptor imaging by linking MR contrast agents such as Gd^{3+}-DOTA or iron oxide nanoparticles to targeting molecules (e.g.. to assess the hepatic asialoglycoprotein receptors' Weissleder *et al.* 1990b), pancreatic secretin receptors (>200,000 receptors/cell) (Shen *et al.*, 1996), HER-*2/neu*-expressing tumors (Artemov *et al.*, 2003a; Funovics *et al.*, 2004), thrombosis (Botnar *et al.*, 2004b), the transferrin receptor (Weissleder *et al.*, 2000), and apoptosis (>Schellenberger *et al.*, 2002; Zhao *et al.*, 2001). This has been accomplished by thoughtful choice of over-expressed or tissue-specific targets and by selection or development of target-specific ligands, such as monoclonal antibodies (Anderson *et al.*, 2000). For recent reviews of this subject, see Allen and Meade (2004), Artemov (2003), and Weinmann *et al.* (2003).

VIII. Targeted Imaging

A variety of MR probes have been developed to study various types of disease processess (e.g., oncology, cardiovascular, neurology, diabetes) by targeting a spectrum of molecular markers. Descriptions of some of these tools are summarized in Table II.

A. MRI of HER-2/*neu*

As noted previously, the HER-2/*neu* receptor may serve as a target for tumor identification (Artemov, 2003; Artemov *et al.*, 2003a,b). This receptor is overexpressed in certain forms of cancer, including approximately 25% of human breast cancers. Its expression is correlated with poor prognosis (Kim *et al.*, 2001). Artemov *et al.* (2003b) developed a two-component gadolinium-based MR contrast agent to image the HER-2/*neu* receptor. Tumor cells are first prelabeled with a biotinylated anti–HER-2/*neu* antibody, then administered Gd-labeled avidin binds with high affinity to the biotinylated monoclonal antibody. They suggested that potential applications for this type of agent may include determination of HER-2/*neu* status for prognosis and for selecting tumors for monoclonal antibody therapy.

B. Magnetic Resonance Imaging of Fibrin

Imaging of fibrin has potential clinical applications for diagnosis of acute coronary syndromes, deep venous thrombosis, and pulmonary emboli. In recent studies by Yu, Flacke, and by our group (Botner *et al.*, 2004a,b; Flacke *et al.*, 2001; Yu *et al.*, 2000), Gd-labeled fibrin-binding nanoparticles and small peptides have been successfully used to image thrombus in the jugular vein, coronary arteries, and aorta. Figure 8 demonstrates imaging of acute thrombus in an animal model of plaque rupture. Gadolinium concentrations as low as >50 μM (R1 $\cong$20 mM^{-1} $\times$ s^{-1}) were sufficient for ready visualization of mural and lumen encroaching thrombus (Botnar *et al.*, 2004b).

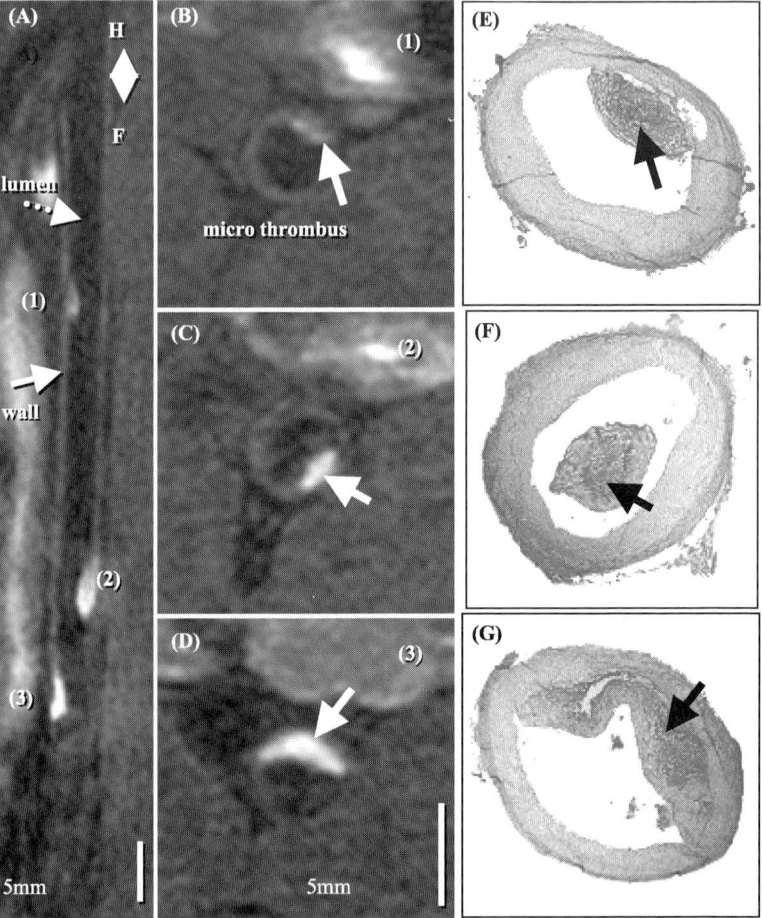

Figure 8 Fibrin imaging in a rabbit model of plaque rupture. (A) Reformatted view of a coronal 3D data set showing the subrenal aorta approximately 20 hours after EP-1873 administration. Three well-delineated mural thrombi (arrows) can be observed with good contrast between thrombus (numbers), arterial blood (dotted arrow), and the vessel wall (dashed arrow). The in-plane view of the aorta allows simultaneous display of all thrombi showing head, tail, length, and relative location. (B–D) Corresponding cross-sectional views show good agreement with histopathology (E–G). From Botnar *et al.*, 2004b. (See Color Insert.)

C. Magnetic Resonance Imaging of Integrins

Integrins, such as $\alpha v\beta 3$, are overexpressed in activated neovascular endothelial cells, which are believed to play an integral role in tumor growth and the initiation and development of atherosclerosis. Wickline, Lanza, and co-workers have developed perfluoro-nanoparticles that can carry as many as 90,000 paramagnetic Gd chelates and can be targeted against various biomarkers by attaching appropriate ligands (Yu *et al.*, 2000). In a recent study, they directed such nanoparticles to the $\alpha v\beta 3$ integrin by attaching a peptidomimetic vitronectin antagonist. With this approach they were able to image angiogenesis in nascent Vx-2 rabbit tumors (Winter *et al.*, 2003a) and in early-stage atherosclerosis (Fig. 9; Winter *et al.*, 2003b).

The use of Gd-labeled nanoparticles is another means (in contrast to iron oxides) of delivering high concentrations of an imaging agent to low-concentration targets.

D. Magnetic Resonance Imaging of Transgene Expression (ETR Imaging)

Basilion and co-workers have used dynamic amplification to image transgene expression by MRI (Fig. 10). In this strategy an internalizing receptor transgene, the engineered transferring receptor (ETR), was imaged by

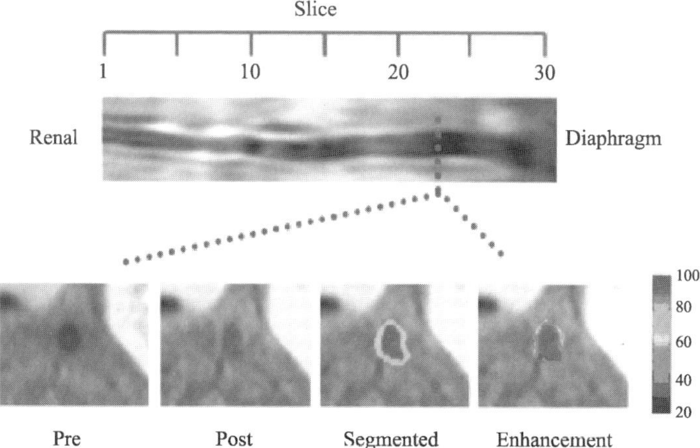

Figure 9 Imaging of $\alpha v\beta 3$ in early-stage atherosclerosis. *In vivo* spin echo image reformatted to display long axis of aorta from renal arteries to diaphragm of one cholesterol-fed rabbit (top) and at single transverse level (bottom) before (Pre) and after (Post) treatment, after semiautomated segmentation (Segmented, grayish ring; see text for description), and with color-coded signal enhancement (Enhancement) above baseline (in percent). From Winter *et al.*, 2003b. (See Color Insert.)

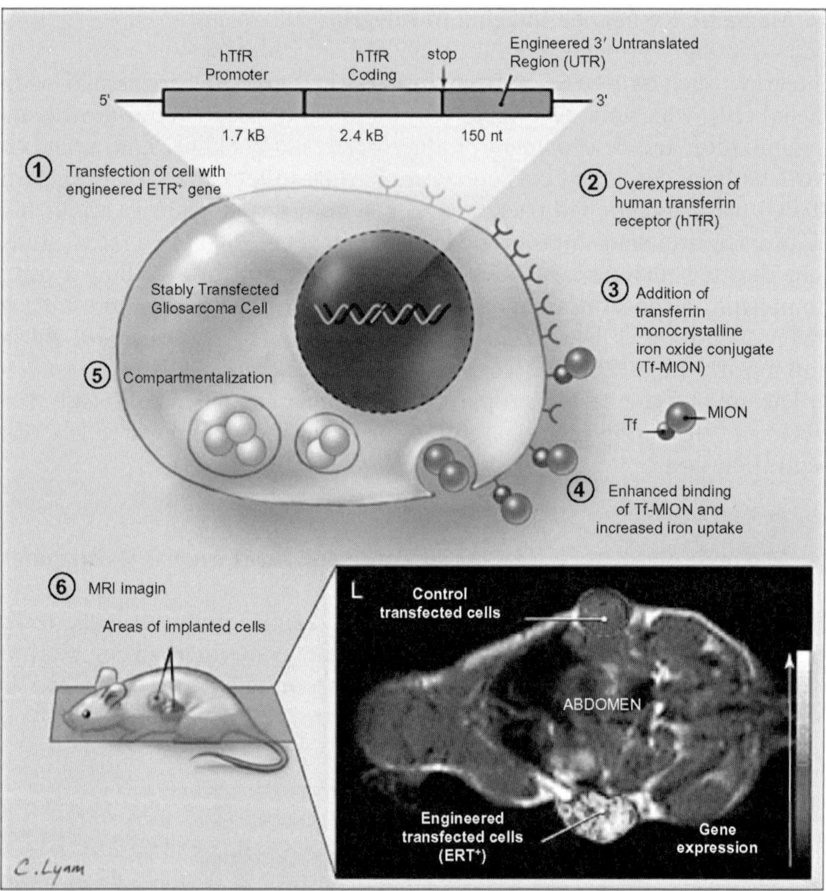

Figure 10 Imaging ETR expression. Several synergistic steps were used to reveal transgene expression in cells by MR imaging. (A) Overexpression of ETR expression results in an approximately fivefold higher cell uptake. (B) During each ETR-mediated internalization event, several thousand iron atoms enter the cell (MION contains an average of 2064 Fe per 3-nm particle core). (C) Upon cellular internalization and compaction in endosomes, the R2 and R2* relaxivities of superparamagnetic MION further increase approximately fourfold (depicted as a change from red to orange), increasing MR detectability. (D) Cellular internalization of iron does not downregulate the level of ETR expression. The ETR cDNA sequence consists of the hTfR promoter, the coding sequence, and the engineered 3' UTR-regulatory sequence (top inset). From Tempany and McNeil, 2001. (See Color Insert.)

targeting it with iron oxide nanoparticles (Fig. 10; Hogemann-Savellano *et al.*, 2003; Tempany and McNeil, 2001; Weissleder *et al.*, 2000). As noted previously, iron oxide nanoparticles are superparamagnetic and have a prominent T2* effect, so they can be detected at very low tissue concentrations (<20 μM iron in tissue; Shen *et al.*, 1993). To generate a targeted contrast agent, iron oxide particles were covalently bound to *holo*-transferrin (Tf), the natural ligand for ETR. In this strategy, two amplification events occur: (1) upon internalization of superparamagnetic particles into cells, R2 relaxivity increases up to fivefold due to compaction in lysosomes (Hogemann-Savellano *et al.*, 2003; Shibata *et al.*, 1992; Weissleder *et al.*, 1997), and (2) receptor recycling resulting in internalization of many probes by one cycling receptor and high cellular accumulation and compactation of probe (Hogemann *et al.*, 2000; Hogemann-Savellano *et al.*, 2003; Ichikawa *et al.*, 2002; Moore *et al.*, 1998; Weissleder *et al.*, 2000). These studies have shown that it is possible to detect receptor expression and regulation in cell lines (Hogemann *et al.*, 2000; Moore *et al.*, 1998) and have demonstrated the utility of this method for detecting *relative* differences in the levels of receptor expression *in vivo* with no measurable cellular or systemic toxicity (Weissleder *et al.*, 2000).

This system has been developed to the point that it is feasible to detect small changes in ETR levels *in vitro* and *in vivo* (Hogemann *et al.*, 2000; Moore *et al.*, 1998; Weissleder *et al.*, 2000). The inset in Fig. 10 shows that modest overexpression of the ETR (approximately 5- to 10-fold) results in significant changes in T2-weighted MR signals. MRI studies of animals harboring two flank tumors, one overexpressing the ETR and one not expressing ETR, revealed no significant differences in tumor signal intensity using either T1- or T2-weighted pulse sequences prior to Tf–iron oxide injection. These results indicate that sources of endogenous ferric iron do not alter image contrast and demonstrate the power of multicomponent amplification schemes to identify gene expression *in vivo*.

To determine if the ETR transgene administered via gene therapy vectors could be monitored by MRI, herpes amplicon vectors harboring the ETR as well as beta-galactosidase and a therapeutic transgene have been developed (Ichikawa *et al.*, 2002). Infection of cells *in vitro* clearly demonstrated the ability of MR to monitor viral transduction of the cells with ETR imaging with respect to probe concentration (Fig. 11), and shows good correlation between ETR expression and expression and efficacy of the simultaneously transferred therapeutic transgene (Hogemann *et al.*, 2000; Hogemann-Savellano *et al.*, 2003; Ichikawa *et al.*, 2002). Exploitation of this method for imaging the transferrin receptor *in vivo* thus shows promise for imaging the effects of gene transfer as well as imaging the effects of promoter activity by utilizing the ETR or other internalizing receptors as MRI reporter genes.

E. Informative Targets: Distribution of Detected Targets

Assuming a strong statistical correlation between a marker and therapeutic design, it may be important to establish not only the presence or absence of a molecular target but also the distribution of the target in a tissue or organ.

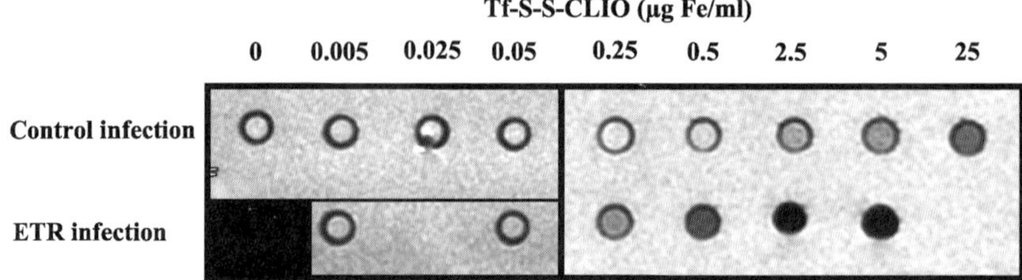

Figure 11 Detection of increased cellular uptake of Tf–iron oxide particles by cells infected with ETR-amplicon. Twenty-four hours after infection the cells were incubated 1 hour with increasing concentrations of Tf-S-S-CLIO contrast agent, washed, pelleted into tissue culture tubes, and imaged in a clinical General Electric 1.5-Tesla MRI. T2-weighted MRI of wells containing cell pellets treated with Tf-S-S-CLIO in culture show that ETR-infected cells show a significant signal decrease at 0.5 mg/ml of iron compared to cells infected with the ETR-negative control vector. From Ichikawa *et al.*, 2002.

For example, it may be useful to produce an image of a cancer mass demonstrating the presence or absence of an informative target and thereby informative therapeutic design. With appropriate technology the topography and heterogeneity of expression could also be measured. This would allow the utility of specific therapies to be assessed. For example, if biopsy of a tumor shows expression of a marker that indicates a specific therapy, this result does not necessarily mean that all the cells within the tumor express the gene product. There may be regional patterns of target expression within the cancer that were not sampled and that may have characteristics potentially requiring therapies other than those predicted by the biopsy results. We have tested the feasibility for imaging topography of gene expression using the ETR and have demonstrated that expression heterogeneities within the tumor can be detected (Weissleder *et al.*, 2000). Sections from tumors overexpressing the ETR were subjected to microscopic MRI to reveal the pattern of iron uptake within the tumors and, presumably, the level of ETR expression (Fig. 12). These results verify that patterns of gene expression can be mapped using MRI *in vitro* and as technology progresses, imaging times and resolutions will improve to make this approach viable for *in vivo* imaging of expression patterns.

IX. Cautions and Conclusions

When contrast enhancement with a targeted probe is observed in or around the specified region, the results must be initially interpreted with caution since targeted contrast agents may nonspecifically accumulate in areas of interest (e.g., nonspecific accumulation of Gd-based agents in areas of

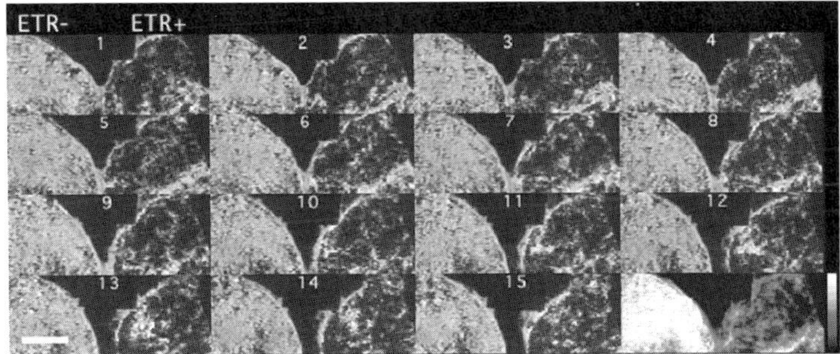

Figure 12   MR microscopy of excised tumor specimen. ETR− (left) and ETR+ (right) tumor specimen obtained from the same animal depicted in Figure 4 following intravenous injection of Tf–iron oxides. The contiguous MR sections are numbered from 1–15. The ETR− tumor (left half of each image) appears of higher signal intensity (absence of Tf-MION uptake) and is homogenous. The ETR+ tumor (right half of each image) has lower signal intensity (Tf-MION uptake) and appears heterogeneous because of regional differences in Tf–iron oxide uptake. The false color image in the lower right represents a projection image through sections 1–15 to show the marked difference in MR signal intensity between the tumors. The scale bar in the lower left represents 1000 μm. Serial 58-μm thick MR images (gradient echo pulse sequence TR/TE/α: 150/3.7/35 degrees; 7.1 T; 58-μm^3 isotropic voxel resolution). Image and legend from reference Weissleder *et al.*, 2000. (See Color Insert.)

atherosclerotic plaques; Kramer *et al.*, 2004; Sirol *et al.*, 2004; Weinmann *et al.*, 2003; Yuan *et al.*, 2002. That is, contrast may accumulate at sites of interest due to the effects of kinetics rather than molecular targeting. Furthermore, when using negative contrast agents, there is a potential for a host of dark band artifacts caused by susceptibility changes or water–fat-shift at tissue borders. Therefore, careful use of controls such as precontrast images is necessary to help distinguish contrast agent effects (signal attenuation due to T2* effect) from potential artifacts (signal voids).

In choosing molecular imaging targets it is important, for all modalities, to probe targets that are highly relevant to the biology or disease of interest. This information can be obtained via proteomic or genomic analysis of diseased and normal tissues. As described earlier, often the information derived from genomic analysis of disease does not identify single markers but rather identifies groups of markers whose variable expression as a cohort is diagnostic. Therefore, in the future molecular imaging probes and technologies will have to be designed to survey the expression of several targets simultaneously in order to meaningfully report on the biological state.

Molecular MRI has great promise as a tool to improve understanding of biologic processes and to aid in clinical diagnosis and monitoring of response to treatment. Although progress in the field must cope with numerous technical challenges to develop novel contrast agents and optimize imaging methods, a multidisciplinary approach to these problems should lead to further advances in this field.

References

Aime, S., Cabella, C., Colombatto, S., Geninatti Crich, S., Gianolio, E., and Maggioni, F. (2002). Insights into the use of paramagnetic Gd(III) complexes in MR-molecular imaging investigations. *J. Magn. Reson. Imaging* **16**, 394–406.

Alizadeh, A. A., Eisen, M. B., Davis, R. E., Ma, C., Lossos, I. S., Rosenwald, A., Boldrick, J. C., Sabet, H., Tran, T., Yu, X., Powell, J. I., Yang, L., Marti, G. E., Moore, T., Hudson, J., Jr., Lu, L., Lewis, D. B., Tibshirani, R., Sherlock, G., Chan, W. C., Greiner, T. C., Weisenburger, D. D., Armitage, J. O., Warnke, R., and Staudt, L. M. (2000). Distinct types of diffuse large B-cell lymphoma identified by gene expression profiling. *Nature* **403**, 503–511.

Allen, M. J., and Meade, T. J. (2004). Magnetic resonance contrast agents for medical and molecular imaging. *Met. Ions Biol. Syst.* **42**, 1–38.

Anderson, S. A., Glod, J., Arbab, A. S., Noel, M., Ashari, P., Fine, H. A., and Frank, J. A. (2004). Noninvasive MR imaging of magnetically labeled stem cells to directly identify neovasculature in a glioma model. *Blood* **105**, 420–425.

Anderson, S. A., Rader, R. K., Westlin, W. F., Null, C., Jackson, D., Lanza, G. M., Wickline, S. A., and Kotyk, J. J. (2000). Magnetic resonance contrast enhancement of neovasculature with alpha(v)beta(3)-targeted nanoparticles. *Magn. Reson. Med.* **44**, 433–439.

Artemov, D. (2003). Molecular magnetic resonance imaging with targeted contrast agents. *J. Cell. Biochem.* **90**, 518–524.

Artemov, D., Mori, N., Okollie, B., and Bhujwalla, Z. M. (2003a). MR molecular imaging of the Her-2/neu receptor in breast cancer cells using targeted iron oxide nanoparticles. *Magn. Reson. Med.* **49**, 403–408.

Artemov, D., Mori, N., Ravi, R., and Bhujwalla, Z. M. (2003b). Magnetic resonance molecular imaging of the HER-2/neu receptor. *Cancer Res.* **63**, 2723–2727.

Ben-Dor, A., Bruhn, L., Friedman, N., Nachman, I., Schummer, M., and Yakhini, Z. (2000). Tissue classification with gene expression profiles. *J. Comput. Biol.* **7**, 559–583.

Bogdanov, A., Jr., Matuszewski, L., Bremer, C., Petrovsky, A., and Weissleder, R. (2002). Oligomerization of paramagnetic substrates result in signal amplification and can be used for MR imaging of molecular targets. *Mol. Imaging* **1**, 16–23.

Botnar, R. M., Buecker, A., Wiethoff, A. J., Parsons, E. C., Jr., Katoh, M., Katsimaglis, G., Weisskoff, R. M., Lauffer, R. B., Graham, P. B., Gunther, R. W., Manning, W. J., and Spuentrup, E. (2004b). *In vivo* molecular imaging of acute and subacute thrombosis using a fibrin-binding magnetic resonance imaging contrast agent. *Circulation* **110**, 1463–1466.

Botnar, R. M., Perez, A. S., Witte, S., Wiethoff, A. J., Laredo, J., Hamilton, J., Quist, W., Parsons, E. C., Jr., Vaidya, A., Kolodziej, A., Barrett, J. A., Graham, P. B., Weisskoff, R. M., Manning, W. J., and Johnstone, M. T. (2004c). *In vivo* molecular imaging of acute and subacute thrombosis using a fibrin-binding magnetic resonance imaging contrast agent. *Circulation* **109**, 2023–2029.

Butte, A. J., Tamayo, P., Slonim, D., Golub, T. R., and Kohane, I. S. (2000). Discovering functional relationships between RNA expression and chemotherapeutic susceptibility using relevance networks. *Proc. Natl. Acad. Sci. USA* **97**, 12182–12186.

Caravan, P., Ellison, J. J., McMurry, T. J., and Lauffer, R. B. (1999). Gadolinium(III) Chelates as MRI Contrast Agents: Structure, Dynamics, and Applications. *Chem. Rev.* **99**, 2293–2352.

Caravan, P., Greenwood, J. M., Welch, J. T., and Franklin, S. J. (2003). Gadolinium-binding helix-turn-helix peptides: DNA-dependent MRI contrast agents. *Chem. Commun. (Camb.)* **20,** 2574–2575.

Chen, J. W., Pham, W., Weissleder, R., and Bogdanov, A., Jr. (2004). Human myeloperoxidase: A potential target for molecular MR imaging in atherosclerosis. *Magn. Reson. Med.* **52,** 1021–1028.

Clare, S., Francis, S., Morris, P. G., and Bowtell, R. (2001). Single-shot T2(*) measurement to establish optimum echo time for fMRI: Studies of the visual, motor, and auditory cortices at 3.0 T. *Magn. Reson. Med.* **45,** 930–933.

Dahnke, H., Weiss, S., and Schaeffter, T. (2004). Simultaneous T2* mapping and anatomical imaging using a fast radial multi-gradient-echo acquisition, Paper presented at: ISMRM (Kyoto: International Society for Magnetic Resonance Imaging).

Dousset, V., Delalande, C., Ballarino, L., Quesson, B., Seilhan, D., Coussemacq, M., Thiaudiere, E., Brochet, B., Canioni, P., and Caille, J. M. (1999). *In vivo* macrophage activity imaging in the central nervous system detected by magnetic resonance. *Magn. Reson. Med.* **41,** 329–333.

Enochs, W. S., Petherick, P., Bogdanova, A., Mohr, U., and Weissleder, R. (1997). Paramagnetic metal scavenging by melanin: MR imaging. *Radiology* **204,** 417–423.

Ferrucci, J. T., and Stark, D. D. (1990). Iron oxide-enhanced MR imaging of the liver and spleen: Review of the first 5 years. *AJR Am. J. Roentgenol.* **155,** 943–950.

Fischer, S. E., Wickline, S. A., and Lorenz, C. H. (1999). Novel real-time R-wave detection algorithm based on the vectorcardiogram for accurate gated magnetic resonance acquisitions. *Magn. Reson. Med.* **42,** 361–370.

Flacke, S., Fischer, S., Scott, M. J., Fuhrhop, R. J., Allen, J. S., McLean, M., Winter, P., Sicard, G. A., Gaffney, P. J., Wickline, S. A., and Lanza, G. M. (2001). Novel MRI contrast agent for molecular imaging of fibrin: Implications for detecting vulnerable plaques. *Circulation* **104,** 1280–1285.

Funovics, M. A., Kapeller, B., Hoeller, C., Su, H. S., Kunstfeld, R., Puig, S., and Macfelda, K. (2004). MR imaging of the her2/neu and 9.2.27 tumor antigens using immunospecific contrast agents. *Magn. Reson. Imaging* **22,** 843–850.

Gayed, I., Vu, T., Iyer, R., Johnson, M., Macapinlac, H., Swanston, N., and Podoloff, D. (2004). The role of 18F-FDG PET in staging and early prediction of response to therapy of recurrent gastrointestinal stromal tumors. *J. Nucl. Med.* **45,** 17–21.

Goerres, G. W., Stupp, R., Barghouth, G., Hany, T. F., Pestalozzi, B., Dizendorf, E., Schnyder, P., Luthi, F., Von Schulthess, G. K., and Leyvraz, S. (2004). The value of PET, CT and in-line PET/CT in patients with gastrointestinal stromal tumours: Long-term outcome of treatment with imatinib mesylate. *Eur. J. Nucl. Med. Mol. Imaging* **32,** 153–162.

Gohr-Rosenthal, S., Schmitt-Willich, H., Ebert, W., and Conrad, J. (1993). The demonstration of human tumors on nude mice using gadolinium-labelled monoclonal antibodies for magnetic resonance imaging. Invest. Radiol. **28,** 789–795.

Golub, T. R. (2001). Genome-wide views of cancer. *N. Engl. J. Med.* **344,** 601–602.

Golub, T. R., Slonim, D. K., Tamayo, P., Huard, C., Gaasenbeek, M., Mesirov, J. P., Coller, H., Loh, M. L., Downing, J. R., Caligiuri, M. A., Bloomfield, C. D., and Lander, E. S. (1999). Molecular classification of cancer: Class discovery and class prediction by gene expression monitoring. *Science* **286,** 531–537.

Grimm, J., Perez, J. M., Josephson, L., and Weissleder, R. (2004). Novel nanosensors for rapid analysis of telomerase activity. *Cancer Res.* **64,** 639–643.

Harisinghani, M. G., Barentsz, J., Hahn, P. F., Deserno, W. M., Tabatabaei, S., van de Kaa, C. H., de la Rosette, J., and Weissleder, R. (2003). Noninvasive detection of clinically occult lymph-node metastases in prostate cancer. *N. Engl. J. Med.* **348,** 2491–2499.

Henderson, E., McKinnon, G., Lee, T. Y., and Rutt, B. K. (1999). A fast 3D look-locker method for volumetric T1 mapping. *Magn. Reson. Imaging* **17,** 1163–1171.

Hogemann, D., and Basilion, J. P. (2002). "Seeing inside the body": MR Imaging of gene expression. *Eur. J. Nucl. Med. Mol. Imaging* **29,** 400–408.

Hogemann, D., Josephson, L., Weissleder, R., and Basilion, J. P. (2000). Improvement of MRI probes to allow efficient detection of gene expression. *Bioconjug. Chem.* **11,** 941–946.

Hogemann, D., Ntziachristos, V., Josephson, L., and Weissleder, R. (2002). High throughput magnetic resonance imaging for evaluating targeted nanoparticle probes. *Bioconjug. Chem.* **13,** 116–121.

Hogemann-Savellano, D., Bos, E., Blondet, C., Sato, F., Abe, T., Josephson, L., Weissleder, R., Gaudet, J., Sgroi, D., Peters, P. J., and Basilion, J. P. (2003). The transferrin receptor: A potential molecular imaging marker for human cancer. *Neoplasia* **5,** 495–506.

Ichikawa, T., Hogemann, D., Saeki, Y., Tyminski, E., Terada, K., Weissleder, R., Chiocca, E. A., and Basilion, J. P. (2002). MRI of transgene expression: Correlation to therapeutic gene expression. *Neoplasia* **4,** 523–530.

Jacobs, A., Tjuvajev, J. G., Dubrovin, M., Akhurst, T., Balatoni, J., Beattie, B., Joshi, R., Finn, R., Larson, S. M., Herrlinger, U., Pechan, P. A., Chiocca, E. A., Breakefield, X. O., and Blasberg, R. G. (2001a). Positron emission tomography-based imaging of transgene expression mediated by replication-conditional, oncolytic herpes simplex virus type 1 mutant vectors *in vivo. Cancer Res.* **61,** 2983–2995.

Jacobs, A., Voges, J., Reszka, R., Lercher, M., Gossmann, A., Kracht, L., Kaestle, C., Wagner, R., Wienhard, K., and Heiss, W. D. (2001b). Positron-emission tomography of vector-mediated gene expression in gene therapy for gliomas. *Lancet* **358,** 727–729.

Joensuu, H. (2002). Treatment of inoperable gastrointestinal stromal tumor (GIST) with Imatinib (Glivec, Gleevec). *Med. Klin (Munich)* **97**(Suppl. 1), 28–30.

Johansson, L. O., Bjornerud, A., Ahlstrom, H. K., Ladd, D. L., and Fujii, D. K. (2001). A targeted contrast agent for magnetic resonance imaging of thrombus: Implications of spatial resolution. *J. Magn. Reson. Imaging* **13,** 615–618.

Johnson, G. A., Benveniste, H., Black, R. D., Hedlund, L. W., Maronpot, R. R., and Smith, B. R. (1993). Histology by magnetic resonance microscopy. *Magn. Reson. Quarterly* **9,** 1–30.

Kang, H. W., Josephson, L., Petrovsky, A., Weissleder, R., and Bogdanov, A., Jr. (2002). Magnetic resonance imaging of inducible E-selectin expression in human endothelial cell culture. *Bioconjug. Chem.* **13,** 122–127.

Kim, Y. S., Konoplev, S. N., Montemurro, F., Hoy, E., Smith, T. L., Rondon, G., Champlin, R. E., Sahin, A. A., and Ueno, N. T. (2001). HER-2/neu overexpression as a poor prognostic factor for patients with metastatic breast cancer undergoing high-dose chemotherapy with autologous stem cell transplantation. *Clin. Cancer Res.* **7,** 4008–4012.

Kramer, C. M., Cerilli, L. A., Hagspiel, K., Di Maria, J. M., Epstein, F. H., and Kern, J. A. (2004). Magnetic resonance imaging identifies the fibrous cap in atherosclerotic abdominal aortic aneurysm. *Circulation* **109,** 1016–1021.

Laniado, M., Weinmann, H. J., Schorner, W., Felix, R., and Speck, U. (1984). First use of GdDTPA/dimeglumine in man. *Physiol. Chem. Phys. Med. NMR.* **16,** 157–165.

Lanza, G. M., Wallace, K. D., Scott, M. J., Cacheris, W. P., Abendschein, D. R., Christy, D. H., Sharkey, A. M., Miller, J. G., Gaffney, P. G., and Wickline, S. A. (1996). A novel site-targeted ultrasonic contrast agent with broad biomedical application. *Circulation* **94,** 3334–40.

Laxman, B., Hall, D. E., Bhojani, M. S., Hamstra, D. A., Chenevert, T. L., Ross, B. D., and Rehemtulla, A. (2002). Noninvasive real-time imaging of apoptosis. *Proc. Natl. Acad. Sci. USA* **99,** 16551–16555.

Lewin, M., Carlesso, N., Tung, C. H., Tang, X. W., Cory, D., Scadden, D. T., and Weissleder, R. (2000). Tat peptide-derivatized magnetic nanoparticles allow *in vivo* tracking and recovery of progenitor cells. *Nat. Biotechnol.* **18,** 410–414.

Look, D. C., and Locker, D. R. (1970). Time saving in measurement of NMR and EPR relaxation times. *Rev. Sci. Instrum.* **41,** 250–251.

Louie, A. Y., Huber, M. M., Ahrens, E. T., Rothbacher, U., Moats, R., Jacobs, R. E., Fraser, S. E., and Meade, T. J. (2000). *In vivo* visualization of gene expression using magnetic resonance imaging. *Nat. Biotechnol.* **18,** 321–325.

Mahmood, U., and Weissleder, R. (2003). Near-infrared optical imaging of proteases in cancer. *Mol. Cancer Ther.* **2,** 489–496.

Messroghli, D. R., Radjenovic, A., Kozerke, S., Higgins, D. M., Sivananthan, M. U., and Ridgway, J. P. (2004). Modified Look-Locker inversion recovery (MOLLI) for high-resolution T1 mapping of the heart. *Magn. Reson. Med.* **52,** 141–146.

Moore, A., Basilion, J. P., Chiocca, E. A., and Weissleder, R. (1998). Measuring transferrin receptor gene expression by NMR imaging. *Biochim. Biophys. Acta* **1402,** 239–249.

Perez, J. M., Josephson, L., O'Loughlin, T., Hogemann, D., and Weissleder, R. (2002a). Magnetic relaxation switches capable of sensing molecular interactions. *Nat. Biotechnol.* **20,** 816–820.

Perez, J. M., Josephson, L., and Weissleder, R. (2004a). Use of magnetic nanoparticles as nanosensors to probe for molecular interactions. *Chembiochem.* **5,** 261–264.

Perez, J. M., O'Loughin, T., Simeone, F. J., Weissleder, R., and Josephson, L. (2002b). DNA-based magnetic nanoparticle assembly acts as a magnetic relaxation nanoswitch allowing screening of DNA-cleaving agents. *J. Am. Chem. Soc.* **124,** 2856–2857.

Perez, J. M., Simeone, F. J., Saeki, Y., Josephson, L., and Weissleder, R. (2003). Viral-induced self-assembly of magnetic nanoparticles allows the detection of viral particles in biological media. *J. Am. Chem. Soc.* **125,** 10192–10193.

Perez, J. M., Simeone, F. J., Tsourkas, A., Josephson, L., and Weissleder, R. (2004b). Peroxidase Substrate Nanosensors fo MR Imaging. *Nano Letters* **4,** 119–122.

Pirko, I., Johnson, A., Ciric, B., Gamez, J., Macura, S. I., Pease, L. R., and Rodriguez, M. (2004). *In vivo* magnetic resonance imaging of immune cells in the central nervous system with superparamagnetic antibodies. *FASEB J.* **18,** 179–182.

Reddy, M. P., Reddy, P., and Lilien, D. L. (2003). F-18 FDG PET imaging in gastrointestinal stromal tumor. *Clin. Nucl. Med.* **28,** 677–679.

Reeder, S. B., Faranesh, A. Z., Boxerman, J. L., and McVeigh, E. R. (1998). *In vivo* measurement of T*2 and field inhomogeneity maps in the human heart at 1.5 T. *Magn. Reson. Med.* **39,** 988–998.

Rogers, J., Lewis, J., and Josephson, L. (1994). Use of AMI-227 as an oral MR contrast agent. *Magn. Reson. Imaging* **12,** 631–639.

Schellenberger, E. A., Hogemann, D., Josephson, L., and Weissleder, R. (2002). Annexin V-CLIO: A nanoparticle for detecting apoptosis by MRI. *Acad Radiol.* **9**(Suppl.2), S310–S311.

Schellenberger, E. A., Sosnovik, D., Weissleder, R., and Josephson, L. (2004). Magneto/Optical annexin v, a multimodal protein. *Bioconjug. Chem.* **15,** 1062–1067.

Shen, T. T., Bogdanov, A., Jr., Bogdanova, A., Poss, K., Brady, T. J., and Weissleder, R. (1996). *Bioconjug. Chem.* **7,** 311–316.

Shen, T., Weissleder, R., Papisov, M., Bogdanov, A., Jr., and Brady, T. J. (1993). Magnetically labeled secretin retains receptor affinity to pancreas acinar cells. Monocrystalline iron oxide nanocompounds (MION): Physicochemical properties. *Magn. Reson. Med.* **29,** 599–604.

Shibata, T., Weissleder, R., Schaeffer, B., Shen, T., Papisov, M., and Brady, T. (1992). Tissue relaxivities and detectability of monocrystalline iron oxides. *In* "Society of Magnetic Resonance in Medicine Annual Meeting," p. 1416, Berlin.

Sirol, M., Itskovich, V. V., Mani, V., Aguinaldo, J. G., Fallon, J. T., Misselwitz, B., Weinmann, H. J., Fuster, V., Toussaint, J. F., and Fayad, Z. A. (2004). Lipid-rich atherosclerotic plaques detected by gadofluorine-enhanced *in vivo* magnetic resonance imaging. *Circulation* **109**, 2890–2896.

Small, W. C., Nelson, R. C., and Bernardino, M. E. (1993). Dual contrast enhancement of both T1- and T2-weighted sequences using ultrasmall superparamagnetic iron oxide. *Magn. Reson. Imaging* **11**, 645–654.

Smith, B., Johnson, G., Groman, E., and Linney, E. (1994). Magnetic resonance microscopy of mouse embryos. *Proc. Natl. Acad. Sci. USA* **91**, 3530–3533.

Stroobants, S., Goeminne, J., Seegers, M., Dimitrijevic, S., Dupont, P., Nuyts, J., Martens, M., van den Borne, B., Cole, P., Sciot, R., Dumez, H., Silberman, S., Mortelmans, L., and van Oosterom, A. (2003). 18FDG-Positron emission tomography for the early prediction of response in advanced soft tissue sarcoma treated with imatinib mesylate (Glivec). *Eur. J. Cancer* **39**, 2012–2020.

Tempany, C. M., and McNeil, B. J. (2001). Advances in biomedical imaging. *JAMA* **285**, 562–567.

Van den Abbeele, A. D., and Badawi, R. D. (2002). Use of positron emission tomography in oncology and its potential role to assess response to imatinib mesylate therapy in gastrointestinal stromal tumors (GISTs). *Eur. J. Cancer* **38**(Suppl. 5), S60–S65.

Wei, J. S., Greer, B. T., Westermann, F., Steinberg, S. M., Son, C. G., Chen, Q. R., Whiteford, C. C., Bilke, S., Krasnoselsky, A. L., Cenacchi, N., Catchpoole, D., Berthold, F., Schwab, M., and Khan, J. (2004). Prediction of clinical outcome using gene expression profiling and artificial neural networks for patients with neuroblastoma. *Cancer Res.* **64**, 6883–6891.

Weinmann, H. J., Ebert, W., Misselwitz, B., and Schmitt-Willich, H. (2003). Tissue-specific MR contrast agents. *Eur. J. Radiol.* **46**, 33–44.

Weissleder, R., Cheng, H., Bogdanova, A., and Bogdanov, A., Jr. (1997). Magnetically labeled cells can be detected by MR imaging. *J. Magn. Reson. Imaging* **7**, 258–263.

Weissleder, R., Elizondo, G., Wittenberg, J., Rabito, C. A., Bengele, H. H., and Josephson, L. (1990a). Ultrasmall superparamagnetic iron oxide: Characterization of a new class of contrast agents for MR imaging. *Radiology* **175**, 489–493.

Weissleder, R., Moore, A., Mahmood, U., Bhorade, R., Benveniste, H., Chiocca, E. A., and Basilion, J. P. (2000). *In vivo* magnetic resonance imaging of transgene expression. *Nat. Med.* **6**, 351–355.

Weissleder, R., and Ntziachristos, V. (2003). Shedding light onto live molecular targets. *Nat. Med.* **9**, 123–128.

Weissleder, R., Reimer, P., Lee, A. S., Wittenberg, J., and Brady, T. J. (1990b). MR receptor imaging: Ultrasmall iron oxide particles targeted to asialoglycoprotein receptors. *AJR Am. J. Roentgenol.* **155**, 1161–1167.

Winter, P. M., Caruthers, S. D., Kassner, A., Harris, T. D., Chinen, L. K., Allen, J. S., Lacy, E. K., Zhang, H., Robertson, J. D., Wickline, S. A., and Lanza, G. M. (2003a). Molecular imaging of angiogenesis in nascent Vx-2 rabbit tumors using a novel alpha(nu)beta3-targeted nanoparticle and 1.5 tesla magnetic resonance imaging. *Cancer Res.* **63**, 5838–5843.

Winter, P. M., Morawski, A. M., Caruthers, S. D., Fuhrhop, R. W., Zhang, H., Williams, T. A., Allen, J. S., Lacy, E. K., Robertson, J. D., Lanza, G. M., and Wickline, S. A. (2003b). Molecular imaging of angiogenesis in early-stage atherosclerosis with alpha(v)beta3-integrin-targeted nanoparticles. *Circulation* **108**, 2270–2274.

Yeang, C. H., Ramaswamy, S., Tamayo, P., Mukherjee, S., Rifkin, R. M., Angelo, M., Reich, M., Lander, E., Mesirov, J., and Golub, T. (2001). Molecular classification of multiple tumor types. *Bioinformatics* **17**, S316–S322.

Yu, X., Song, S. K., Chen, J., Scott, M. J., Fuhrhop, R. J., Hall, C. S., Gaffney, P. J., Wickline, S. A., and Lanza, E. (2000). High-resolution MRI characterization of human thrombus using a novel fibrin-targeted paramagnetic nanoparticle contrast agent. *Magn. Reson. Med.* **44,** 867–72.

Yuan, C., Kerwin, W. S., Ferguson, M. S., Polissar, N., Zhang, S., Cai, J., and Hatsukami, T. S. (2002). Contrast-enhanced high resolution MRI for atherosclerotic carotid artery tissue characterization. *J. Magn. Reson. Imaging* **15,** 62–67.

Zhao, M., Beauregard, D. A., Loizou, L., Davletov, B., and Brindle, K. M. (2001). Non-invasive detection of apoptosis using magnetic resonance imaging and a targeted contrast agent. *Nat. Med.* **7,** 1241–1244.

2

Magnetic Resonance Imaging Contrast Agents in the Study of Development

Angelique Louie
Department of Biomedical Engineering
University of California Davis
Davis, California 95616

The use of magnetic resonance imaging (MRI) as a research tool in the study of development has increased in recent years due in part to improvements in spatial resolution, new imaging agents, and increased availability of MRI scanners. This chapter describes how contrast agent–enhanced MRI can be used to study development. In addition, we highlight some novel applications of contrast agent–enhanced MRI in biology that may be useful as tools for the study of development. © 2005, Elsevier Inc.

I. Introduction

Much of what it known about the intricate patterns and timings of cell movements in development has been revealed by optical microscopy. Optical imaging techniques are limited by the opacity of the system to the passage of light; thus, much is known about development in model embryos that are relatively transparent, such as zebrafish, chick and *Xenopus,* but mammalian systems are less well understood. MRI offers attractive possibilities for the study of development in that it allows noninvasive interrogation of deep tissues that are refractory to optical techniques. Although other techniques such as x-ray computed- or positron emission tomography are also capable of noninvasively imaging deep tissues, MRI may be more suitable for developing systems as MRI avoids the use of ionizing radiation. Improvements in instrumentation, imaging probes, and image acquisition programs

Current Topics in Developmental Biology, Vol. 70
Copyright 2005, Elsevier Inc. All rights reserved.

0070-2153/05 $35.00
DOI: 10.1016/S0070-2153(05)70002-8

have now allowed investigators to push the limits of resolution to the cellular (micron) range. While still expensive instruments, MR scanners are much more common on university campuses, and high-field instruments are available at numerous multiuser centers across the country, putting MRI within reach of many researchers. This review highlights some of the most recent applications of MRI and contrast agents as they have been applied to study developing systems, and introduces uses for contrast agents in imaging *in vivo* that demonstrate their utility as tools for research.

II. MRI Contrast Agents

Biomedical visualization methodologies yield images in a multitude of ways. In fluorescence microscopy, the signal comes from fluorophores that emit light. In positron emission tomography, the signal comes from decay products of radioactive isotopes. In computed tomography, the signal comes from X-rays that pass through the specimen. In conventional ^{1}H MRI, the signal comes from water protons. While differences in relaxation times (i.e., T1 and T2) from tissue to tissue can provide sufficient contrast in an image, quite often tissues do not demonstrate distinguishable differences. In this case, a contrast-enhancing agent can be introduced to modulate signal intensity, much as fluorophores modulate intensity in fluorescence images.

Most commercially available contrast agents, and those approved for clinical use, are relaxation agents. Relaxation agents affect image contrast by strongly influencing relaxation times of local water protons. The agents fall into two classes, those that more strongly affect T1 and those that more strongly affect T2. Typical T1 agents such as ProHance and Magnevist, both in clinical use, consist of gadolinium ions bound by a chelating ligand (Lauffer, 1987; Tweedle, 1992). Note that in the context of inorganic chemistry, ligand refers to groups bound to a metal ion. The chelator is a single molecule that occupies several binding sites of gadolinium and due to this "chelate effect" forms an extremely high stability complex with the metal. A metal complexed with a chelator is much more thermodynamically stable than a complex formed with several comparable monodentate ligands. The name *chelator* is derived from the Latin *chela-* for the pincher claws of crustaceans, and came about from the observation of clawlike structures that are formed as associating groups on the chelator fasten around the central metal ion (Fig. 1). Gd(III) has nine coordination, or binding, sites and the chelators in ProHance and Magnevist occupy eight of those sites. This degree of chelation makes for a very stable complex with little chance of release of gadolinium from the complex *in vivo* and reduces the risk for toxicity as bound gadolinium cannot substitute for calcium as free Gd^{3+} ions can. High stability of the complex ensures that gadolinium will remain

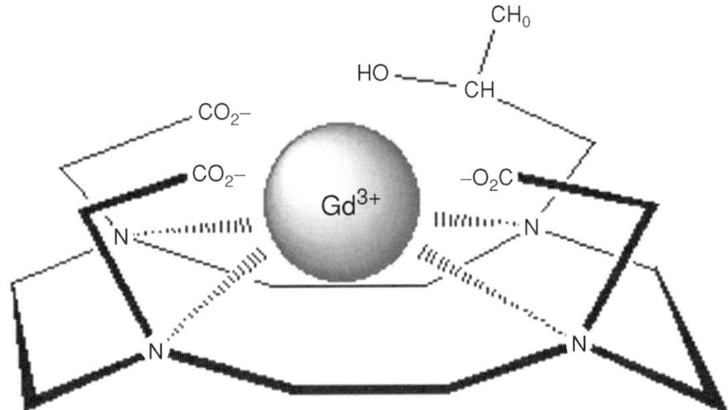

Figure 1 Structure of the commercially available contrast agent ProHance. A ringlike macrocycle (blue) surrounds and binds a central metal ion (magenta) at several positions (eight, in the example shown). (See Color Insert.)

complexed over the long term. Gadolinium has an affinity for nitrogen and oxygen groups and typically binds to chelators through the nitrogen and pendant carboxyl groups.

Gadolinium(III) is the most widely used paramagnetic ion in T1 agents. It has a very high magnetic moment and seven unpaired electrons, which is the most of any stable ion. The effect of gadolinium on relaxation times requires direct interaction between unpaired electrons in the ion and water protons. Thus, contrast agents are designed with open sites for coordination of water to gadolinium. Both Prohance and Magnevist leave one site open for water access. Although more sites for interaction with water would increase the effect on signal intensity, it is a delicate tradeoff between more sites for water access and decreased complex stability. In general, chemists have opted on the side of safety and maximized for stability. Other paramagnetic ions such as manganese may be used as T1 agents, but these can be less effective because manganese has fewer unpaired electrons (five) and a smaller magnetic moment.

T2 agents comprise another class of relaxation agents. As the name suggests, they affect T2 relaxation times. These agents affect contrast by introducing a large regional magnetic field gradient in the vicinity of the agent. Thus, their effect is more long range than that for gadolinium, which requires direct interaction with water protons. Typical T2 agents are composed of superparamagnetic species such as iron in the form of an iron oxide core, coated by a water-soluble shell such as dextran (Kawaguchi and Hasegawa, 2000; Landfester and Ramirez, 2003). The shortening of T2 relaxation time caused by the agents creates a decrease of signal intensity in tissues containing the agents, which results in contrast enhancement compared to neighboring tissues. A consequence of the high number of iron atoms in each core is that less molar amounts of contrast agents need to be delivered compared to T1 agents. However, because the contrast enhancement comes from a signal decrease rather than a signal increase, T1 agents are considered to provide images with larger dynamic range. Superparamagnetic iron oxide particles (SPIO) are prepared in a range of sizes from small (>50 nm) to ultrasmall (<50 nm). Ultrasmall superparamagnetic iron oxide particles (USPIO) have a longer residence time in the blood and thus are less subject to removal by the reticuloendothelial system and lymphatic system. The larger SPIO are preferred where accumulation in the liver is desirable, for example, to image hepatocytes. The larger particles also experience higher uptake in phagocytic cell types (Bowen *et al.*, 2002; Hauger *et al.*, 2000; Kehagias *et al.*, 2001; Paul *et al.*, 2004).

Although nongadolinium lanthanides are generally not efficient relaxation agents, they show promise as contrast agents for magnetization transfer (or chemical exchange saturation transfer, CEST) experiments. Magnetization transfer is another type of acquisition in MRI which takes advantage of the fact that there are two different pools of protons found *in vivo* with distinct nuclear magnetic resonance (NMR) spectral properties (i.e., mobile protons and bound protons) (Henkelman *et al.*, 2001). The majority of protons are mobile protons in liquid water. The bound protons are those that are associated with macromolecules and membranes; these generally have T2 values that are too short to be detected in typical relaxation experiments. Because the bound protons have a very broad NMR spectrum compared to the mobile protons, they can be selectively excited to saturation and via coupling can transfer this saturation to the mobile population of protons with the rate of transfer dependent on the rate of exchange between the two populations. This method can provide additional contrast in some applications; for example, it has shown utility in imaging white matter diseases of the brain, as the resonance of the bound protons changes with progressive disease (Henkelman *et al.*, 2001).

The nature of contrast agents for CEST varies widely and falls into three major classes: endogenous molecules, exogenous nonmetal agents, and lanthanide-based agents. Chemical groups such as –NH and –OH groups have been identified as exchange sites for CEST and these can be exploited in endogenous molecules or exogenous nonmetal–based agents. For example, amide protons of intracellular proteins and peptides have been used to detect pH *in vivo* (Zhou *et al.*, 2003). The concentration of these molecules is in the millimolar range, and large numbers of amide groups on intracellular macromolecules supply a sufficient amount of sites for observable exchange between amide protons and water protons. However, the absolute effects of amide proton transfer on signal are difficult to evaluate as other effects, including blood oxygen level–dependent MRI effects, direct saturation of water, and transfer between solidlike macromolecules to cellular water, also contribute to the signal. In order to determine absolute pH, these confounding factors must be removed. Investigators were able to detect pH changes on the order of 0.2 pH units or larger and identified regions of focal ischemic damage in the rat brain. Figure 2 shows MR images of a rat brain following mid-cerebral arterial occlusion (MCAO). MCAO typically affects the caudate nucleus and this is confirmed in the images (arrow). Intracellular pH was independently verified by ^{31}P spectroscopy; however, these images were only calibrated against postmortem water exchange transfer spectroscopy results; this makes it difficult to assess the accuracy of the pH measurements *in vivo*, but the enhancement of image contrast is readily apparent.

Exogenous agents possessing suitable exchange groups can also serve as CEST agents (Ward and Balaban, 2000; Ward *et al.*, 2000). In an investigation of various macromolecules as pH indicators, 5,6 dihydrouracil (5-DH), 5-hydroxytryptophan (5-HT), and 2-imidizolidinethione (2-IL, in combination with 5-HT) were found to be sensitive to pH in useful range (Ward *et al.*, 2000; Ward and Balaban, 2000). The authors note that high concentrations, >40 mM, are required to generate large effects and that this may limit the utility of the probes. Among these types of agents the largest effects have been observed for macromolecules carrying large numbers of exchange groups such as polypeptides, oligonucleotides, and dendrimers (Snoussi *et al.*, 2003).

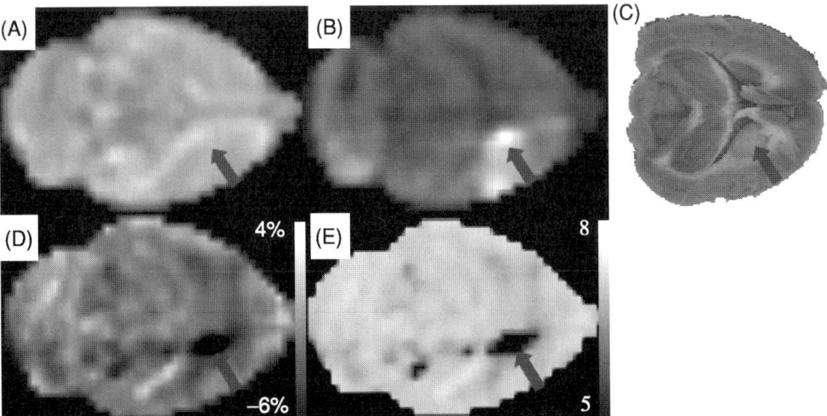

Figure 2 Ischemic areas in rat brain revealed by magnetization transfer ratio (MTR) imaging. Comparison of T2-weighted (A), isotropic diffusion weighted (B), absolute MTR$_{asym}$ (C), 2,3,5 triphenyl tetrazodium chloride (TTC) stained (D), and absolute pH (E) images for an ischemic rat brain. Focal ischemia was introduced by middle cerebral artery occlusion in seven rats. These injuries typically produce ischemia in the caudate nucleus and the pH-sensitive images show a clear decrease of signal intensity in the caudate nucleus (arrow) that is not evident in the T2-weighted image. Contrast is highest for absolute pH images (E) that are calibrated against intracellular pH values obtained by ^{31}P. From Zhou *et al.*, 2003, with permission.

Recently much interest has been directed toward the use of nongadolinium lanthanides as useful centers for CEST agents (Aime *et al.*, 2002a,b; Zhang and Sherry, 2003; Zhang *et al.*, 1999, 2001, 2002, 2003b). These agents bind water and create a bound water pool that exchanges with free water protons; protons bound to the lanthanide ions are chemically shifted away from the free pool. The shift enhances the effect on contrast compared to endogenous macromolecules, which typically have resonances quite close to that of bulk water, by reducing the probability of unintentional saturation of bulk water with the off-resonance pulse. If bulk water is itself already saturated, this decreases the amount of detectable exchange events and thus, signals. The primary desirable factor for lanthanide CEST contrast agents appears to be a slow rate of water exchange. Most Ln^{3+} ions meet the conditions for slow exchange at 4.7 T; however, at typical clinical field strengths (1.5 T) only Eu^{3+}, Tb^{3+}, $Dy,^{3+}$ and Ho^{3+} have $\Delta\omega T_M > 1$, where $T_M =$ lifetime for water exchange and $\Delta\omega =$ difference in frequency between bound and free pools of protons (Zhang and Sherry, 2003; Zhang *et al.*, 2003a). Nongadolinium lanthanides can be stabilized by the same types of chelators used for gadolinium ions and many new CEST agents are composed of DOTA-based ligands.

Lanthanide-based CEST pH reporters have also been described (Aime *et al.*, 2002a,b). Aime and colleagues investigated a series of Ln(III) chelates of tetraglycineamide-derivatized DOTA (DOTAM-Gly). The researchers report both a ratiometric method using a combination of Eu(III) and Yb (IIII) chelates and "single-molecule" pH-responsive agents of Pr(III), Nd (III), and Eu(III) chelates of DOTAM-Gly. The mixture of chelates was chosen based on the degree of the saturation transfer effect between two chemically different exchanging pools, namely, the amide protons on the Yb (III) complex and metal-coordinated water on the Eu(III) complex. These agents are still under development, but there is great interest in using ratiometric methods as the efficiency of saturation transfer can be made independent of the concentration of agent (Terreno *et al.*, 2004). The advantage of the single ratiometric agent is that lower concentrations can be used (Aime *et al.*, 2002b). While none of these pH-sensitive agents has yet been used for developmental studies, they represent a new class of agents whose utility for MRI is only recently being explored and that offer the ability to probe the biochemical environment in a living system.

III. Emerging Contrast Agent Applications in Model Systems

A. Anatomical Mapping

One of the earliest applications of MRI in the study of development in model systems was for generating anatomical atlases of developing embryos. Anatomical atlases provide detailed structural information about developing systems over time. In conjunction with molecular and biochemical studies, atlases allow correlation of structural changes with changes in gene or protein expression during development. As noted earlier, MRI offers the ability to probe structural information in living, thick specimens that are refractory to optical methods. The benefits of noninvasive imaging by MRI have been exploited to document mouse and human embryonic development (Dhenain *et al.*, 2001; Jacobs *et al.*, 2003; Smith *et al.*, 1996, 1999). Although these studies do not require contrast agent inside cells, iron-containing agents are sometimes used in the solution bathing the embryos to reduce ^{1}H background signal. Highly detailed atlases can be generated and the data analyzed and presented in a number of ways. For example, a site developed by researchers at the California Institute of Technology (http://mouseatlas. caltech.edu) presents a digital atlas of the mouse. Figure 3 illustrates views of a reconstructed mouse embryo from this atlas. A similar site has been developed for sharing information about human embryo (http://embryo. soad.umich.edu/). In this study, 10 stages of developing human embryos were imaged by MRI; slices through the embryos can be selectively viewed and completed rendering and movies can be viewed.

Contrast agents have come into play in developmental anatomical imaging primarily for imaging vasculature. Perfusion of contrast agent has been used to image microvasculature in various animal systems such as the chick (Smith *et al.*, 1992; Zhang *et al.*, 2003c), mice (Chapon *et al.*, 2002), and rats (Hamilton *et al.*, 1994). Some studies rely on perfusion of a contrast agent that cannot diffuse out of the blood vessel, such as Gd-DTPA-BSA (Gd–diethyle-netriaminepentaacetic acid–bovine serum albumin) delivered at the same time as a fixative (Zhang *et al.*, 2003c); others have used $GdCl_3$ salts (in warmed 10% gelatin solution) (Smith *et al.*, 1992). Both methods allow detailed imaging of microvascular and cardiac structures in the embryo systems. Figure 4 gives an example of an MRI image compared to the corresponding histological section. The boxed area in parts (A) and (C), showing the single heart chamber in a day 4 chick embryo, is shown in close-up view in (B) and (D). Parts (A) and (B) give histological sections, while (C) and (D) show the corresponding MRI in the same embryo. Details such as the presumptive atrium (a), atrioventricular endocardial cushions (ec arrows), and the presumptive ventricle (v) can be identified quite well in the MR images.

In an interesting application of perfusion imaging, Chapon *et al.* (2002) have examined *in utero* development of mice over time by perfusing contrast agent through the mother's tail vein. Dextran iron oxide nanoparticles were used for these studies revealing structural detail in developing organs such as heart and skeletal system. The authors note that this is far from the quality of traditional histological slices, but at the resolution of the MRI (195 μm), one could envision using MRI to phenotype embryos *in utero* to characterize trangenics.

Vascular imaging has also been harnessed to image embryo implantation as a possible diagnostic tool for fertility programs (Hamilton *et al.*, 1994).

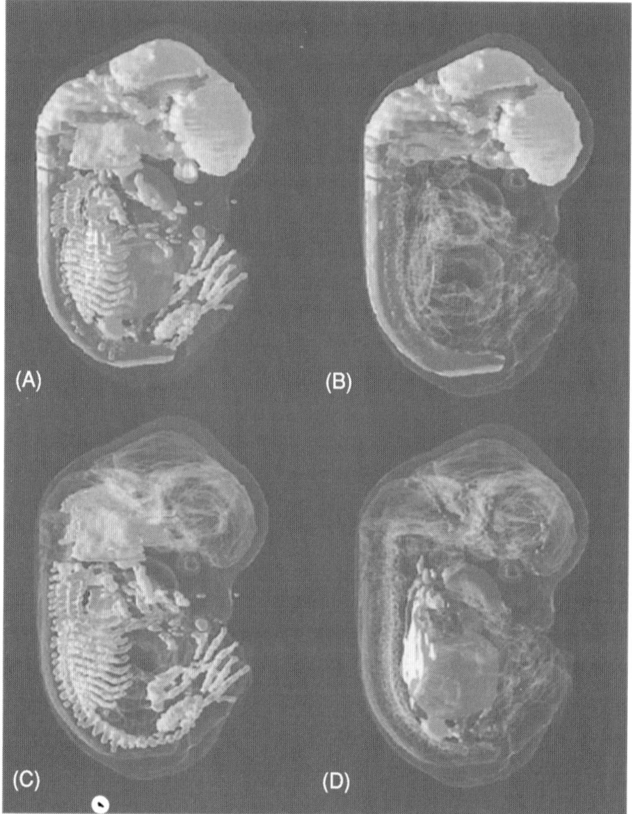

Figure 3 Customizable views of 3D models from the Caltech mouse atlas. The Web-accessible mouse atlas developed by researchers from Caltech allows the users to view specific organ systems. For example the user can view the entire reconstructed embryo with all organ systems (A), or request specific views of the central nervous system (B), skeletal system, (C) or combinations of systems (D). The enteric, pulmonary, and circulatory systems are shown. From Dhenain *et al.*, 2001, with permission.

Using the rat model, investigators imaged the uterine horns before and after introduction of gadopentate meglumine through a femoral venous catheter. Vascular changes induced by embryo implantation resulted in distinct punctate patterns of enhancement in MR images resembling "peas in a pod" and correlated with histological staining for embryo implantation.

B. Cell/Lineage Tracking

The observation of cell movements and cell lineages is key to morphological studies of embryogenesis. Optical imaging in conjunction with the use of fluorescent probes has revealed many key developmental phenomena, such

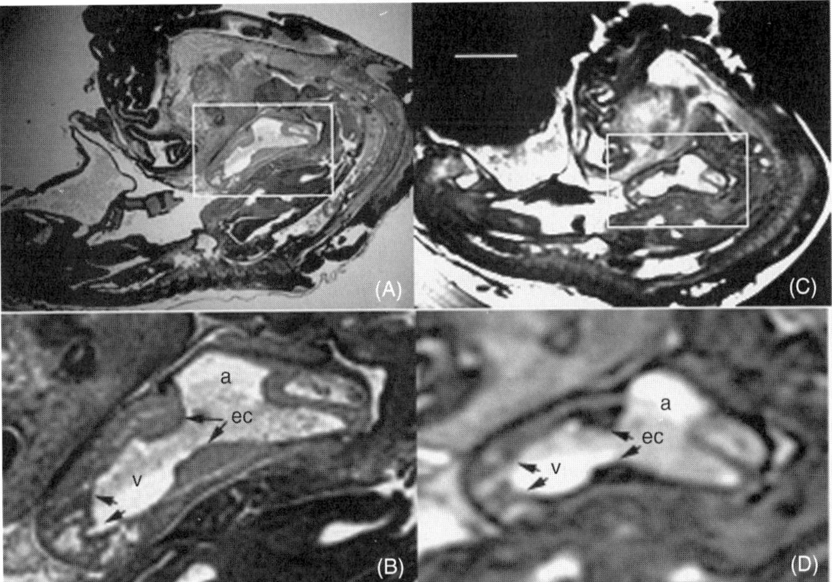

Figure 4 Comparison of histological sections with MR images from same chick embryo. Histological and MRI sections from a 1-day-old chick embryo (parts (A) and (C), respectively) with close-up view of the heart ([B] and [D]). MR microscopy view of the heart (D) allows identification of presumptive ventricle (v), presumptive atrium (a), and atrioventricular endocardial cushions (ec) with good resolution. From Zhang *et al.*, 2003c, with permission.

as gastrulation. However, optical methods are largely limited to early events, or surface events, because light cannot access the interior reaches of most embryos. As the resolution of MRI has improved to cellular levels, the use of MRI to noninvasively track cell movements and lineages in larger, opaque embryos has been explored.

Jacobs and Fraser first proposed tracking of cell lineages using MRI in 1994 (Jacobs and Cherry, 2001). Using a 7-T MRI system and magnetic field gradients several orders of magnitude larger than those used for clinical imaging, they were able to achieve of order of 10-μm scale resolution. In this work, they injected Gd(III)-based MRI tracers in a single blastomere of 16 cell *Xenopus laevis* embryos, then imaged the embryo over time to follow the descendants of the injected cell during subsequent gastrulation and neurulation to 98 hours after fertilization. Images from these studies are shown in Fig. 5. Labeled cells have the highest opacity, while cavities are transparent and unlabeled cells are rendered semi-transparent. Their results revealed that the mesoderm and surface ectoderm do not move simultaneously during convergent-extension movements during formation of the embryonic axis. By being able to observe movements of the underlying

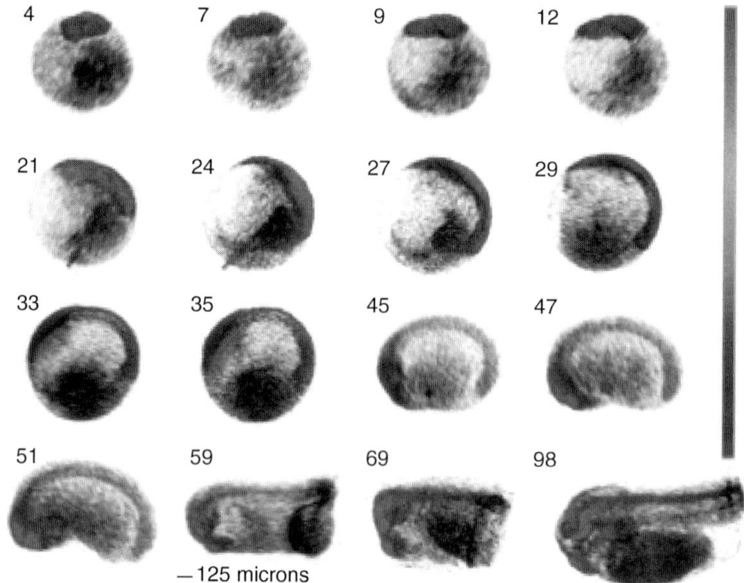

Figure 5 MRI of developing *Xenopus* embryo. Images of a developing *Xenopus* embryo were taken at 4, 7, 9, 12, 21, 24, 27, 29, 33, 35, 45, 47, 51, 59, 69, and 98 hours after fertilization. A single cell of a 16-cell blastomere was injected with contrast agent. 4–12, Development of blastula; 21–35, gastrulation; 36–98, neurula to early tailbud stage. Descendants of the injected cell are visible as higher-intensity regions. The opacity of each voxel was adjusted so that higher-intensity voxels are of exponentially higher opacity; this allows visualization of labeled cells against background that is more transparent by this adjustment. From Jacobs *et al.*, 1994, with permission.

mesoderm, these researchers found that the timing of the mesoderm and surface ectoderm movements differed from what was previously assumed to be extension in concert at this stage of development. These observations demonstrated the power of MRI to probe deeper tissues than could be seen by light microscopy. Other work from the same laboratory used a contrast agent that was both fluorescent and magnetic, allowing corroboration with fluorescence microscopy for early-stage embryos (Hueber *et al.*, 1998).

One of the challenges of imaging cell trafficking is to load cells with sufficient amounts of contrast agent so that very few, or even single cells can be detected in the MR image. The above studies require injection of tracer directly into the observed cell. Alternatively, exogenous cells may be labeled *ex vivo* and transplanted into the system of interest. The first report of MRI tracking of transplanted cells was described by Yeh *et al.* (1995). In that early work, rat T cells were labeled with USPIO and introduced into a rat model in which tissue inflammation had been induced in the testes. MR images were able to identify migration of labeled T cells to the site of inflammation.

Many studies in recent literature reports are applying *ex vivo* labeling methods, followed by transplantation, to track stem cell migration. For example, MRI has been used to track stem cell movements in cardiac applications (Hill *et al.*, 2003; Kraitchman *et al.*, 2003; Rickers *et al.*, 2004), in the brain (Bulte *et al.*, 2003; Modo *et al.*, 2002, 2003), and in muscle (Walter *et al.*, 2004). Figure 6 illustrates tracking of neural precursor cells that were labeled with iron oxide nanoparticles *ex vivo* and stereotactically injected into ventricles of Lewis EAE (experimental allergic encephalomyelitis) rats. Figure 6 shows the distribution of labeled cells in various MR slice views in the fixed rat brain. The cells were labeled after 8–9 days in culture using either dextran-coated iron oxide particles that were conjugated to anti-rat transferrin monoclonal antibodies, or dendrimer-coated iron oxide particles (nonspecific). Animals were imaged 1 week after transplantation. Results indicated that both types of agents worked equally well. The majority of stem cell labeling in the literature is done with iron oxide nanoparticle agents, but dual-labeled polymers carrying magnetic

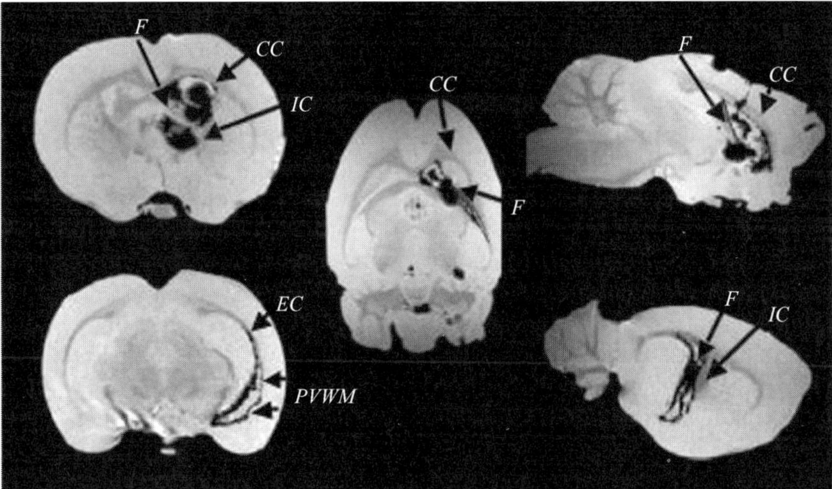

Figure 6 MRI tracking of transplanted neural precursor cells in experimental autoimmune encephalitis (EAE) rat model. Neural precursor cells were labeled with iron oxide nanoparticles stereotactically injected to the lateral ventricle of EAE Lewis rats at the peak of disease (and normal and nondiseased animals as control; only diseased animals are shown in this figure). Slices from MR images show that labeled cells are localized to the corpus callosum (CC), external capsule (EC), fimbria (F), internal capsule (IC), and periventricular white matter (PVWM). The distribution pattern revealed by MRI is similar to that obtained by traditional histological methods. From Bulte *et al.*, 2003, with permission.

and fluorescent moieties have also been described (Modo *et al.*, 2002, 2003). It is beyond the scope of this review to analyze the numerous nanoparticle variants that have been reported in the literature, but readers should be aware of the many choices of surface-modified nanoparticles available for cell tracking experiments.

One recent report was able to distinguish single large iron oxide particles ($\sim$1 μm) in developing mouse embryos (Shapiro *et al.*, 2004). In this approach only a single large particle is needed to render the cell magnetically distinct in the MR image; In contrast, cell tracking with the smaller USPIO or SPIO particles requires large numbers of particles per cell, which are diluted with each cell division. In these investigations micron-sized particles were injected to the pronucleus or cytoplasm of collected transgenic mouse embryos, and then the embryos were transferred to pseudopregnant recipients for further development. Figure 7 clearly shows isolated punctuate spots, which are presumably single particles, within the fixed E11.5 embryos. The embryo studies, and those of cultured cells, indicate that resolution and tracking of single cells is feasible. Advancements in imaging probes and imaging techniques will continue to push the limits of MRI to track single cells *in vivo*.

C. Neuronal Tract Tracing

Recently an exciting field of MRI has developed around the discovery that $MnCl_2$ is axonally transported along neurons. It is believed that Mn^{2+} enters excitable cells through voltage-gated Ca^{2+} channels (Lin and Koretsky, 1997). The Mn^{2+} tends to accumulate along the signaling pathway that responds to the applied stimulus (Pautler and Fraser, 2003; Tjalve *et al.*, 1995). Mn^{2+} *E*nhanced *M*agnetic *R*esonance *I*maging (MEMRI) takes advantage of the paramagnetism of Mn(II) to allow T1-weighted imaging of the labeled neuronal tracts. Using this technique, investigators have been able to generate three-dimensional maps of a number of neural pathways in mammalian systems. In an early MEMRI study, Pautler *et al.* (1998) delivered $MnCl_2$ intranasally and was able to image the olfactory pathway from the olfactory epithelium to the olfactory bulbs and cortex in a rodent model. This work demonstrated the feasibility of using Mn^{2+} to label axons emanating from one sensory tissue up to its termination in the brain. Exposure to odorant with aerosoled Mn^{2+} resulted in localized accumulation of Mn^{2+} as seen in Figure 8. The location of accumulation correlates with previous data obtained using blood oxygen level–dependent (BOLD) functional magnetic resonance imaging (fMRI) (Pautler and Koretsky, 2002).

Mn^{2+} also can be introduced by direct injection. For example, investigators have used MEMRI to image tracing of brain pathways after focal

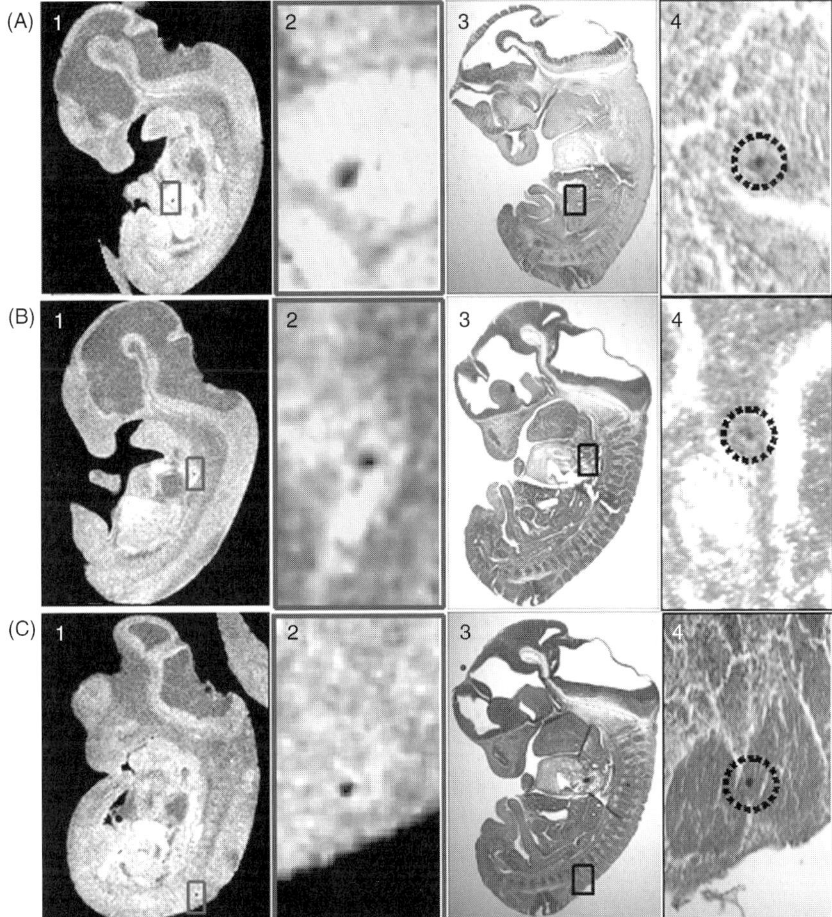

Figure 7 MRI and histology of large Bangs particles in E11.5 mouse embryos. Bangs particles (micron-diameter iron oxide particles) were injected into the pronucleus or cytoplasm of collected transgenic mouse embryos. Approximately 50–100 particles were injected per embryo. Individual particles are visible in E11.5 embryos. (A1–C1) Slices from MR data sets; (A2–C2) expansions of boxed regions in (A1–C1) showing contrast enhancement; (A3–C3) matching histological sections from same plane as MR images; (A4–C4) close-up views of histological sections corresponding to (A2–C2). From Shapiro *et al.*, 2004, with permission.

cortical injections (Allegrini and Weissner, 2003; Leergaard *et al.*, 2003), to inject specific brain regions such as the striatum and amygdala (Pautler *et al.*, 2003), or as intraocular injections (Watanabe *et al.*, 2001). Stereotaxic injection to the dorsal striatum in the mouse model showed enhancement of the globus pallidus and prefrontal cortex, regions of the brain recognized to be

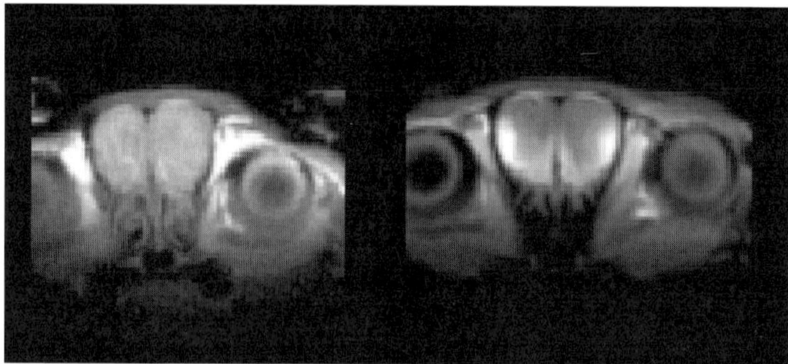

Figure 8 Enhancement of accessory olfactory bulb after exposure to amyl acetate and Mn^{2+}. The 6- to 8-week-old FVB mice were exposed to amyl acetate and Mn^{2+} delivered through a humidifier. Images show mouse exposed to Mn^{2+} alone (left) and Mn^{2+} plus amyl acetate. Co-exposure with amyl acetate reveals localized accumulation of Mn^{2+} in the accessory olfactory bulb (regions of positive contrast enhancement). From Pautler *et al.*, 2002, with permission.

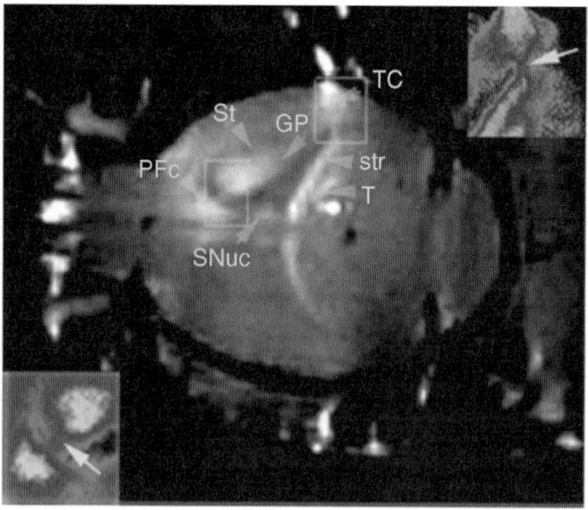

Figure 9 MRI section from mouse injected with Mn^{2+} to dorsal striatum. Tracings from the striatum show projections to the globus pallidus (GP), prefrontal cortex (PFc), the stria terminalis (str), and part of the thalamus (T) and temporal cortex (TC). From Pautler *et al.*, 2003, with permission.

connected to the striatum (Fig. 9); tracings could be followed for up to five synapses. Intracortical injections to the somatosensory cortex (Leergaard *et al.*, 2003) and sensorimotor cortex (Allegrini and Weissner, 2003) in healthy and lesioned rats visualized major projection pathways. MRI also allows noninvasive mapping of neuronal connections of higher-level animal models such as primates (Saleem *et al.*, 2002). Similar to studies in rodent models, injections of $MnCl_2$ to the caudate and putamen resulted in signal enhancement in the globus pallidus and substantia nigra (Fig. 10), which are regions of the brain known to receive projections from the injected areas. Intraocular injections of Mn^{2+} (to vitreous body) enhanced elements of the visual pathway in the rat (Watanabe *et al.*, 2001). In Fig. 11 the injected left eye shows enhancement of the retina (1), optic nerve (2), optic chiasm (3), right optic tract (4), right geniculate nucleus (5), right brachium of the superior colliculus (6), right pretectal region (7), and the right superior colliculus (8), very clearly illustrating contralateral projection enhancement with submillimeter resolution on a 2.35-T MRI system. Although the sensitivity of MRI is not as high as for histological techniques, the ability to study connections in living animals is of enormous value. In addition, studies indicate that Mn^{2+} washes out after 10 days, suggesting that connections can be studied up to a week and then probed over extended time periods by repeat injection without accumulation of Mn^{2+} to toxic levels.

Systemic introduction of $MnCl_2$ by intravenous, subcutaneous, or intraperitoneal injection demonstrated the ability of Mn^{2+} to cross the blood–brain barrier (BBB) and accumulate in active regions of the brain. Early studies by Lin and Koretsky (1997) demonstrated accumulation in the brain after intravenous injection. Subcutaneous injection resulted in pronounced signal enhancement in the brain 24 hours after injection of $MnCl_2$ to axillary adipose tissues in mice (Watanabe *et al.*, 2002). Intraperitoneal injections to neonatal mice resulted in signal enhancement in ventricles, olfactory bulb, hippocampus, and cerebellum (Wadghiri *et al.*, 2004). These methods show the promise for using MEMRI to study behavioral studies to map brain function in awake animals or for analyzing brain development. While the BOLD contrast mechanism can give functional information, BOLD effects only occur as tissues are actively stimulated while the animal is constrained in the MR instrument; this makes the observation of some behaviors difficult. The persistence of the Mn^{2+} along tracts after uptake would allow imaging *after* the behavioral event. While there are artifactual concerns to consider while designing experiments, MEMRI represents a promising methodology to map behavioral centers in the brain that cannot currently be observed otherwise.

MEMRI can also be applied to observe dynamic activity much like BOLD fMRI. Known as dynamic manganese-dependent contrast, time-resolved MEMRI has been applied to correlate behavior with specific brain regions in several animal systems. Observations of canaries exposed to conspecific song revealed that projection neurons from two specific regions of the brain are differentially activated in response to auditory stimuli (Tindemans *et al.*, 2003). Dynamics were followed by repeated injections of Mn^{2+} (through a stereotaxic cannula) in the presence or absence of song over the course of 8 hours. Like BOLD, this method requires confinement of the subject in

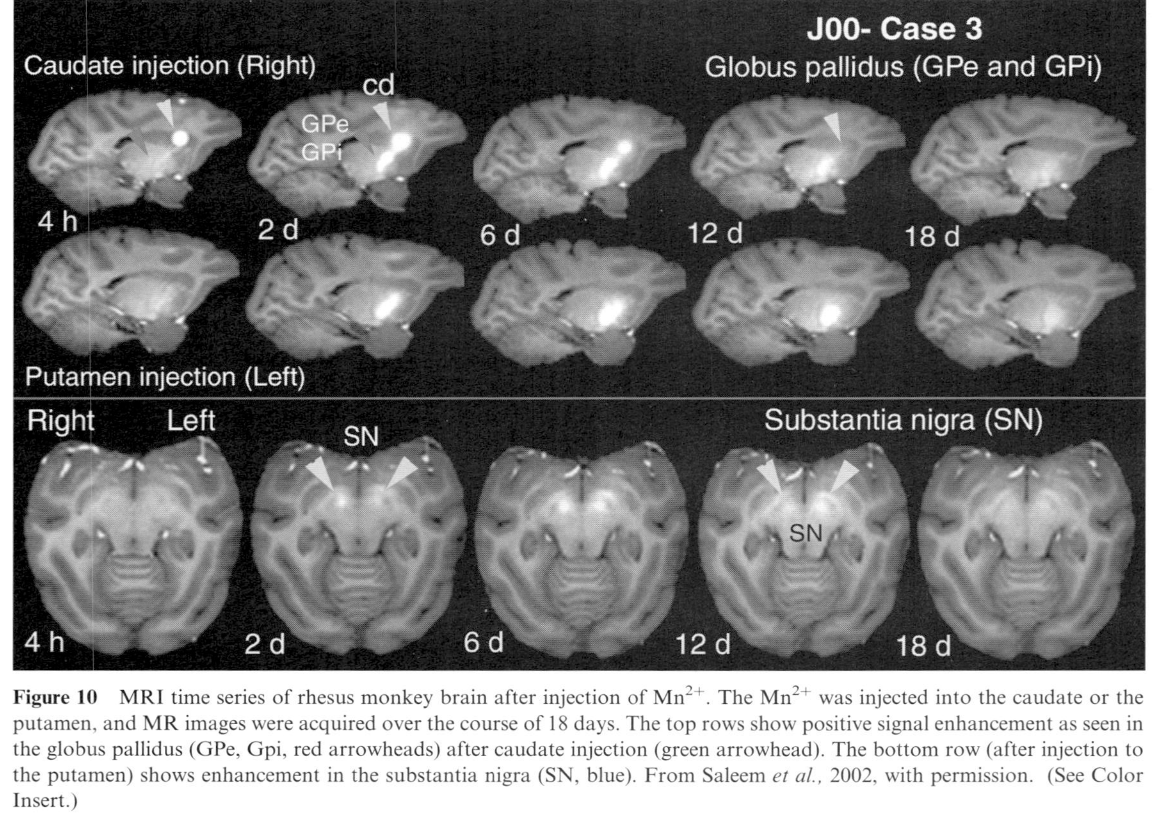

Figure 10 MRI time series of rhesus monkey brain after injection of Mn^{2+}. The Mn^{2+} was injected into the caudate or the putamen, and MR images were acquired over the course of 18 days. The top rows show positive signal enhancement as seen in the globus pallidus (GPe, Gpi, red arrowheads) after caudate injection (green arrowhead). The bottom row (after injection to the putamen) shows enhancement in the substantia nigra (SN, blue). From Saleem *et al.*, 2002, with permission. (See Color Insert.)

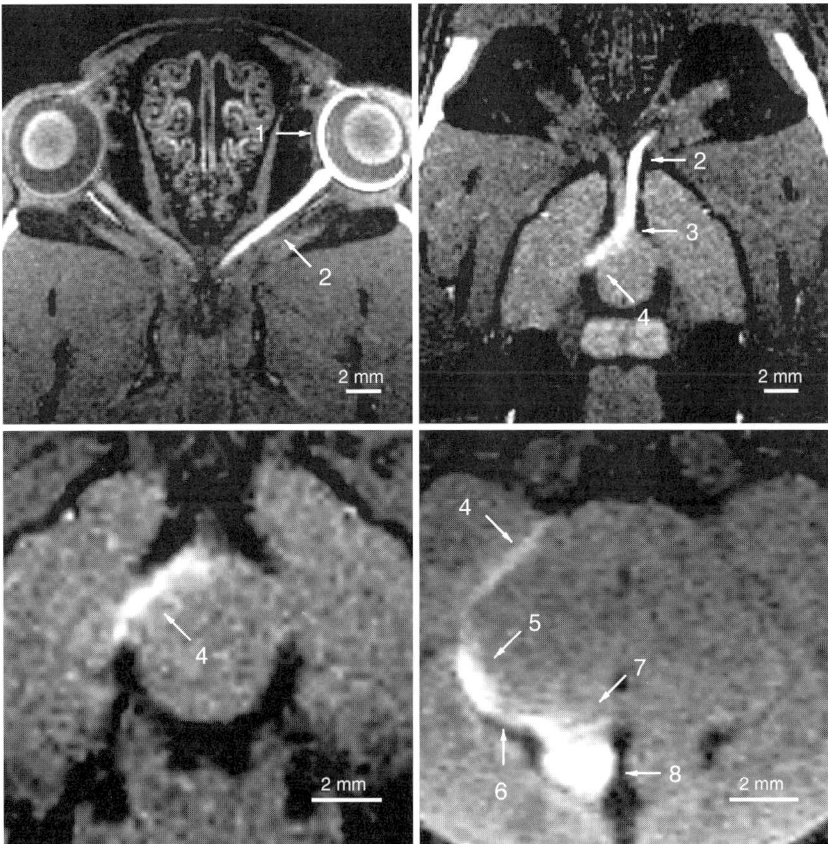

Figure 11 Signal enhancement of the rat visual pathway. Intraocular injection of Mn^{2+} to the vitreous of the left eye of rats was performed under microscopic guidance. Oblique sections of MR images in wide and close up-view show enhancement of (1) left retina, (2) left optic nerve, (3) optic chiasm, (4) right optic tract, (5) right lateral geniculate nucleus, (6) right brachium of the superior colliculus, (7) right pretectal region, and (8) right superior colliculus. From Watenabe *et al.*, 2001, with permission.

the magnet during the behavioral activity and is therefore limited to certain types of stimulation. In addition, long-term cannula placement, while effective, is less than desirable for many applications. Other methods for delivery of Mn^{2+} in dynamic studies have been investigated, including infusion of Mn^{2+} into the cerebrospinal space without breaking the BBB (Liu *et al.*, 2004) and intra-arterially after the BBB had been disrupted with a hyperosmolar agent. After administration of Mn^{2+} animals are awake during most of the uptake period for behavioral stimulation. The Mn^{2+} remained primarily in cerebrospinal fluid spaces and was observed to wash out after about 24 hours (Liu *et al.*, 2004). In other studies, intra-arterially delivered Mn^{2+} was infused before, during, and after functional stimulation (Aoki *et al.*, 2002). Functional maps were corrected against images obtained during nonspecific stimulation. Corrected maps were found to better delineate specific regions of the brain that were activated by stimulation.

D. Molecular Imaging

The most recent advancement in MRI has been the development of novel contrast agents that are targeted, activatable, or multimodal. There are numerous extensive reviews on these topics available in the literature (Aime *et al.*, 2003; Bogdanov and Weissleder, 1998; Glogard *et al.*, 2003; Hanaoka *et al.*, 2002; Hogemann and Basilion, 2002; Jacques and Desreux, 2002; Louie *et al.*, 2002; Louie and Meade, 2000; Modo and Williams, 2003; Zhang *et al.*, 1999) (also see Chapter 1 of this volume). These reviews describe agents that are turned on by specific biochemical parameters such as pH, temperature, oxygen level, metal ion concentration, or enzyme activity. These novel agents extend the utility of MRI beyond anatomical mapping and will increase the use of MRI for developmental biological research.

IV. Conclusion

The studies highlighted here serve as examples of the many applications for MRI and the use of contrast agents in the study of biology. These contrast agents represent tools that could be applied to developing systems to probe cell movement, tissue rearrangements, and establishment of neural connections. As the resolution of MRI improves, we may be able to probe the movement of single cells in developing mammalian embryos *in utero*. This would be a powerful tool for revealing the mysteries of mammalian development in real time over extended periods. Traditional optical methods have illuminated developmental processes in externally developing, transparent embryos but fail when the embryos become opaque to light due to size or pigmentation. MRI does not suffer from this limitation. The studies here hint at the future applications that will be possible as MRI contrast agents, imaging systems, and methodologies continue to improve.

References

Aime, S., Barge, A., Castelli, D., Fedeli, F., Mortillaro, A., Nielsen, F., and Terreno, E. (2002a). Paramagnetic lanthanide(III) complexes as pH-sensitive chemical exchange saturation transfer (CEST) contrast agents for MRI applications. *Magn. Reson. Med.* **47,** 639–648.

Aime, S., Castelli, D., and Terreno, E. (2002b). Novel pH-reporter MRI contrast agents. *Angew Chem. Int. Ed. Engl.* **41,** 4334–4336.

Aime, S., Dastru, W., Crich, S., Gianolio, E., and Mainero, V. (2003). Innovative magnetic resonance imaging diagnostic agents based on paramagnetic Gd(III) complexes. *Biopolymers* **66,** 419–428.

Allegrini, P., and Weissner, C. (2003). Three-dimensional MRI of cerebral projections in the rat brain *in vivo* after intracortical injection of MnCl2. *NMR Biomed.* **16,** 252–256.

Aoki, I., Tanaka, C., Takegami, T., Ebisu, T., Umeda, M., Fukunaga, M., Fukuda, K., Silva, A., Koretsky, A., and Naruse, S. (2002). Dynamic activity-induced manganese-dependent contrast magnetic resonance imaging (DAIM MRI). *Magn. Reson. Med.* **48,** 927–933.

Bogdanov, A., and Weissleder, R. (1998). The development of *in vivo* imaging systems to study gene expression. *TIBTECH* **16,** 5–10.

Bowen, C., Zhang, X., Saab, G., Gareau, P., and Rutt, B. (2002). Application of the static dephasing regime theory to superparamagnetic iron-oxide loaded cells. *Magn. Reson. Med.* **48,** 52–61.

Bulte, J., BenHur, T., Miller, B., Mizrachi Kol, R., Einstein, O., Reinhartz, E., Zywicke, H., and Douglas, T. (2003). MR microscopy of magnetically labeled neurospheres transplanted into the Lewis EAE rat brain. *Magn. Reson. Med.* **50,** 201–205.

Chapon, C., Franconi, F., Marescaux, L., Le Jeune, J., and Lemaire, L. (2002). *In utero* time-course assessment of mouse embryo development using high resolution magnetic resonance imaging. *Anat. Embryol.* **206,** 131–137.

Dhenain, M., Ruffins, S., and Jacobs, R. (2001). Three-dimensional digital mouse atlas using high-resolution MRI. *Dev. Biol.* **232,** 458–470.

Glogard, C., Stensrud, G., and Aime, S. (2003). Novel radical-responsive MRI contrast agent based on paramagnetic liposomes. *Magn. Reson. Chem.* **41,** 585–588.

Hamilton, G., Kennedy, T., and Karlik, S. (1994). Early identification of sites of embryo implantation in rats by means of gadolinium-enhanced MR imaging. *J. Magn. Reson. Imaging* **4,** 481–484.

Hanaoka, K., Kikuchi, K., Urano, Y., Narazaki, M., Yokawa, T., Sakamoto, S., Yamaguchi, K., and Nagano, T. (2002). Design and synthesis of a novel magnetic resonance imaging contrast agent for selective sensing of zinc ion. *Chem. Biol.* **9,** 1027–1032.

Hauger, O., Delalande, C., Deminiere, C., Fouqueray, B., Ohayon, C., Garcia, S., Trillaud, H., Combe, C., and Grenier, N. (2000). Nephrotoxic nephritis and obstructive nephropathy: Evaluation with MR imaging enhanced with ultrasmall superparamagnetic iron oxide–preliminary findings in a rat model. *Radiology* **217,** 819–826.

Henkelman, R., Stanisz, G., and Graham, S. (2001). Magnetization transfer in MRI: A review. *NMR Biomed.* **14,** 57–64.

Hill, J., Dick, A., Raman, V., Thompson, R., Yu, Z., Hinds, A., Pessanha, B., Guttman, M., Varney, T., Martin, B., *et al.* (2003). Serial cardian magnetic resonance imaging of injected mesenchymal stem cells. *Circulation* **108,** 1009–1014.

Hogemann, D., and Basilion, J. P. (2002). Seeing inside the body: MR imaging of gene expression. *Eur. J. Nuc. Med.* **29,** 400–408.

Hueber, M., Staubli, A., Kustedjo, K., Gray, M., Shih, J., Fraser, S., Jacobs, R., and Meade, T. (1998). Fluorescently detectable magnetic resonance imaging agents. *Biocong. Chem.* **9,** 242–249.

Jacobs, R., and Cherry, S. (2001). Complementary emerging techniques: High resolution PET and MRI. *Curr. Opin. Neurobiol.* **11**, 621–629.

Jacobs, R., Papan, C., Ruffins, S., Tyszka, J., and Fraser, S. (2003). MRI: Volumetric imaging for vital imaging and atlas construction. *Imag. Cell Biol., (Nature special supplement)*, SS10–SS16.

Jacques, V., and Desreux, J. (2002). New classes of MRI contrast agents. *Top. Curr. Chem.* **221**, 123–164.

Kawaguchi, T., and Hasegawa, M. (2000). Structure of dextran-magnetite complex: Relation between conformation of dextran chains covering core and its molecular weight. *J. Matl. Sci. Matl. Med.* **11**, 31–35.

Kehagias, D., Gouliamos, A., Smyrniotis, V., and Vlahos, L. (2001). Diagnostic efficacy and safety of MRI of the liver with superparamagnetic iron oxide particles (SH U 555 A). *J. Mag. Res. Med.* **14**, 595–601.

Kraitchman, D., Heldman, A., Atalar, E., Amado, L., Martin, B., Pittenger, M., Hare, J., and Bulte, J. (2003). *In vivo* magnetic resonance imaging of mesenchymal stem cells in myocardial infarction. *Circulation* **107**, 2290–2293.

Landfester, K., and Ramirez, L. (2003). Encapsulation of magnetite particles for biomedical application. *J. Phys. Condens Matter* **15**, S1345–S1361.

Lauffer, R. (1987). Paramagnetic metal complexes as water proton relaxation agents for NMR imaging: Theory and Design. *Chem. Rev.* **87**, 901–927.

Leergaard, T., Bjaalie, J., Devor, A., Wald, L., and Dale, A. (2003). *In vivo* tracing of major rat brain pathways using manganese-enhanced magnetic resonance imaging and three-dimensional digital atlasing. *Neuroimage* **20**, 1591–1600.

Liu, C., DArceuil, H., and deCrespigny, A. (2004). Direct CSF injection of MnCl2 for dynamic manganese-enhanced MRI. *Mag. Res. Med.* **51**, 978–987.

Louie, A., Duimstra, J., and Meade, T. (2002). Mapping gene expression by MRI. *In* "Brain Mapping" (A. a. M. Toga JC, Ed.), pp. 819–828. Elsevier, London, UK.

Louie, A., and Meade, T. (2000). Recent advances in MRI:Novel contrast agents shed light on *in vivo* biochemistry. *New Technologies in the Life Sciences: A Trends Guide,* 7–11.

Modo, M., Cash, D., Mellodew, K., Williams, S., Fraser, S., Meade, T., Price, J., and Hodges, H. (2002). Tracking transplanted stem cell migration using bifunctional, contrast agent-enhanced, magnetic resonance imaging. *Neuroimage* **17**, 803–811.

Modo, M., Mellodew, K., Cash, D., Fraser, S., Meade, T., Price, J., and Williams, S. (2003). Mapping transplanted stem cell migration after a stroke: A serial, *in vivo* magnetic resonance imaging study. *Neuroimage* **21**, 311–317.

Modo, M., and Williams, S. (2003). MRI and novel contrast agents for molecular imaging. *In* "Biomedical Imaging in experimental neuroscience" (M. a. R. T. VanBruggen, Ed.), pp. 293–322. CRC Press, Boca Raton, FL.

Paul, K., Frigo, T., Groman, J., and Groman, E. (2004). Synthesis of ultrasmall super-paramagnetic iron oxides using reduced polysaccharides. *Bioconj. Chem.* **15**, 394–401.

Pautler, R., and Fraser, S. (2003). The year(s) of the contrast agent-microMRI in the new millenium. *Curr. Opin. Immunol.* **15**, 385–392.

Pautler, R., and Koretsky, A. (2002). Tracing odor-induced activation in the olfactory bulbs of mice using manganese-enhanced magnetic resonance imaging. *Neuroimage* **16**, 441–448.

Pautler, R., Mongeau, R., and Jacobs, R. (2003). *In vivo* trans-synaptic tract tracing from the murine striatum and amygdala utlizing manganese enhanced MRI (MEMRI). *Mag. Res. Med.* **50**, 33–39.

Pautler, R., Silva, A., and Koretsky, A. (1998). *In vivo* neuronal tract tracing using manganese-enhanced magnetic resonance imaging. *Mag. Res. Med.* **40**, 740–748.

Rickers, C., Gallegos, R., Seethamraju, R., Wang, X., Swingen, C., Jayaswal, A., Rahrmann, E. P., KAstenberg, Z., Clarkson, C., Bianco, R., *et al.* (2004). Applications of magnetic resonance imaging for cardiac cell therapy. *J. Interv. Cardiol.* **17,** 37–46.

Saleem, K., Pauls, J., Augath, M., Trinath, R., Prause, B., Hashikawa, T., and Logothesis, N. (2002). Magnetic resonance imaging of neuronal connections in the macaque monkey. *Neurotechnique* **34,** 685–700.

Smith, B., Elffmann, E., and Johnson, G. (1992). MR Microscopy of chick embryo vasculature. *JMRI* **2,** 237–240.

Smith, B., Huff, D., and Johnson, G. (1999). Magnetic resonance imaging of embryos: And Internet resource for the study of embryonic development. *Comput. Medi. Imag. Graph.* **23,** 33–40.

Smith, B., Linney, E., Huff, D., and Johnson, G. (1996). Magnetic Resonance Microscopy of Embryos. *Comput. Med. Imag. Graph.* **20,** 482–490.

Snoussi, K., Bulte, J., Gueron, M., and vanZijl, P. (2003). Sensitive CEST agents based on nucleic acid imino proton exchange: Detection of poly(rU) and of a dendrimer-poly(rU) model for nucleic acid delivery and pharmacology. *Magn. Reson. Med.* **49,** 998–1005.

Terreno, E., Castelli, D., Cravotto, G., Milone, L., and Aime, S. (2004). Ln(III)-DOTAMGly complexes: A versatile series to assess the determinants of efficacy for paramagnetic chemical exchange saturation transfer agents for magnetic resonance imaging applications. *Invest. Radiol.* **39,** 235–243.

Tindemans, I., Verhoye, M., Balthazart, J., and Vander Linden, A. (2003). *In vivo* dynamic ME-MRI reveals differential functional responses of RA- and area X-projecting neurons in the HVC of canaries exposed to conspecific song. *Eur. J. Neurosci.* **18,** 3352–3360.

Tweedle, M. F. (1992). Physicochemical properties of gadoteridol and other magnetic resonance contrast agents. *Invest. Radiol.* **27,** S2–S6.

Walter, G., Cahill, K., Huard, J., Feng, H., Douglas, T., Sweeney, H., and Bulte, J. (2004). Noninvasive monitoring of stem cell transfer for muscle disorders. *Magn. Reson. Med.* **51,** 273–277.

Ward, D., Aletras, A., and Balaban, R. (2000). A new class of contrast agents for MRI based on proton chemical exchange dependent saturation transfer (CEST). *J. Mag. Res.* **143,** 79–87.

Ward, K., and Balaban, R. (2000). Determination of pH using water protons and chemical exchange dependent saturation transfer (CEST). *Mag. Res. Med.* **44,** 799–802.

Watanabe, T., Michaelis, T., and Frahm, J. (2001). Mapping of retinal projections in the living rat using high-resolution 3D gradient-echo MRI with $Mn2^{+}$-induced contrast. *Mag. Res. Med.* **46,** 424–429.

Watanabe, T., Natt, O., Boretius, S., Frahm, J., and Michaelis, T. (2002). *In vivo* 3D MRI staining of mouse brain after subcutaneous application of $MnCl2$. *Mag. Res. Med.* **48,** 852–859.

Yeh, T., Zhang, W., Ildstad, S., and Ho, C. (1995). *In vivo* dynamic MRI tracking of rat T-cells labeled with superparamagnetic iron-oxide particles. *Mag. Res. Med.* **33,** 200–208.

Zhang, S., Merritt, M., Woessner, D., Lenkinski, R. E., and Sherry, A. (2003a). PARACEST agents:Modulating MRI contrast via water proton exchange. *Acc. Chem. Res.* **36,** 783–790.

Zhang, S., Michaudet, L., Burgess, S., and Sherry, A. (2002). The amine protons of an ytterbium(III)dota tetraamide complex act as efficient antennae for transfer of magnetization to bulk water. *Angew. Chem. Int. Ed. Engl.* **41,** 1919–1921.

Zhang, S., and Sherry, A. (2003). Physical characteristics of lanthanide complexes that act as magnetization transfer (MT) contrast agents. *J. Solid State Chem.* **171,** 38–43.

Zhang, S., Trokowski, R., and Sherry, A. (2003b). A paramagnetic CEST agent for imaging glucose by MRI. *JACS* **50,** 15288–15289.

Zhang, S., Winter, P., Wu, K., and Sherry, A. (2001). A novel europium(III)-based MRI contrast agent. *JACS* **123,** 1517–1518.

Zhang, S., Wu, K., and Sherry, A. (1999). A novel pH sensitive MRI contrast agent. *Angew. Chem. Int. Ed. Engl.* **38,** 3192–3194.

Zhang, X., Yelbuz, M., Cofer, G., Chorna, M., Kirby, M., and Johnson, G. (2003c). Improved preparation of chick embryonic samples for magnetic resonance microscopy. *Mag. Res. Med.* **49,** 1192–1195.

Zhou, J., Payen, J., Wilson, D., Traystman, R., and vanZijl, P. (2003). Using the amide proton signals of intracellular proteins and peptides to detect pH effects in MRI. *Nat. Med.* **9,** 1085–1090.

Further Reading

Arbab, A., Yocum, G., Kalish, H., Jordan, E., Anderson, S., Read, E., and Frank, J. (2004). Efficient magnetic cell labeling with protamine sulfate complexed to ferumoxides for cellular MRI. *Blood* **104,** 1217–1223.

Chew, S., Ahmadi, A., Goh, P., and Foong, L. (2001). The effects of 1.5T magnetic resonance imaing on early murine *in-vitro* embryo development. *J. Mag. Res. Imaging* **13,** 417–420.

DaldrupLink, H., Rudelius, M., Metz, S., Piontek, G., Pichler, B., Settles, M., Heinzmann, U., Schlegel, J., Oostendorp, R., and Rummeny, E. (2004). Cell tracking with gadophrin-2: A bifunctional contrast agent for MR imaging, optical imaging, and fluorescence microscopy. *Eur. J. Nuc. Med. Mol. Imaging* **31,** 1313–1321.

Fleige, G., Seeberger, F., Laux, D., Kresse, M., Taupitz, M., Pilgrimm, H., and Zimmer, C. (2002). *In vitro* characterization of two different ultrasmall iron oxide particles for magnetic resonance cell tracking. *Invest. Radiol.* **37,** 482–488.

Halpern, M. (1987). The organization and function of the vomeronasal system. *Ann. Rev. Neurosci.* **10,** 325–362.

Hinds, K., Hill, J., Laukkanen, M., Silva, A., Combs, C., Varney, T., Balaban, R., Koretsky, A., and Dunbar, C. (2003). Highly efficient edosomal labeling of progenitor and stem cells with large magnetic particles allows magnetic resonance imaging of single cells. *Blood* **102,** 867–872.

Josephson, L., Tung, C., Moore, A., and Weissleder, R. (1999). High-efficiency intracellular magnetic labeling with novel superparamagnetic-Tat peptide conjugates. *Bioconj. Chem.* **10,** 186–191.

Kircher, M., Allport, J., Graves, E., Love, V., Josephson, L., Lichtman, A., and Weissleder, R. (2003). *In vivo* high resolution three-dimensional imaging of antigen-specific cytotoxic T-cell lymphocyte tracking to tumors. *Cancer Res.* **63,** 6838–6846.

Lee, S., Kim, K., Kim, J., Lee, S., Yi, J., Kim, S., Ha, K., and Cheong, C. (2001). One micrometer resolution NMR microscopy. *J. Mag. Res.* **150,** 207–213.

Lu, Y., Yin, Y., Mayers, B., and Xia, Y. (2002). Modifying the surface properties of superparamagnetic iron oxide nanoparticles through a sol-gel approach. *Nano Lett.* **2,** 183–186.

3

^{1}H/^{19}F Magnetic Resonance Molecular Imaging with Perfluorocarbon Nanoparticles

Gregory M. Lanza,[*,†] *Patrick M. Winter,*[*] *Anne M. Neubauer,*[*,†]
Shelton D. Caruthers,[*,‡] *Franklin D. Hockett,*[*] *and*
Samuel A. Wickline[*,†]
*Division of Cardiology, Washington University Medical School, St. Louis
Missouri 63110
[†]Division of Bioengineering, Washington University Medical School, St. Louis
Missouri 63110
[‡]Philips Medical Systems, Cleveland, Ohio 44143

Developments in genomics, proteomics, and cell biology are leading a trend toward individualized segmentation and treatment of patients based on early, noninvasive recognition of unique biosignatures. Although developments in molecular imaging have been dominated by nuclear medicine agents in the past, the advent of nanotechnology in the 1990s has led to magnetic resonance (MR) molecular agents that allow detection of sparse biomarkers with a high-resolution imaging modality that can provide both physiological and functional agents.

A wide variety of nanoparticulate MR contrast agents have emerged, most of which are superparamagnetic iron oxide–based constructs. However, this chapter focuses on a diagnostic and therapeutic perfluorocarbon (PFC) nanoparticulate platform that is not only effective as a T1-weighted agent, but also supports ^{19}F MR spectroscopy and imaging. The unique capability of ^{19}F permits confirmation and segmentation of MR contrast images as well as direct quantification of nanoparticle concentrations within a voxel.

Current Topics in Developmental Biology, Vol. 70
Copyright 2005, Elsevier Inc. All rights reserved.

0070-2153/05 $35.00
DOI: 10.1016/S0070-2153(05)70003-X

PFC nanoparticles have the capability to effectively deliver therapeutic agents to target sites by a novel mechanism termed "contact-facilitated drug delivery." Combined with MR spectroscopy, the concentration of drug delivered to the target site can be determined and the expected response predicted. Moreover, mixtures of nanoparticles with different perfluorocarbon cores can provide a quantitative, multispectral signal, which can be used to simultaneously distinguish the relative concentrations of several important epitopes within a region of interest. In conjunction with rapid improvements in MR imaging, the prospects for personalized medicine and early recognition and treatment of disease have never been better. © 2005, Elsevier Inc.

I. Introduction

Until recently, medical diagnosis and therapy has been a one-size-fits-all strategy. However, a paradigm shift to individualize treatment is emerging from technology developments in genomics, proteomics, molecular imaging, and targeted drug delivery, which provide new capabilities to understand, recognize, and characterize early pathology based on unique biosignatures. Although the concept of targeted imaging agents and therapies dates back three decades to the discovery of monoclonal antibody production by Kohler and Milstein (1975), the great promise of molecular imaging agents produced by coupling radionuclide or paramagnetic chelates to immunoglobulins has gone essentially unfulfilled in clinical practice. But the dream and motivation to develop such agents has persisted, almost like a "holy grail" in medicine. From this common beginning, a variety of formulations and imaging techniques have emerged that are tailored to the needs of each specific imaging modality. For instance, in nuclear medicine contrast agents have trended away from large immunoglobulin-based molecules toward peptides, peptidomimetics, and other essentially massless probes, which retain high target specificity and display rapid systemic clearance (Krenning *et al.*, 1992). Conversely, in the field of magnetic resonance imaging (MRI), most new formulations have gravitated toward functionalized nanoparticles, which are visualized with very high spatial resolution (Lanza *et al.*, 1998; Sipkins *et al.*, 1998).

In contradistinction to typical gadolinium-based MR contrast agents (e.g., gadolinium–diethylenetriaminepentaacetate [GD-DTPA]), which highlight vascular contours, targeted nanoparticulate contrast agents, particularly the paramagnetic forms, typically provide little or no blood-pool contrast after injection. Nanoparticulate agents may function to *passively* accumulate within the reticuloendothelial system and provide contrast essentially by creation of a signal void (Hahn and Saini, 1998; Harisinghani *et al.*, 2003; Kim and Harisinghani, 2004; Kooi *et al.*, 2003; Muhler *et al.*, 1995; Ruehm *et al.*, 2002; Saini *et al.*, 1995; Vassallo *et al.*, 1994; Weissleder *et al.*, 1990), or they may be preferentially targeted to unique biochemical signatures of disease via homing ligands, such as antibodies, peptides, aptamers, or peptidomimetics (Anderson *et al.*, 2000; Flacke *et al.*, 2001; Johansson *et al.*, 2001, 2002; Kaufman *et al.*, 2003; Kircher *et al.*, 2003; Sipkins *et al.*, 1998; Winter *et al.*, 2003b,c). Recently superparamagnetic nanoparticles have been incorporated into stem cells or lymphocytes *in vitro* and MR has been used to track the migration and integration of iron-tagged cells into target tissues following implantation or systemic injection (Bulte *et al.*, 2002; Daldrup-Link *et al.*, 2003; Moore *et al.*, 2004). In this chapter, a liquid perfluorocarbon (PFC)-based nanoparticle (Lanza *et al.*, 1996), which is a successful platform technology for molecular imaging and drug delivery, is discussed with significant attention to the important opportunities presented by its high internal concentrations of fluorine.

II. PFC Nanoparticles for MRI

A. A T1-Weighted "Ultraparamagnetic" Contrast Agent

PFC emulsion nanoparticles may be functionalized for targeted MR molecular imaging by the surface incorporation of paramagnetic chelates and homing ligands into the outer phospholipid monolayer (Flacke *et al.*, 2001; Lanza *et al.*, 1998). With a nominal diameter of ~250 nm, PFC nanoparticles present an enormous surface area to transport and concentrate paramagnetic metal to important vascular biomarkers sites. Rather than the 2 to 4 gadolinium ions that could be bound to and delivered with an antibody or peptide, paramagnetic nanoparticles can deliver 50,000 to 90,000+ gadolinium ions each, and in the case of PFC nanoparticles, all of the paramagnetic ions are presented to the outer aqueous phase for maximum relaxivity (Flacke *et al.*, 2001; Lanza *et al.*, 1998; Winter *et al.*, 2003a).

Proton relaxivity of paramagnetic nanoparticulates is described in both ionic and macromolecular contexts. "Ionic relaxivity," universally applied to blood pool agents, is calculated with respect to absolute Gd–chelate concentration. However, for targeted nanoparticulate agents the effective T1 relaxivity of each construct is better reflected as the "particulate" or "molecular relaxivity." PFC nanoparticle molecular relaxivities at 1.5 T are between 1,000,000 $(mM \bullet s)^{-1}$ and greater than 2,000,000 $(mM \bullet s)^{-1}$, dependent on the lipophilic chelate utilized and the paramagnetic loading of the lipid surfactant (Flacke *et al.*, 2001; Winter *et al.*, 2003a).

B. Unique Chemistry of PFC Nanoparticles

In addition to the surface payload of paramagnetic chelates, PFC nanoparticles are 98% PFC by volume, which equates for perfluorooctylbromide (1.98 g/ml, 498 daltons) to approximately 100 M concentration of fluorine within a nanoparticle. PFC nanoparticles are distinctly different from other oil-based emulsions by virtue of the physicochemical properties of fluorine, the most electronegative of all elements. Fluorine has a high ionization potential and very low polarizability (Krafft, 2001). Larger than hydrogen, fluorine creates bulkier, stiffer compounds that typically adopt a helical conformation. The C–F bond is chemically and thermally stable and essentially biologically inert. The dense electron cloud of fluorine atoms creates a barrier to encroachment on the perfluorinated chain by other chemical reagents. The large surface area combined with the low polarizability presented by the fluorinated chains enhances hydrophobicity. Uniquely, perfluorinated chains are extremely hydrophobic and lipophobic simultaneously.

The biocompatibility of liquid fluorocarbons is well documented. Even at large doses, most fluorocarbons are innocuous and physiologically inactive. No toxicity, carcinogenicity, mutagenicity, or teratogenic effects have been reported for pure fluorocarbons within the 460–520-MW range. PFCs have tissue half-life residencies ranging from 4 days for perfluorooctylbromide up to 65 days for perfluorotripropylamine, and are not metabolized, but rather slowly reintroduced to the circulation in dissolved form by lipid carriers and expelled through the lungs. Increased pulmonary residual volumes with blood transfusion level dosages of PFC emulsions have been reported in rabbits, swine, and macaque but not in mouse, dog, or humans (Krafft, 2001).

C. PFC Nanoparticles for ^{19}F Magnetic Resonance Spectroscopy and Imaging

Fluorine is an excellent element for MR spectroscopy and imaging because of the following characteristics:

- ^{19}F is has a gyromagnetic ratio nearly equivalent to proton
- ^{19}F has a spin 1/2 nucleus
- ^{19}F has 100% natural abundance
- ^{19}F has essentially no detectable background concentration.

In addition, since fluorine has seven outer shell electrons, rather than a single electron as with hydrogen, the range and the sensitivity of fluorine chemical shifts to the details of the local environment are much higher for fluorine than hydrogen. As a consequence, ^{19}F MRI has been applied to study tumor metabolism (Ikehira *et al.*, 1999; Schlemmer *et al.*, 1999; Wolf *et al.*, 2000), to map physiologic Po$_2$ tension (Fan *et al.*, 2002; Hunjan *et al.*, 2001; Noth *et al.*, 1999), and to characterize liquid ventilation (Huang *et al.*, 2002; Laukemper-Ostendorf *et al.*, 2002). Unfortunately, most of these studies have required high magnetic field strengths—4.7 T or greater—and/or direct infusion of ^{19}F constructs to compensate for the relatively low fluorine concentrations available to detect.

Others have suggested that choosing a PFC with most of the signal at a single frequency, such as perfluoro-2.2.2'.2'-tetramethyl–4.4'-bis(1.3-dioxolane (Sotak *et al.*, 1993) or perfluoro-15-crown-5-ether (Dardzinski and Sotak, 1994), as well as a prolonged T2 time and no homonuclear fluorine–fluorine coupling can significantly increase detectability of the ^{19}F. A variety of rapid imaging techniques for proton imaging have been adapted for ^{19}F to reduce the total data acquisition time relative to standard spin echo imaging sequences and to address the additional challenges posed by multiple chemical shifts, for example, slice selection and frequency-encoding artifacts.

In the past, investigators using PFC emulsion droplets designed for artificial blood substitutes were plagued by rapid reticuloendothelial clearance by virtue of their micron size and low circulatory concentration. However, the prolonged systemic half-life of PFC nanoparticles in conjunction with the local concentrating effect produced by ligand-directed binding now permit ^{19}F spectroscopy and imaging studies at clinically relevant magnetic field strengths (1.5 T) (Morawski *et al.*, 2004a).

III. Applications in Molecular Imaging

A. Fibrin-Imaging for Detection of Unstable Plaque and Thrombus

PFC nanoparticles have been targeted to a variety of molecular epitopes using biotinylated ligands sandwiched to a biotinylated nanoparticle through avidin–biotin interactions (Anderson *et al.*, 2000) or direct, covalent conjugation of ligands—monoclonal antibodies, F(ab) fragments (Flacke *et al.*, 2001), and peptidomimetics—to the surfactant surface (Winter *et al.*, 2003b,c). Fibrin-targeted nanoparticles densely and specifically adhere to fibrin fibrils along the clot surface, delivering tens of thousands of gadolinium atoms with each bound particle. Using a typical low-resolution clinical ^{1}H imaging protocol, the fibrin clots targeted with nanoparticles provide homogeneous T1-weighted contrast enhancement. The gadolinium-rich nanoparticles overcome the partial volume dilution effect of the low-resolution voxel, and the targeted clot appears completely filled with signal despite restriction

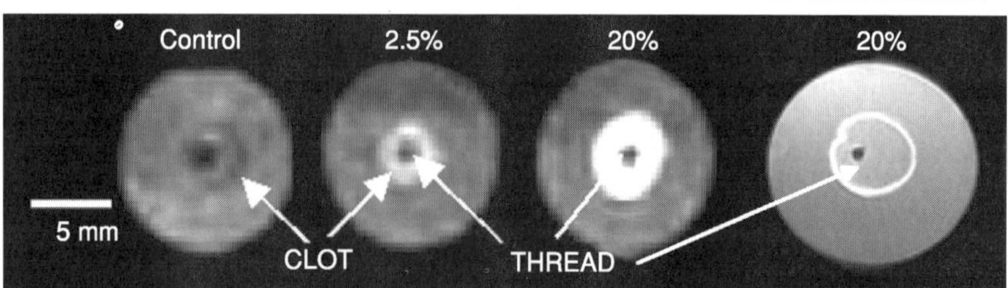

Figure 1   Left, fibrin clots targeted with nanoparticles presenting a homogeneous, T1-weighted enhancement (0.7 mm × 0.7 mm in-plane, 3D gradient spin echo) that improves with increasing gadolinium level (0, 2.5, and 20 mole%). Far right, higher-resolution scan of fibrin clot (3D T1-weighed gradient recalled echo sequence) reveals thin layer of nanoparticles along surface (20 mole% gadolinium) and a decreased voxel size (0.1 mm × 0.1 mm). Reproduced and modified with permission from Flacke *et al.,* 2001.

of the nanoparticles to the surface, which is revealed with higher in-plane resolution images (Fig. 1).

In dogs, gradient echo images of thrombus targeted with anti-fibrin paramagnetic nanoparticles produce high signal intensity (1780 ± 327), whereas, the contralateral control clot had a signal intensity of 815 ± 41, which was similar to that of the adjacent muscle (768 ± 47) (Fig. 2). The contrast-to-noise ratio (CNR) between the targeted clot and blood measured with this sequence was approximately 118 ± 21, whereas the CNR between the targeted clot and the control clot was 131 ± 37. The concept of detecting human ruptured plaque was illustrated *in vitro* using carotid artery endarterectomy specimens from a symptomatic patient in which microscopic fibrin deposits within the ruptured "shoulders" of the plaque were readily apparent in contradistinction to control specimens (Fig. 3).

We have exploited the high fluorine content of fibrin-targeted nanoparticles and the lack of background for ^{19}F imaging and spectroscopy. Fibrin-targeted, paramagnetic perfluoro–crown-ether nanoparticles were mixed in titrated ratios with fibrin-targeted nanoparticles containing safflower oil and bound to plasma clots *in vitro*. A linear and parallel decrease in ^{19}F and the gadolinium signal was measured proportionate to the amount of competing safflower oil agent in the targeted blend. As expected, the number of bound paramagnetic fluorinated nanoparticles, as calculated from the normalized ^{19}F spectroscopic signal, was directly proportional to the measured gadolinium content of the clots (Fig. 4).

In Fig. 5, treatment of a fibrin clot with the crown ether emulsion provided a high number of bound nanoparticles, each composed of a large amount of

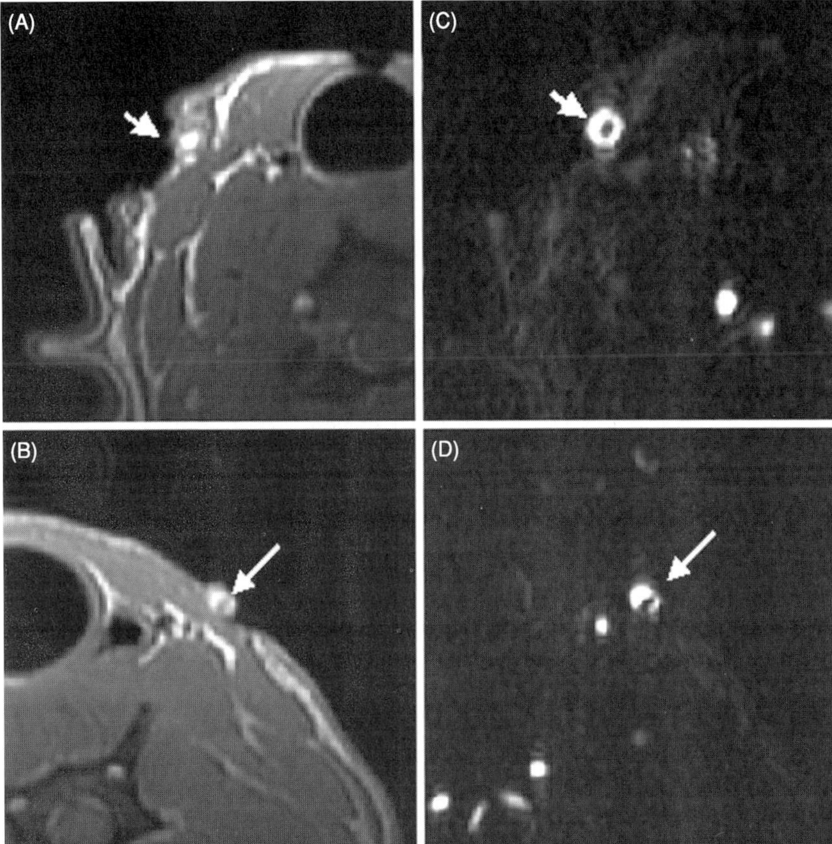

Figure 2 Thrombus imaged with a 3D T1-weighted, fat suppression, fast gradient echo sequence revealing a marked improvement in detectability of targeted clot (A) by the fibrin-specific paramagnetic nanoparticles relative to control thrombus (B). Phase-contrast angiography revealed the clots as flow deficits in the external jugular veins for the targeted (C) and control thrombi (D). Reproduced and modified with permission from Flacke *et al.*, 2001.

PFC, allowing acquisition of high signal-to-noise ratio (20.8) fluorine images at 4.7 T in less than 5 minutes. The corresponding ^{1}H image of the same slice showed that the ^{19}F signal from the bound nanoparticles originated from the clot surface.

Human carotid endarterectomy samples have complex atherosclerotic lesions with several plaques and areas of calcification distributed throughout the vessel (Fig. 6). Multislice ^{1}H images showed high levels of signal enhancement along the luminal surface due to binding of targeted paramagnetic nanoparticles to fibrin deposits. A ^{19}F projection image of the artery,

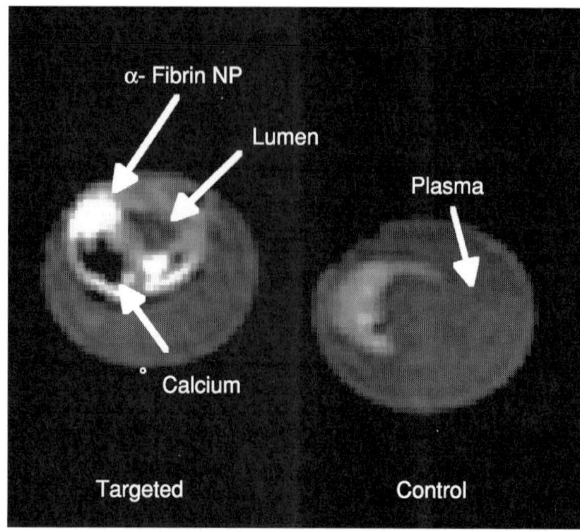

Figure 3 Color-enhanced MRI images of fibrin-targeted and control carotid endarterectomy specimens revealing contrast enhancement (white) of a small fibrin deposit on a symptomatic ruptured plaque. Calcium deposit (black). This image is a 3D, fat-suppressed, T1-weighted fast gradient echo. NP, nanoparticle. Reproduced with permission from Flacke *et al.*, 2001.

acquired in less than 5 minutes, shows an asymmetrical distribution of fibrin-targeted nanoparticles around the vessel wall corroborating the signal enhancement observed with ^{1}H MRI. Spectroscopic quantification of nanoparticle binding allowed calibration of the ^{19}F MRI signal intensity. Co-registration of the quantitative nanoparticle map with the ^{1}H image permits visualization of anatomical and pathological information in a single image. Indeed, combination of the ^{1}H and ^{19}F signals in "real time" could synergistically increase information content, helping to identify which regions bright on T1-weighted proton images are targeted thrombus. In principle, estimating the exposed microthrombi surface area may predict the probability of subsequent occlusion or distal embolization. These risk data could support rational guidelines for decisions to acutely intervene for plaque stabilization or to follow a more expectant course of medical therapy.

B. Detection and Treatment of Angiogenesis in Cancer and Early Atherosclerosis

Paramagnetic PFC nanoparticles have been used to detect the sparse expression of the $\alpha_v\beta_3$ integrin on neovasculature and to deliver anti-angiogenic therapy, which is an important goal for biologists and physicians involved

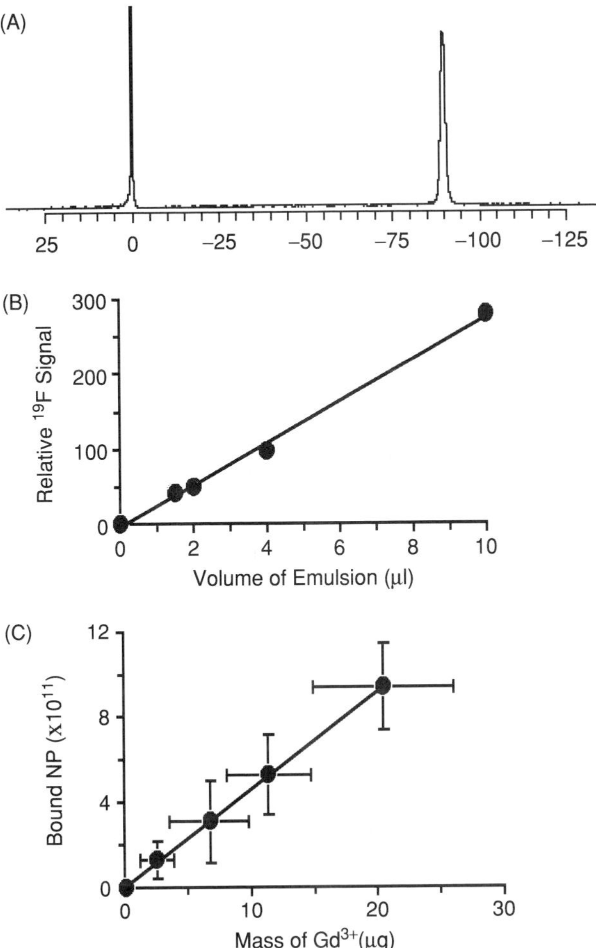

Figure 4   (A) Representative spectrum taken at 4.7 T of crown ether emulsion (−90 ppm) and trichlorofluormethane (0 ppm) used as a reference. (B) The calibration curve for crown ether emulsion has a slope of 28.06 with an r^2 of 0.9968. (C) The calculated number of bound nanoparticles (mean standard error) as calculated from ^{19}F spectroscopy versus the mass of total gadolinium (Gd^{3+}) in the sample as determined by neutron activation analysis shows excellent agreement as independent measures of fibrin-targeted nanoparticle binding to clots. The linear regression line has an r^2 of 0.9997. Reproduced with permission from Morawski *et al.*, 2004b.

with the care of patients with cardiovascular, oncologic, and rheumatologic disease. The $\alpha_v\beta_3$ integrin has garnered prominent early attention for angiogenic targeting applications because it is expressed on the luminal surface of activated endothelial cells but not on mature quiescent cells.

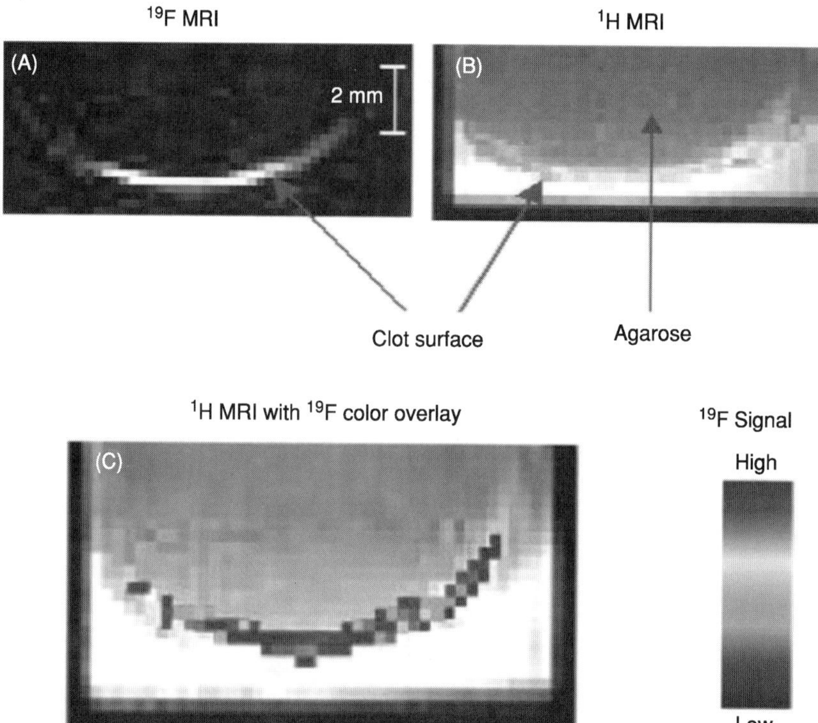

Figure 5 (A) ^{19}F image (4.7 T) of a single slice through a clot treated with crown ether emulsion. High signal is observed at the clot surface due to bound fluorinated nanoparticles. (B) ^{1}H image (4.7 T) of the same slice showing the anatomy of the clot. (C) False color overlay of ^{19}F signal onto ^{1}H image clearly showing localization of ^{19}F signal to the clot surface. Reproduced with permission from Morawski *et al.*, 2004b.

$\alpha_v\beta_3$ integrin–targeted nanoparticles can sensitively detect angiogenic endothelium at 1.5 T in New Zealand White rabbits bearing Vx-2 tumors (<1.0 cm) (Winter *et al.*, 2003b), melanoma tumors ($\sim$33 mm^3) in nude mice (Schmieder *et al.*, 2005), and in hyperlipidemic New Zealand White rabbits with early atherosclerotic disease (Winter *et al.*, 2003c).

In cancers, $\alpha_v\beta_3$ integrin expression within the neovasculature is observed by MR and corroborated by histology as an asymmetrical feature distributed along portions of the outer tumor capsule. In addition, angiogenic sprouts incited by vascular endothelial growth factor (VEGF) and fibroblast growth factor (FGF) secreted by the tumor are identified within the adventitia of neighboring vessels as well as adjoining connective tissue interfaces

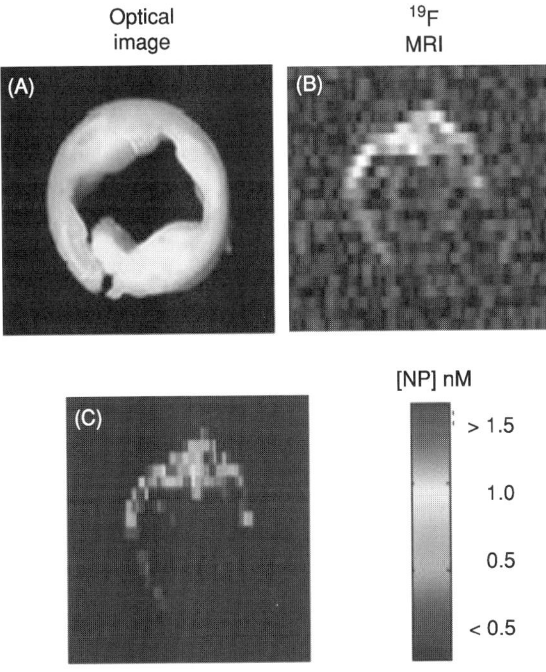

Figure 6 (A) Optical image of a 5-mm cross-section of a human carotid endarterectomy sample with moderate luminal narrowing and several atherosclerotic lesions. (B) A ^{19}F projection image acquired at 4.7 T through the entire carotid artery sample shows high signal along the lumen due to nanoparticles bound to fibrin. (C) Concentration map of bound nanoparticles in the carotid sample. NP, nanoparticle. Reproduced with permission from Morawski *et al.*, 2004b.

between tumor and muscle. Within 2 hours of injection, $\alpha_v\beta_3$-targeted nanoparticles enhanced MRI signal from tumor neovasculature by 126% relative to baseline. Moreover, *in vivo* competition-blockade studies support the specificity of the $\alpha_v\beta_3$-targeted paramagnetic agent as revealed by 50% decreases in $\alpha_v\beta_3$ integrin–targeted signal enhancement by pretreatment receptor blockade (Winter *et al.*, 2003b).

In an analogous study conducted in athymic mice subcutaneously implanted with human melanoma xenografts (C-32, ATCC), $\alpha_v\beta_3$ integrin–targeted paramagnetic nanoparticles produced MR signal enhancement at 1.5 T from the targeted angiogenic vasculature in 0.5 hours (Fig. 7), which became progressively more prominent over 2 hours (177%; Fig. 8). Again, competition studies markedly diminished the signal from $\alpha_v\beta_3$-targeted

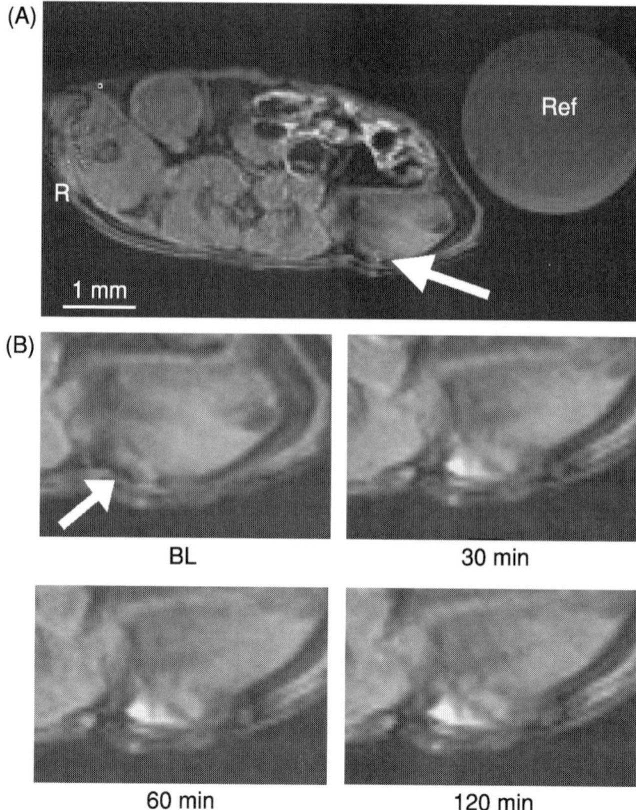

Figure 7 (A) Full-slice T1-weighted MR image (axial view) of athymic nude mouse before injection. The arrow indicates difficult to detect C32 tumor (reference = gadolinium in 10-ml syringe). (B) Enlarged section of MR image showing T1-weighted signal enhancement of angiogenic vasculature of early tumors over 2 hours as detected by $\alpha_v\beta_3$-targeted paramagnetic nanoparticles. BL, baseline. Reproduced with permission from Schmieder *et al.*, 2005.

nanoparticles and histology corroborated the MR images. Importantly, the tumors imaged in nude mice were only 33 mm^3 nominally, which demonstrates the progressive improvements of integrin-targeted paramagnetic agent sensitivity since the first report by Sipkins *et al.* (1998).

Angiogenesis is also an essential component of atherosclerotic plaque development (Moulton, 2001) and affects plaque formation in the following manner:

- By permitting expansion of plaques beyond a wall thickness that limits nutrient diffusion
- By conducting inflammatory cells into lesions

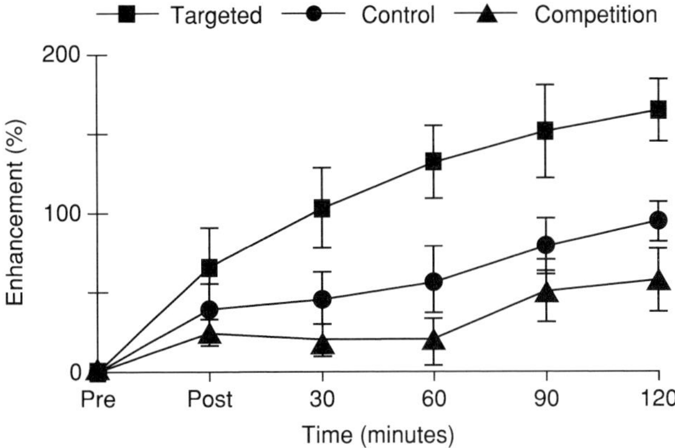

Figure 8 Time course of MRI signal enhancement in tumor bearing mice treated with $\alpha_v\beta_3$-targeted contrast agent (■), nontargeted contrast agent (●), or competition procedure (▲). T1-weighted signal enhancement in the targeted group was nearly twice that of the control animals given the untargeted agent ($p < 0.05$). At the later time points, both the percent enhancement and the rate of increase in signal intensity are greater in the targeted animals. Competitive blockage of $\alpha_v\beta_3$ integrin sites greatly diminished contrast signal enhancement compared to that of targeted particles at the 2-hour time point ($p < 0.05$), confirming the specificity of the targeted nanoparticles. Reproduced with permission from Schmieder *et al.*, 2005.

- By undermining thinning intimal caps and instigating plaque instability and rupture.

Noninvasive and specific recognition of molecular events associated with early vascular disease is not possible with traditional medical imaging techniques and relies on indirect correlation with circulating inflammatory markers, such as C-reactive peptide (CRP) or lipoprotein markers such as low-density lipoprotein. In contradistinction to these correlative measures, $\alpha_v\beta_3$-targeted paramagnetic nanoparticles have been demonstrated to spatially localize and quantify early atherosclerotic progression directly within the aorta of hyperlipidemic New Zealand White rabbits (Winter *et al.*, 2003c), before the gross manifestations of lumen encroachment or wall thickening are detectable by MR imaging.

$\alpha_v\beta_3$-Targeted nanoparticles can also specifically and locally deliver potent pharmaceutical agents to elicit a marked anti-angiogenic effect with even a single dose. Fumagillin and other drugs (Lanza *et al.*, 2002) may be delivered to targeted cells through "contact facilitated drug delivery," a mechanism in which ligand-directed binding promotes the direct transfer of lipids and drug from the nanoparticle's surfactant monolayer to the targeted cell membrane. For example, after 80 days of diet, early atherosclerotic disease in hyperlipidemic New Zealand White rabbits were injected with $\alpha_v\beta_3$-targeted paramagnetic nanoparticles with and without fumagillin incorporated into the surfactant. MRI signal enhancement averaged over all imaged slices from the renal artery to the diaphragm provided an integrated assessment of the atherosclerotic burden, which was identical at baseline for animals given $\alpha_v\beta_3$-targeted nanoparticles with ($16.7 \pm 1.1\%$) or without ($16.7 \pm 1.6\%$) fumagillin. One week after nanoparticle treatment, the residual angiogenic activity within the aortic wall was reassessed by injection of $\alpha_v\beta_3$-targeted paramagnetic nanoparticles (no drug). MRI signal enhancement of angiogenic vasculature 1 week after $\alpha_v\beta_3$-targeted fumagillin nanoparticle treatment was markedly reduced ($2.9 \pm 1.6\%$; $p < 0.05$) in both spatial distribution as well as intensity while the signal from the neovasculature of rabbits after $\alpha_v\beta_3$-targeted nanoparticles *lacking fumagillin* was undiminished ($18.1 \pm 2.1\%$).

D. Personalized Medicine with PFC Nanoparticles

Combining drug delivery and MR molecular imaging permits patients to be phenotypically characterized and individually matched to an appropriate therapy. High-resolution ^{1}H imaging spatially delineates regions of delivery and permits the local concentration of drug to be qualitatively estimated (i.e., *rational drug dosing*), as we have demonstrated in the assessment of early atherosclerosis. However, the ^{1}H signal received can be greatly influenced by the local environment as well as artifacts unrelated to the nanoparticles *per se*. ^{19}F provides both quantitative and multispectral features that are more linear and insensitive to local field inhomogeneity than is T1-weighted imaging. As aforementioned, the total fluorine spectrum, which is quantitatively reflective of the mass of PFC deposited within a selected voxel, can be used to confirm and segment the proton images for easier recognition of contrast enhancement with reduced confusion induced by artifacts. ^{19}F imaging multispectral capability will permit PFC nanoparticles containing spectrally distinct fluorocarbons to be simultaneously targeted and independently measured through deconvolution of their composite spectral signal, as illustrated in Fig. 9. This approach may be extended to several nanoparticle formulations to provide a multispectral palette with which to phenotypically characterize the biochemical nature or therapeutic sensitivity of a lesion. Such information may have prognostic utility when establishing a treatment strategy as well as retrospective value when reinterrogating local response to therapy. Together, the use of both ^{1}H MRI and ^{19}F imaging and spectroscopy offers many synergistic possibilities for enhancing clinical treatment strategies.

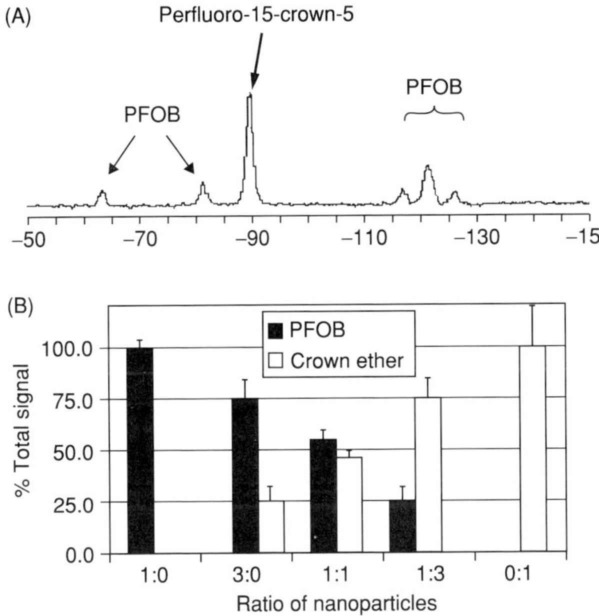

Figure 9 (A) Representative ^{19}F spectrum acquired at 4.7 T of a clot treated with a mixture of fibrin-targeted crown ether and perfluorooctylbromide (PFOB) emulsions. The crown ether peak and five discernible perfluorooctylbromide (PFOB) peaks are easily detected and individually resolved. (B) Percentage of total ^{19}F signal attributed to crown ether or PFOB for the clots treated with different emulsion mixtures (n = 3; plotted as mean ± standard error). Emulsion mixtures are listed as the ratio of PFOB to crown ether. Spectral discrimination of crown ether and PFOB allows quantification of the two nanoparticle species within a single sample. Reproduced with permission from Morawski *et al.*, 2004b.

IV. Challenges for ^{19}F MR spectroscopy and ^{19}F/^{1}H imaging

Certain challenges exist for the clinical application of ^{19}F spectroscopy and imaging. First, the fluorine signal is intrinsically weak compared with that of protons in view of the limited concentration. To be clinically applicable, ^{19}F imaging of targeted nanoparticles must be possible at 1.5 or 3.0 T. Previous studies have reported the minimum detectable limit of ^{19}F at 1.5 T is 30 μM. This concentration is difficult to achieve with a circulating agent that might passively accumulate at a selected site *in vivo*. Targeted PFC nanoparticles, however, can overcome this limitation because the specific binding permits local accumulation of nanoparticles at the site of pathology and each individual nanoparticle carries an extremely high payload of fluorine ($\sim$100 M). These factors allow ^{19}F accumulation at the target surface to a concentration well above the reported minimum detection limit. Therefore, we suggest that molecular imaging and quantitative ^{19}F spectroscopy and imaging should be clinically feasible at 1.5 T and greater field strengths.

The practical effectiveness of molecular imaging depends on both the chemistry of the contrast agents and the engineering of the associated MR equipment. Optimized coils are one important key element for ^{19}F and dual ^{19}F/^{1}H imaging into which we have exerted significant effort. RF coils are the "antenna" of the MRI system that excites the MR signal within the patient and/or receives the return signal. RF coils can be receive only, in which case the "built-in" body coil is used as a transmitter, or transmit and receive (i.e., transceive).

^{19}F imaging in small animals is most easily accomplished with surface coils, which are, simply put, a loop of wire, commonly circular or rectangular, placed over the region of interest. The sensitivity of a surface coil decreases with distance and is generally limited to about one radius of the coil. We have developed circular and rectangular frame surface coils, which are used in the transmit-and-receive mode (Fig. 10A). With these coils, the required radiofrequency (RF) power requirement is reduced by factors of 2 to 100 with respect to typical head and body coils. These loop coils are

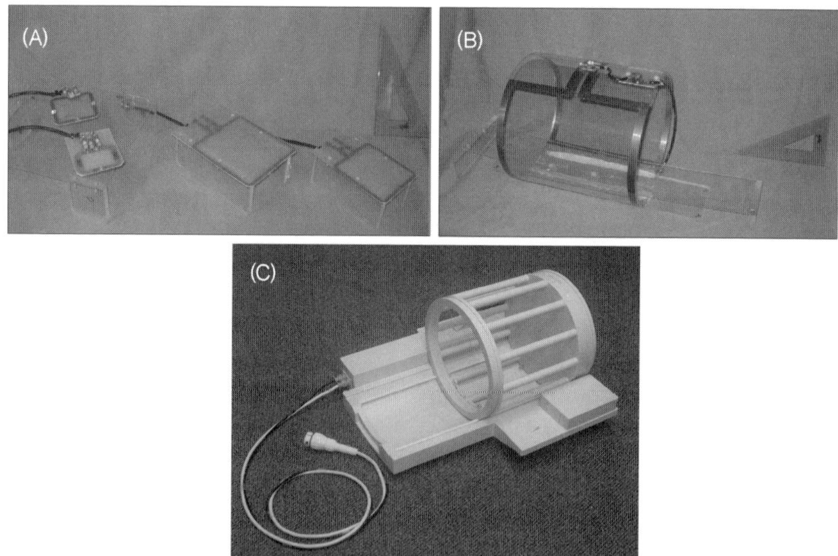

Figure 10 (A) Various ^{19}F surface tranceiver coils from 3 to 16 cm. (B) Helmholtz volume ^{19}F transceiver coils (~20-cm diameter) with a Plexiglass sliding tray for removing the coil from the field of view for ^{1}H image acquisition while leaving the sample in place. (C) A modified Philips quadrature birdcage transmit/receive (Proton) head coil retuned to ^{19}F displaying the "sliding" principle but also capable of dual ^{19}F/^{1}H MR imaging without coil displacement.

tuned for ^{19}F and allow high-resolution images over a small region of interest without foldover and motion artifacts originating outside the field of interest.

We have also designed and constructed Helmholtz volume coils named after the German scientist, Hermann von Helmholtz (1821–1894). A Helmholtz coil is a pair of circular coils separated by a distance equal to the coil radius, which optimizes the homogeneity of the magnetic field that is produced by equal current flowing through each coil (Fig. 10B). The improved field homogeneity of these coils makes them ideal for small animal imaging studies but, in practice, more complex quadrature coil arrangements that provide excellent homogeneity and sensitivity as well as dual ^{19}F/^{1}H applications will be best suited for clinical applications (Fig. 10C).

RF coil development efforts have focused on phased-array coil development for four-, and more, channel MRI clinical and research MRI systems. A phased-array coil is simply an "array" of surface or volume coil elements each connected to a separate receiver channel of the MR system. When the signals from individual receive elements are combined with a certain "phase," the signal-to-noise ratio (SNR) of the smaller elements is maintained over a larger field of view. Also, if an array of surface coils is placed around the surface of a cylinder forming a volume head array, the composite array image will have peripheral SNR equal to that of the surface coil elements, and the SNR in the center of the "volume array" will be the same as that obtained with a quadrature volume coil of the same size. Moreover, new array coils in development will further minimize inductive and resistive coupling effects to provide superior SNR with parallel imaging (e.g., SENSE) in comparison to current phased-array coils. Continued surface and volume phased-array coil development will likely play a major role in the success of ^{19}F/^{1}H imaging in a clinical setting.

V. Conclusion

Molecular imaging and targeted drug delivery are emerging medical tools that promise to change the traditional paradigms of medicine. Efforts to diagnose, quantify, and treat illnesses in patients with early disease will be greatly bolstered by the ongoing development of these new agents and complementary imaging hardware/software. To date, the concept of molecular imaging has generally been focused on simple detection of pathology, but the future of clinical management requires quantification. PFC nanoparticles offer excellent assessments of disease distribution and severity with ^{1}H imaging; however, the addition of fluorine spectroscopy and imaging greatly expands the quantitative opportunities as well as diagnostic precision. In particular, the multispectral capability of PFC nanoparticles could permit several prognostically important biomarkers to be assessed simultaneously for rational stratification of patients into individualized treatment regimens with focused, noninvasive reassessment of patient response.

References

Anderson, S. A., Rader, R. K., Westlin, W. F., Null, C., Jackson, D., Lanza, G. M., Wickline, S. A., and Kotyk, J. J. (2000). Magnetic resonance contrast enhancement of neovasculature with $\alpha_v\beta_3$-targeted nanoparticles. *Magn. Reson. Med.* **44,** 433–439.

Bulte, J. W., Duncan, I. D., and Frank, J. A. (2002). *In vivo* magnetic resonance tracking of magnetically labeled cells after transplantation. *J. Cereb. Blood Flow Metab.* **22,** 899–907.

Daldrup-Link, H. E., Rudelius, M., Oostendorp, R. A., Settles, M., Piontek, G., Metz, S., Rosenbrock, H., Keller, U., Heinzmann, U., Rummeny, E. J., Schlegel, J., and Link, T. M. (2003). Targeting of hematopoietic progenitor cells with MR contrast agents. *Radiology* **228,** 760–767.

Dardzinski, B. J., and Sotak, C. H. (1994). Rapid tissue oxygen tension mapping using 19F inversion-recovery echo-planar imaging of perfluoro-15-crown-5-ether. *Magn. Reson. Med.* **32,** 88–97.

Fan, X., River, J., Zamora, M., Al-Hallaq, H., and Karczmar, G. (2002). Effect of carbogen on tumor oxygenation: Combined fluorine-19 and proton MRI measurements. *Int. J. Radiat. Oncol. Biol. Phys.* **54,** 1202–1209.

Flacke, S., Fischer, S., Scott, M., Fuhrhop, R., Allen, J., Mc Lean, M., Winter, P., Sicard, G., Gaffney, P., Wickline, S., and Lanza, G. (2001). A novel MRI contrast agent for molecular imaging of fibrin:implications for detecting vulnerable plaques. *Circulation* **104,** 1280–1285.

Hahn, P. F., and Saini, S. (1998). Liver-specific MR imaging contrast agents. *Radiol. Clin. North Am.* **36,** 287–297.

Harisinghani, M. G., Barentsz, J., Hahn, P. F., Deserno, W. M., Tabatabaei, S., van de Kaa, C. H., de la Rosette, J., and Weissleder, R. (2003). Noninvasive detection of clinically occult lymph-node metastases in prostate cancer. *N. Engl. J. Med.* **348,** 2491–2499.

Huang, M., Ye, Q., Williams, D., and Ho, C. (2002). MRI of lungs using partial liquid ventilation with water-in-perfluorocarbon emulsions. *Magn. Reson. Med.* **48,** 487–492.

Hunjan, S., Zhao, D., Canstandtinescu, A., Hahan, E., Antich, P., and Mason, R. (2001). Tumor oximetry: Demonstration of an enhanced dynamic mapping procedure using fluorine-19 echo planar magnetic resonance imaging the Dunning prostate R3327-Atl rat tumor. *Int. J. Radiat. Oncol. Biol. Phys.* **49,** 1097–1108.

Ikehira, H., Girard, F., Obata, T., Ito, H., Yoshitomi, H., Miyazaki, M., Nakajiman, N., Kamei, H., Kanazawa, Y., Takano, H., Tanada, S., and Sasaki, Y. (1999). A preliminary study for clinical pharmacokinetics of oral fluorine anticancer medicines using the commercial MRI system 19F-MRS. *Br. J. Radiol.* **72,** 584–589.

Johansson, L. O., Bjornerud, A., Ahlstrom, H. K., Ladd, D. L., and Fujii, D. K. (2001). A targeted contrast agent for magnetic resonance imaging of thrombus: Implications of spatial resolution. *J. Magn. Reson. Imaging* **13,** 615–618.

Kaufman, C. L., Williams, M., Ryle, L. M., Smith, T. L., Tanner, M., and Ho, C. (2003). Superparamagnetic iron oxide particles transactivator protein-fluorescein isothiocyanate particle labeling for *in vivo* magnetic resonance imaging detection of cell migration: Uptake and durability. *Transplantation* **76,** 1043–1046.

Kim, J. Y., and Harisinghani, M. G. (2004). MR imaging staging of pelvic lymph nodes. *Magn. Reson. Imaging Clin. North Am.* **12**, 581–586.

Kircher, M. F., Mahmood, U., King, R. S., Weissleder, R., and Josephson, L. (2003). A multimodal nanoparticle for preoperative magnetic resonance imaging and intraoperative optical brain tumor delineation. *Cancer Res.* **63**, 8122–8125.

Kohler, G., and Milstein, C. (1975). Continuous cultures of fused cells secreting antibody of predefined specificity. *Nature* **256**, 495–497.

Kooi, M. E., Cappendijk, V. C., Cleutjens, K. B., Kessels, A. G., Kitslaar, P. J., Borgers, M., Frederik, P. M., Daemen, M. J., and van Engelshoven, J. M. (2003). Accumulation of ultrasmall superparamagnetic particles of iron oxide in human atherosclerotic plaques can be detected by *in vivo* magnetic resonance imaging. *Circulation* **107**, 2453–2458.

Krafft, M. (2001). Fluorocarbons and fluorinated amphiphiles in drug delivery and biomedical research. *Adv. Drug Del. Rev.* **47**, 209–228.

Krenning, E. P., Kwekkeboom, D. J., Reubi, J. C., Van Hagen, P. M., van Eijck, C. H., Oei, H. Y., and Lamberts, S. W. (1992). ^{111}In-octreotide scintigraphy in oncology. *Metabolism* **41**, 83–86.

Lanza, G., Lorenz, C., Fischer, S., Scott, M., Cacheris, W., Kaufman, R., Gaffney, P., and Wickline, S. (1998). Enhanced detection of thrombi with a novel fibrin-targeted magnetic resonance imaging agent. *Acad. Radiol.* **5**(Suppl. 1), S173–S176.

Lanza, G., Wallace, K., Scott, M., Cacheris, W., Abendschein, D., Christy, D., Sharkey, A., Miller, J., Gaffney, P., and Wickline, S. (1996). A novel site-targeted ultrasonic contrast agent with broad biomedical application. *Circulation* **94**, 3334–3340.

Lanza, G. M., Yu, X., Winter, P. M., Abendschein, D. R., Karukstis, K. K., Scott, M. J., Chinen, L. K., Fuhrhop, R. W., Scherrer, D. E., and Wickline, S. A. (2002). Targeted antiproliferative drug delivery to vascular smooth muscle cells with a magnetic resonance imaging nanoparticle contrast agent: Implications for rational therapy of restenosis. *Circulation* **106**, 2842–2847.

Laukemper-Ostendorf, S., Scholz, A., Burger, K., Heussel, C., Schmittner, M., Weiler, N., Markstaller, K., Eberle, B., Kauczor, H., Quintel, M., Thelen, M., and Schreiber, W. (2002). 19F-MRI of perflubron for measurement of oxygen partial pressure in porcine lungs during partial liquid ventilation. *Magn. Reson. Med.* **47**, 82–89.

Moore, A., Grimm, J., Han, B., and Santamaria, P. (2004). Tracking the recruitment of diabetogenic CD8+ T-cells to the pancreas in real time. *Diabetes* **53**, 1459–1466.

Morawski, A. M., Winter, P. M., Crowder, K. C., Caruthers, S. D., Fuhrhop, R. W., Scott, M. J., Robertson, J. D., Abendschein, D. R., Lanza, G. M., and Wickline, S. A. (2004a). Targeted nanoparticles for quantitative imaging of sparse molecular epitopes with MRI. *Magn. Reson. Med.* **51**, 480–486.

Morawski, A. M., Winter, P. M., Yu, X., Fuhrhop, R. W., Scott, M. J., Hockett, F., Robertson, J. D., Gaffney, P. J., Lanza, G. M., and Wickline, S. A. (2004b). Quantitative "magnetic resonance immunohistochemistry" with ligand-targeted (19)F nanoparticles *Magn. Reson. Med.* **52**, 1255–1262.

Moulton, K. (2001). Plaque angiogenesis and atherosclerosis. *Curr. Atheroscler. Rep.* **3**, 225–233.

Muhler, A., Zhang, X., Wang, H., Lawaczeck, R., and Weinmann, H. J. (1995). Investigation of mechanisms influencing the accumulation of ultrasmall superparamagnetic iron oxide particles in lymph nodes. *Invest. Radiol.* **30**, 98–103.

Noth, U., Grohn, P., Jork, A., Zimmermann, U., Haase, A., and Lutz, J. (1999). 19F-MRI *in vivo* determination of the partial oxygen pressure in perfluorocarbon-loaded alginate capsules implanted into the peritoneal cavity and different tissues. *Magn. Reson. Med.* **42**, 1039–1047.

Ruehm, S. G., Corot, C., Vogt, P., Cristina, H., and Debatin, J. F. (2002). Ultrasmall superparamagnetic iron oxide-enhanced MR imaging of atherosclerotic plaque in hyperlipidemic rabbits. *Acad. Radiol.* **9**(Suppl. 1), S143–S144.

Saini, S., Edelman, R. R., Sharma, P., Li, W., Mayo-Smith, W., Slater, G. J., Eisenberg, P. J., and Hahn, P. F. (1995). Blood-pool MR contrast material for detection and characterization of focal hepatic lesions: Initial clinical experience with ultrasmall superparamagnetic iron oxide (AMI-227). *AJR Am. J. Roentgenol.* **164**, 1147–1152.

Schlemmer, H., Becker, M., Bachert, P., Dietz, A., Rudat, V., Vanselow, B., Wollensack, P., Zuna, I., Knopp, M., Weidauer, H., Wannenmacher, M., and van Kaick, G. (1999). Alterations of intratumoral pharacokinetics of 5-fluorouracil in head and neck carcinoma during simulataneous radiochemotherapy. *Cancer Res.* **59**, 2363–2369.

Schmieder, A., Winter, P., Caruthers, S., Harris, T., Williams, T., Allen, J., Lacy, E., Zhang, H., MJ, S., Wickline, S., and Lanza, G. (2005). Molecular MR imaging of melanoma angiogenesis with $\alpha_v\beta_3$-targeted paramagnetic nanoparticles. *Magn. Reson. Med.* **53**, 621–627.

Sipkins, D. A., Cheresh, D. A., Kazemi, M. R., Nevin, L. M., Bednarski, M. D., and Li, K. C. (1998). Detection of tumor angiogenesis *in vivo* by $\alpha_v\beta_3$-targeted magnetic resonance imaging. *Nat. Med.* **4**, 623–626.

Sotak, C. H., Hees, P. S., Huang, H. N., Hung, M. H., Krespan, C. G., and Raynolds, S. (1993). A new perfluorocarbon for use in fluorine-19 magnetic resonance imaging and spectroscopy. *Magn. Reson. Med.* **29**, 188–195.

Vassallo, P., Matei, C., Heston, W. D., McLachlan, S. J., Koutcher, J. A., and Castellino, R. A. (1994). AMI-227-enhanced MR lymphography: Usefulness for differentiating reactive from tumor-bearing lymph nodes. *Radiology* **193**, 501–506.

Weissleder, R., Elizondo, G., Wittenberg, J., Lee, A. S., Josephson, L., and Brady, T. J. (1990). Ultrasmall superparamagnetic iron oxide: An intravenous contrast agent for assessing lymph nodes with MR imaging. *Radiology* **175**, 494–498.

Winter, P., Caruthers, S., Yu, X., Song, S., Fuhrhop, R., Chen, J., Miller, B., Bulte, J., Wickline, S., and Lanza, G. (2003a). Improved molecular imaging contrast agent for detection of human thrombus. *Magn. Reson. Med.* **50**, 411–416.

Winter, P. M., Caruthers, S. D., Kassner, A., Harris, T. D., Chinen, L. K., Allen, J. S., Lacy, E. K., Zhang, H., Robertson, J. D., Wickline, S. A., and Lanza, G. M. (2003b). Molecular imaging of angiogenesis in nascent Vx-2 rabbit tumors using a novel $\alpha_v\beta_3$-targeted nanoparticle and 1.5 tesla magnetic resonance imaging. *Cancer Res.* **63**, 5838–5843.

Winter, P. M., Morawski, A. M., Caruthers, S. D., Fuhrhop, R. W., Zhang, H., Williams, T. A., Allen, J. S., Lacy, E. K., Robertson, J. D., Lanza, G. M., and Wickline, S. A. (2003c). Molecular imaging of angiogenesis in early-stage atherosclerosis with $\alpha_v\beta_3$-integrin-targeted nanoparticles. *Circulation* **108**, 2270–2274.

Wolf, W., Presant, C., and Waluch, V. (2000). 19F-MRS studies of fluorinated drugs in humans. *Adv. Drug Deliv. Rev.* **41**, 55–74.

Further Reading

Josephson, L., Kircher, M. F., Mahmood, U., Tang, Y., and Weissleder, R. (2002). Near-infrared fluorescent nanoparticles as combined MR/optical imaging probes. *Bioconjug. Chem.* **13**, 554–560.

4

Loss of Cell Ion Homeostasis and Cell Viability in the Brain: What Sodium MRI Can Tell Us

Fernando E. Boada, George LaVerde, Charles Jungreis,
Edwin Nemoto, Costin Tanase, and Ileana Hancu
Magnetic Resonance Research Center, Department of Radiology
University of Pittsburgh Medical Center, Pittsburgh, Pennsylvania 15213

This chapter demonstrates the use of sodium magnetic resonance imaging (MRI) as a noninvasive, *in vivo* means to assess metabolic changes that ensue from loss of cell ion homeostasis due to cell death in the brain. The chapter is organized in two sections. In the first section, the constraints imposed on the imaging methods by the nuclear magnetic resonance (NMR) properties of the sodium ion are discussed and strategies for avoiding their potential limitations are addressed. The second section illustrates the use of sodium MRI for monitoring focal brain ischemia in permanent and temporary primate models of endovascular middle cerebral artery occlusion.

I. Introduction

The sodium nucleus yields the second strongest nuclear magnetic resonance (NMR) signal among all biologically relevant NMR-active nuclei. The ^{23}Na nucleus has a spin of 3/2 and a gyromagnetic ratio of 1126 Hz/Gauss, leading to an overall NMR sensitivity (relative to water protons) of 0.092. The

Current Topics in Developmental Biology, Vol. 70
Copyright 2005, Elsevier Inc. All rights reserved.

0070-2153/05 $35.00
DOI: 10.1016/S0070-2153(05)70004-1

sodium ion plays a critical role in cell physiology as many physiological conditions lead to large changes in the average concentration of this nucleus in tissue. For example, during brain neoplasia, sustained cell depolarization, a common precursor of cell division, leads to an increase in the intracellular sodium content and a concomitant rise in the average tissue sodium concentration. As these changes in sodium content are directly related to the underlying physiological changes and typically large (>50%), they represent an ideal means to noninvasively monitor cellular processes associated with the development of pathology *in vivo*.

Magnetic resonance imaging (MRI) has evolved into one of the preferred techniques for noninvasive tomographic imaging *in vivo*. During the past 25 years continued advances in radiofrequency (RF) electronics, computing performance, and high-power amplifiers have led to significant improvements in the capabilities of MRI as an imaging technique. By today's standards, 1.5-Tesla images over a field of view (FOV) of 20 cm at 1×1 mm in-plane resolution are routine, and twice the spatial resolution in comparable or shorter imaging times is also commonplace using the widely available main magnetic field strength of 3.0 T. Moreover, methodological improvements have changed the role of MRI from a high-resolution anatomical imaging tool to that of a functional and anatomical one.

Despite the rapid advances in conventional proton MRI, progress in sodium imaging did not follow suit for many years. This was due in part to the stringent constraints imposed by the sodium nucleus on image signal-to-noise ratio (SNR) *in vivo*. These constraints render the imaging approaches commonly used for proton MRI inadequate. Many of the shortcomings of these techniques for dealing with the specific requirements of the sodium nucleus were recognized early by Hilal *et al.* (1985). However, the methodologies developed at that time overcame some of these limitations at the expense of imaging times that were typically too long for imaging scenarios where fast throughput was required. Fortunately, recent developments in the field have led to imaging schemes that have removed the aforementioned limitations and allow sodium MRI to be performed with adequate spatial resolution and SNR in practical imaging times (<10 minutes). These developments have allowed the study of important physiological conditions such as ischemia and neoplasia *in vivo*.

This chapter begins with a discussion of some of the basic facts about the sodium nucleus and its NMR properties. This discussion is followed by a description of the methodological considerations required for the efficient generation of an *in vivo* sodium MRI protocol. These considerations are then applied in the specific case of brain ischemia, a condition where sodium MRI had been recognized as having an important role for its study and characterization *in vivo*.

II. NMR Properties of the Sodium Nucleus

The potential of sodium imaging for probing metabolic and physiologic aspects of disease was recognized over a decade ago (Hilal *et al.*, 1985). Although a significant effort was devoted early toward the development of efficient protocols for sodium MRI, this technique did not meet its much anticipated usefulness. In light of the many developments in MRI, it is now apparent that many of the previous shortcomings of sodium MRI were related in part to the use of nonoptimal strategies for meeting the data acquisition constraints imposed by the NMR properties of the sodium nucleus. Key among such properties is the fast, biexponential nature of sodium's transverse relaxation decay (T2). Such transverse relaxation behavior makes the use of the conventional spin warp scheme for spatial encoding of the NMR signal impractical (Edelstein *et al.*, 1980). Because the short T2 component of the sodium NMR signal is in the 1–3 ms range, the use of spin warp imaging techniques leads to echo times (TEs) that are too long for sodium MRI (>0.5 ms). Such long TEs translate into poor SNR and sensitivity and intrinsic measurement biases. The reduced specificity of sodium MRI using conventional long TEs has been clearly documented in the literature during the study of neoplasia and stroke (Shimizu *et al.*, 1993; Schuierer *et al.*, 1991). In the context of stroke, for example, early reports (Shimizu *et al.*, 1993) clearly documented the inability of sodium MRI with long TEs to demonstrate the elevation in sodium content in ischemic human brains. A contemporaneous report, using a 4.0-T scanner, reported similar shortcomings in the context of human brain tumors (Schuierer *et al.*, 1991). These shortcoming are, as demonstrated in the following text, a direct consequence of the long TEs used for data acquisition and can be easily circumvented using imaging techniques capable of ultra-short TEs (Boada *et al.*, 1994).

The methodological nature of the effects of sodium's fast transverse relaxation time for imaging focal increases in brain tissue sodium concentration (TSC) can be easily illustrated and quantified using computer simulations. The images presented in Fig. 1 represent a computer model for an idealized human brain with a lesion in the right hemisphere. The model assumes the NMR signal from the brain parenchyma and the lesion relaxes biexponentially (3 ms and 15 ms for the short and long components, respectively, with 60% of the signal relaxing through the short components). Other compartments within the model include the cerebrospinal fluid (CSF), the eyes, and the lesion with an average TSC 200% larger than the surrounding brain tissue. The images in the top row (Fig. 1A and B) correspond to an idealized model reconstructed using short and long TEs (0.5 and 13 ms, for 1A and 1B, respectively). Clearly, the use of a long TE leads to a significant

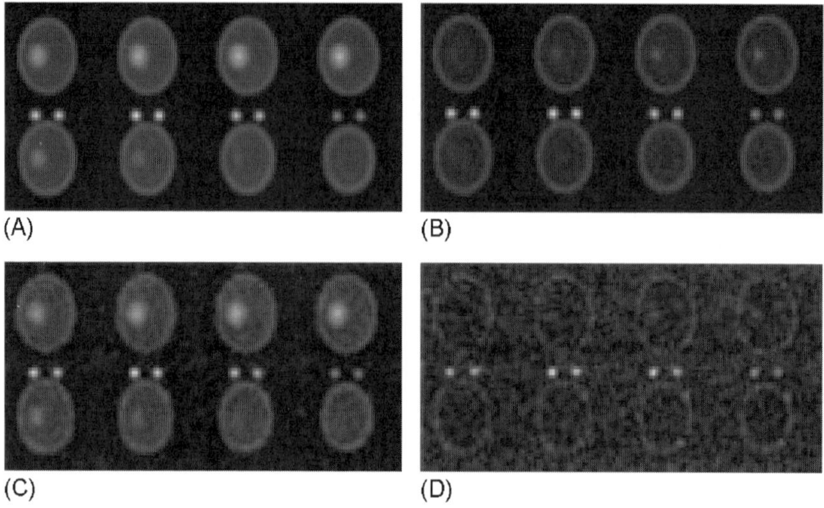

Figure 1 Illustration of fast T_2 decay effects on sodium MRI quality. The images correspond to an idealized model for a human brain with a focal lesion in the right hemisphere. The signal in the simulated brain parenchyma and the region relax biexponentially (3/15 ms, 60/40% contribution for the short/long components, respectively). The top row images correspond to noiseless reconstructions; those on the bottom correspond to the same model after addition of 10% to the raw data before image reconstruction. All the images on the right column correspond to TE = 13 ms, whereas those on the right represent TE = 0.5 ms.

decrease in signal intensity for the lesion and the brain parenchyma. Adding 10% Gaussian noise to the raw data to simulate measurement noise prior to image reconstruction yields the results in the bottom row. In these images the increased variance of the image intensity leads to an almost complete loss of conspicuity for the lesion in the long TE image of Fig. 1D.

Models such as the previous one underscore the need to perform sodium MRI with short TEs and provide a framework for estimating the measurement errors imposed on the imaging methodology by the NMR behavior of the sodium nucleus in the *in vivo* environment. In the *in vivo* environment such NMR behavior can lead to relaxation rates that exacerbate the errors demonstrated above when strong residual quadrupolar interactions exist. Fortunately, such effects are minor in the brain (Reddy *et al.*, 1995) and lead to average relaxation rates that are close to those used in the models above. Consequently, sodium MRI in the brain using short TEs leads to estimated TSCs that agree well with values obtained using invasive biochemical techniques (Boada *et al.*, 2001; Christensen *et al.*, 1996; Wang *et al.*, 2000).

Hilal *et al.* recognized the requirement of ultra-short ($<$0.5 ms) TEs for sodium MRI (Hilal *et al.*, 1992; Ra *et al.*, 1986). However, the methods available for achieving short TEs at the time of his seminal papers in the late

1980s and early 1990s (i.e., projection imaging) rendered the resulting data acquisition protocol impractical because of its exceedingly long data acquisition time (>30 minutes). In projection imaging, the long data acquisition times are a consequence of inefficient sampling of the projection imaging geometry, which is approximately a factor of 3 slower than conventional spin warp techniques. The introduction of the twisted projection imaging (TPI) scheme has provided a fast acquisition scheme, which is approximately 1.6 to 2 times more rapid than conventional spin warp techniques; furthermore, it is capable of ultra-short TEs (Boada *et al.*, 1997). Using the TPI sampling geometry sodium images of adequate quality (SNR >30:1, voxel size <0.2 cc) can be routinely acquired in acceptable imaging times (10 minutes), at commonly available magnetic field strengths (1.5 T and 3.0 T) using commercially available scanner hardware (Boada *et al.*, 1997; Thulborn *et al.*, 1999).

A. Separation of the Intracellular and Extracellular Sodium NMR Signals *In Vivo*

The changes in TSC observed during brain ischemia are a direct consequence of cellular events. It is widely believed that the most sensitive means to study these events *in vivo* is through the use of imaging schemes that can isolate the sodium NMR signal from the intracellular compartment in the brain. From the NMR point of view, this is a challenging requirement as the preferred approach for separating nuclear species in NMR, namely the chemical shift, is not directly applicable to sodium since the intracellular and extracellular sodium pools have the same resonance frequency. Consequently, other approaches have been proposed for the separation of the intracellular and extracellular sodium NMR signal (Gupta and Gupta, 1982; Gelderen *et al.*, 1990; Lee *et al.*, 1990; Pekar and Leigh, 1986; Pekar *et al.*, 1987; Veen *et al.*, 1993). These techniques can be broadly grouped into three categories: shift reagent (SR), diffusion-based, and relaxation-based techniques.

SR techniques (Bansal *et al.*, 1993; Gupta and Gupta, 1982; Navon, 1993) rely on the chemical shift that anionic chelated complexes of paramagnetic lanthanides can introduce on the sodium Larmor frequency. These compounds do not penetrate the cell membrane and therefore create a frequency offset for the sodium nuclei in the extracellular space. Despite their well-known properties and usefulness for studying animal models of disease, SRs are poorly suited for use on humans because of their moderate toxicity. In addition, their use to study central nervous system (CNS) disease is questionable, since they do not cross the blood-brain-barrier.

Diffusion-based techniques (Gelderen *et al.*, 1990; Veen *et al.*, 1993) separate the NMR signal from the intracellular and extracellular compartments based on the differences between the motional properties of the ions in these two compartments. However, the fast transverse relaxation of sodium together with its much lower gyromagnetic ratio require the use of unrealistically large magnetic field gradients to effectively take advantage of these properties using standard clinical MRI scanners.

Relaxation-based techniques have been actively pursued by several investigators (Eliav *et al.*, 1992; Lee *et al.*, 1990; Pekar and Leigh, 1986; Pekar *et al.*, 1987; Wimperis and Wood, 1991). All these approaches take advantage of the significantly different transverse relaxation properties that the sodium nucleus exhibits in the intracellular and extracellular compartments. In the relaxographic imaging approach of Lee *et al.* (1990), an inverse Laplace transformation is used to separate the different exponential components present in the free induction decay signal. Although mathematically concise and well understood, the Laplace transformation is extremely ill conditioned, a fact that imposes serious constraints on the SNR of the input data required to produce acceptable results (Labadie *et al.*, 1994). Multiple-quantum (MQ) filtering techniques (Pekar and Leigh, 1986; Pekar *et al.*, 1987) make explicit use of coherence transfer schemes to generate an NMR signal that is related to the presence of biexponential transverse relaxation. Several reports have demonstrated that the MQ NMR signal from biological tissue arises primarily from the sodium in the intracellular space (Hutchinson *et al.*, 1993; Lyon *et al.*, 1991; Navon, 1993; Schepkin *et al.*, 1996; Seshan *et al.*, 1997; Tauskela *et al.*, 1997). This feature of MQ filtering techniques has been used to study changes in intracellular sodium content in a variety of tissues using spectroscopic techniques (Bansal *et al.*, 1992; Jelicks and Gupta 1989; Knubovets *et al.*, 1998; Lyon *et al.*, 1991; Navon, 1993; Seshan *et al.*, 1997; Tauskela *et al.*, 1997). Imaging extensions of this approach (Bansal and Seshan, 1995; Kalyanapuram *et al.*, 1998; Kemp-Harper *et al.*, 1995; Reddy *et al.*, 1997), however, had not been as common because of the weak nature of the MQ sodium NMR signal. The use of efficient imaging schemes now allows the acquisition of triple-quantum filtered sodium MRI in the *in vivo* human brain (Hancu *et al.*, 1999). This approach currently represents the most suitable candidate to assess the role of intracellular sodium accumulation *in vivo* in the human brain.

B. Sodium MRI Quantification: Data Acquisition Requirements

NMR measurements are inherently quantitative if proper care is taken to eliminate all biases in the conversion of NMR signal to actual nuclear concentrations. This conversion is commonly performed by using calibration standards with known concentration of sodium that are placed inside the image FOV. In this context, for changes in the measured signal to reflect changes in the actual concentration of the tissue metabolites, the effects of transverse relaxation, longitudinal relaxation, and field inhomogeneity must be accounted for or minimized during data acquisition and/or postprocessing.

The transverse relaxation times of sodium in biological tissues can be extremely short (Rooney and Springer, 1991a,b). Therefore, as demonstrated above, the TE used during data acquisition should be kept as short as possible to minimize measurement biases in the MRI signal (Boada *et al.*, 1994). Thus, for typical relaxation rates found in the brain, TEs should be below 0.5 ms (Boada *et al.*, 1994). Such TEs can be efficiently achieved using the TPI scheme mentioned previously with no penalties in the data acquisition time or SNR.

The importance of accounting for errors introduced by the variation of the B_1 field across the FOV has been well documented in the literature (Thulborn and Ackerman, 1983; Tofts and Wray, 1988). This variation is especially severe in MQ sodium MRI due to the strong dependence of the MQ NMR signal on the tip angle (Reddy *et al.*, 1997). Strategies such as the use of adiabatic pulses (Ugurbil *et al.*, 1988) can help minimize RF inhomogeneities. However, their long duration introduces excessive T2 weighting, which leads to an unnecessary bias in the measured NMR signal. Schemes to correct for B_1 inhomogeneities have been proposed in the literature. Two avenues are often used to achieve this goal. In the first, the RF field is mapped over the volume of interest and later used to correct for B_1 inhomogeneities (Lian *et al.*, 1990; Stolberger *et al.*, 1988; Talagala and Gillen, 1991; Thulborn *et al.*, 1993). In the second, retrospective corrections involving some approximations about the spin density distribution or the coil geometry are used to decrease the RF inhomogeneities (Moyher *et al.*, 1995; Meyer *et al.*, 1995). Careful correction in postprocessing requires special consideration of the changes in the B_1 distribution introduced by the electrical loading and RF attenuation of the sample. With these constraints in mind, a suitable solution would be one in which the RF field could be mapped out across the entire volume of interest at the time of the study. Using fast imaging techniques such a solution can be implemented as part of an imaging protocol without significant penalties in the total data acquisition time. Moreover, due to sodium's short T1 in tissue, such an approach can also be implemented in the context of the applications considered here (Boada *et al.*, 1997).

C. Sodium Imaging and Monitoring of Brain Ischemia

Changes in TSC during focal ischemia have been well documented using invasive measurements (Betz *et al.*, 1994; Gotoh *et al.*, 1985b; Hatashita and Hoff, 1990; Ito *et al.*, 1979; Menzies *et al.*, 1993; Minematsu *et al.*, 1992;

O'Brien *et al.*, 1974; Schuier and Hossmann, 1980; Siegel *et al.*, 1973; Wang *et al.*, 2000; Yang *et al.*, 1992; Young *et al.*, 1987). Experimental results in various animal models of middle cerebral artery (MCA) occlusion indicate that TSC rises continuously during the hyperacute stages of ischemia (1–8 hours after stroke onset) (Betz *et al.*, 1994; Gotoh *et al.*, 1985a; Hatashita and Hoff, 1990; Hossman *et al.*, 1977; Ito *et al.*, 1979; Memezawa *et al.*, 1992; Menzies *et al.*, 1993; Minematsu *et al.*, 1992; O'Brien *et al.*, 1974; Schuier and Hossmann, 1980; Siegel *et al.*, 1973; Wang *et al.*, 2000; Yang *et al.*, 1992; Young *et al.*, 1987). The mechanisms responsible for this increase are not known in detail, although the formation of cytotoxic edema seems to be one of the precursory mechanisms (Young *et al.*, 1987). It has been suggested that the intracellular sodium influx that takes place during cytotoxic edema creates an extracellular ionic gradient between the lesion and the plasma compartments (Young *et al.*, 1987.) In the presence of such a gradient, sodium diffuses from the plasma into the lesion, giving rise to a gradual increase in TSC. Studies performed on several animal models have shown that the increase in TSC is reversed by tissue reperfusion after short periods of ischemia (<1 hour) (Ito *et al.*, 1979; Yang *et al.*, 1992). These studies also demonstrate that tissue reperfusion after longer periods of occlusion (>3 hours), corresponding to higher TSC, does not reverse tissue damage but instead accelerates the formation of cerebral edema (Ito *et al.*, 1979). Because TSC increases continuously during the early stages of stroke, the mechanisms giving rise to this increase could have an important predictive value in estimating the ability of the ischemic tissue to recover upon tissue reperfusion. In particular, it may provide an index from which to ascertain whether reperfusion will lead to reduced infarct size.

Tissue plasminogen activator (tPA) has revolutionized the treatment of acute stroke by providing an aggressive means for the reversal of the ischemic insult (Group *et al.*, 1995). The use of this therapeutic option, however, hinges on strict eligibility criteria. The cornerstone of these criteria is the minimization of hemorrhage and cerebral herniation, both of which appear to be exacerbated when the ischemic tissue mass is no longer viable. Because there are no proven, noninvasive means to assess ischemic brain tissue viability *in vivo,* decisions regarding the application of tPA are based on the estimated time of ischemia onset using a critical window of 3 hours after ischemia as the "ideal time" when tPA administration is of substantial benefit. This approach, although successful in most cases (70% of the subjects), does not ensure the minimization of the risks and more importantly might exclude its use on subjects for which the ischemic mass may be viable beyond the 3-hour window of opportunity. Currently there are no proven, noninvasive means for the assessment of tissue viability during acute brain ischemia. Perfusion MR imaging techniques using exogenous or endogenous contrast can delineate perfusion deficits in the brain within seconds of ischemia onset. When combined with diffusion-weighted (DW) MRI, perfusion MRI commonly demonstrates a mismatch between perfusion and diffusion abnormalities in the ischemic brain tissue (Albers, 1999; Baird *et al.*, 1997; Barber *et al.*, 1999; Flacke *et al.*, 1998; Hillis *et al.*, 2000). Because the volume of the perfusion abnormality is usually larger than its diffusion counterpart, the perfusion–diffusion mismatch originally was believed to represent ischemic, but viable, tissue that could potentially be salvaged by thrombolytic therapy (Albers, 1999; Baird *et al.*, 1997; Barber *et al.*, 1999; Flacke *et al.*, 1998; Hillis *et al.*, 2000; Sorensen *et al.*, 1996; Warach *et al.*, 1993, 1996). This interpretation was supported in part by the observation that DW MRI abnormalities in human brain had not been known to be reversed by tPA treatment and, therefore, they could be representative of irreversible ischemic tissue damage. This approach for the assessment of ischemic tissue damage has also been proven to be of limited value as recent studies (Kidwell *et al.*, 2000; Krueger *et al.*, 2000) demonstrate that DW MRI abnormalities can in fact be completely resolved by tPA administration (without any evidence of ischemic damage). The finding that DW MRI does not necessarily represent ischemic tissue damage indicates that the area of ischemic but salvageable tissue is not limited to the volume of mismatch between DW and perfusion MRI; therefore, other means to determine its true extent are required. The use of sodium MRI techniques to assess ion homeostasis and tissue viability may prove to be an important tool for the study and assessment of focal brain ischemia.

III. Methods

A. MR Imaging

A 3-T whole-body scanner (General Electric Medical Systems, Milwaukee, WI) operating under version VH3 of the scanning software was used for all the studies. The NMR data were acquired using a custom-built, dual-tuned, dual-quadrature (^{1}H/^{23}Na) RF birdcage coil that was interfaced with a coil holder, which allowed quick and reproducible positioning of the animal's head at the isocenter of the magnet. A photograph of the RF coil setup is presented in Fig. 3.

The single-quantum (SQ, representing the total sodium content of the voxel) sodium data acquisition scheme relied on the use of the TPI technique. This data acquisition technique allows the collection of the k-space data with ultra-short TEs (<0.5 ms) in a fraction of the time required to acquire a projection imaging of the same spatial resolution and with the same number of averages (Boada *et al.*, 1997). Figure 2 presents a sketch of the trajectory traced by this acquisition scheme in k-space. The k-space

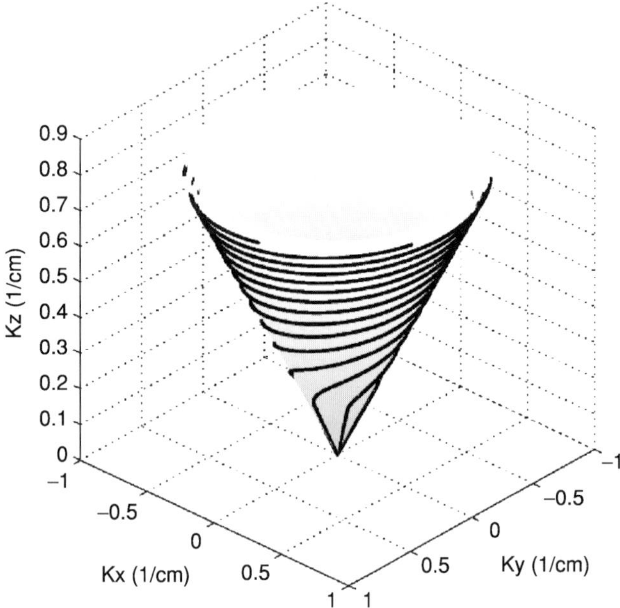

Figure 2 Sketch of the k-space trajectories used for data acquisition. The k-space trajectory lies on the surface of a cone in k-space and changes pitch to maintain constant sample density with spherical symmetry. Use of multiple concentric cones allows the volume of a sphere of radius K_{max} to be sampled efficiently to generate a properly sampled 3D data set.

trajectories lie on the surface of concentric cones with each trajectory starting at the origin and traversing a radial line up to a fraction p (p < 1) of the total radial distance (K_{max}) defining the spherical volume of k-space to be acquired. After pK_{max} is reached, the trajectory evolution is such that the sample density at pK_{max} is preserved throughout k-space. Utilization of this k-space sampling scheme requires customized gradient waveforms that are loaded into waveform memory during imaging.

B_1 inhomogeneities were mapped out over the volume of interest using a version of the TPI sequence modified to automatically acquire a set of sodium images at various RF power levels. Because of the low spatial frequency content of the B_1 fields, a lower-resolution trajectory design was used for mapping the B_1 field. This lower-resolution design used a total of 8 different RF power levels (repetition time [TR] = 100 ms, TE = 0.4 ms).

Triple-quantum (TQ) images representing a signal heavily weighted toward the intracellular sodium content were also collected as part of the experiments using the approach proposed by Hancu *et al.* (1999) in conjunction with the TPI data acquisition scheme discussed above. All sodium data sets were collected using the same imaging coil illustrated in Fig. 3.

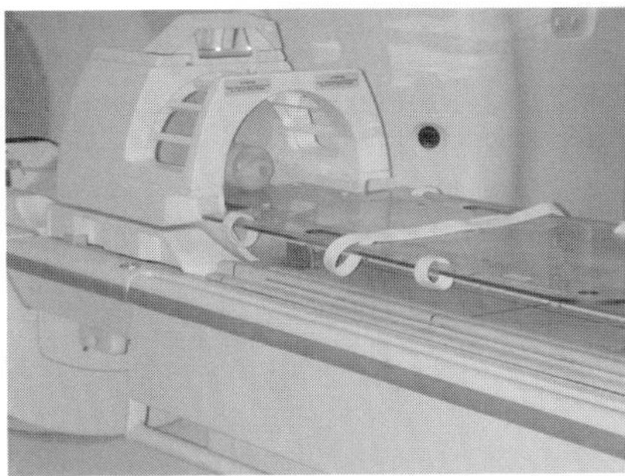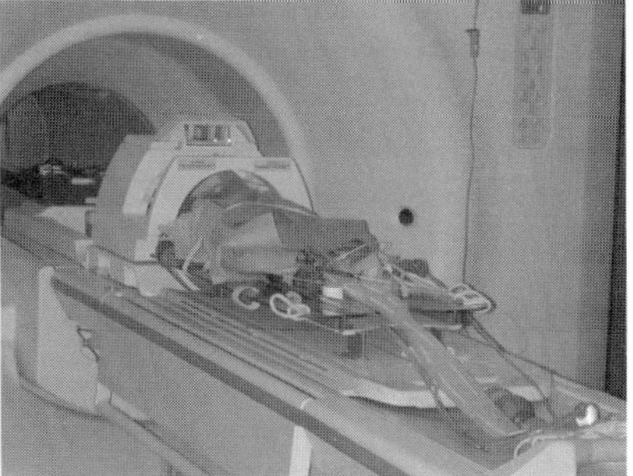

Figure 3 Monkey cradle and RF coil used for the imaging experiments (left). The cradle was built in-house and consists of two removable (and independent) MR-compatible plates that interface with the patient table and the animal, respectively. As shown on the right, the cradle allows easy positioning of the animal's head at the magnet's isocenter while keeping the animal completely stable on the patient table. (See Color Insert.)

B. Middle Cerebral Artery Occlusion in the Monkey Model

All animals were initially anesthetized using ketamine (ketamine, 10 mg/kg intramuscularly), intubated with a cuffed endotracheal tube, and fitted with a peripheral venous catheter for fluid replacement (0.9% NaCl) at 5 ml/kg/hr. Maintenance anesthesia was provided by fentanyl infusion at a rate of 25 μ/kg/hr following an initial loading dose of 25 mg/kg and diazepam (2.5–5.0 mg/hr). All monkeys were mechanically ventilated on 70% nitrous oxide/30% oxygen and chemically paralyzed with Pavulon (0.06 mg/kg/hr). After preparation all animals were strapped in the supine position to the custom-built, MR-compatible cradle with their heads immobilized (Fig. 3). For the permanent middle cerebral artery (MCA) occlusion model, the procedure described by Horowitz *et al.* (2001) was used. This procedure relied on the use of embolization coils in the right posterior cerebral artery (PCA) and the right MCA. Figure 4 presents a schematic description of the inferior view of an animal's neurovasculature (left column) before (top) and after (bottom) the occlusion of the PCA and MCA using embolization coils, alongside angiograms of the animal's head before and after the permanent occlusion procedure.

For the temporary occlusion model (Jungreis *et al.*, 2003) the procedure is modified so that a balloon catheter is used in the MCA instead of an embolization coil. This model also requires the use of systemic anticoagulation with heparin and monitoring of the activated clotting time to ensure that no thrombi occlude the MCA branches upon reperfusion. Figure 5 presents a schematic description of the inferior view of the animal's neurovasculature (top) before (left) and after (right) the occlusion of the PCA and MCA. The actual demonstration of the procedure is also depicted in this figure. Here raw angiograms (left column) and digitally subtracted angiograms of the right

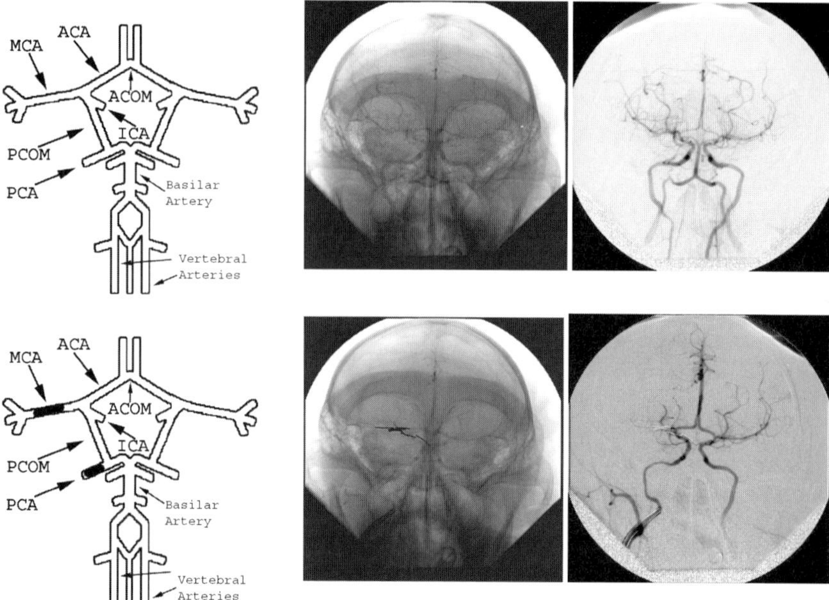

Figure 4 Schematic descriptions (left column), raw angiograms (middle column), and digitally subtracted angiograms (right column) in the monkey brain before (top row) and after (bottom row) occlusion of the right PCA and MCA using embolization coils. The arrows in the schematic diagrams illustrate the location of the MCA, PCA, posterior communicating artery (PCOM), internal carotid artery (ICA), anterior cerebral artery (ACA), the anterior communicating artery (ACOM), and the position of the embolization coils (dark cylinders in the drawing). The actual coils are clearly seen in the raw angiogram (middle column, bottom row) in the posterior aspect of the right orbit. Only the ACA territory remains unaffected in this occlusion model.

internal carotid artery (ICA) are presented before (middle column) and after (right column) the inflation of the balloon. The middle row corresponds to a lateral view while the bottom row corresponds to a frontal projection. These images clearly demonstrate that flow from the right ICA into the right hemisphere has been stopped. Only collateral flow into the territory of the right anterior cerebral artery (ACA) remains through the anterior communicating artery (ACOM) when this procedure is used; this collateral flow is not seen in these angiograms because the contrast is being injected through the right ICA.

IV. Results

Sodium MRI on the permanent occlusion model is demonstrated in Fig. 6 where selected partitions from the SQ (top) and TQ (bottom) sodium images are shown. The TQ image was the first data set collected during the imaging

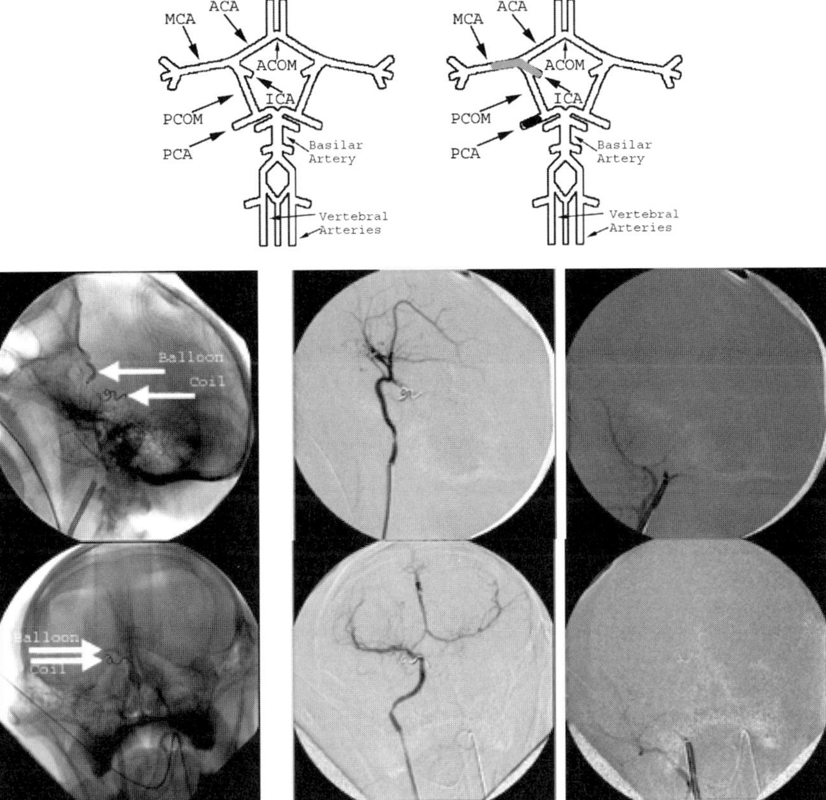

Figure 5 (Top) Schematic diagram for the inferior view of the animal's brain vasculature before (left) and during (right) temporary occlusion of the right MCA using a balloon catheter. The arrows illustrate the location of the MCA, PCA, posterior communicating artery (PCOM), internal carotid artery (ICA), anterior cerebral artery (ACA), the anterior communicating artery (ACOM), and the position of the balloon (gray cylinder) and embolization coils (dark cylinder). (Middle) Digitally subtracted angiograms (right and middle columns) of the right ICA before (middle row) and during (right row) temporary occlusion of the right MCA using a balloon catheter. The actual balloon and coils are clearly seen in the raw angiograms (left column) acquired immediately after balloon inflation. The top and bottom rows correspond to lateral and frontal views, respectively.

session (1.5 hours postischemia). Although the SNR was limited by the large diameter of the coil, the ischemic insult in the right hemisphere has led to a significant increase in the TQ signal intensity (bottom row, first two arrows from left to right). This TQ signal hyperintensity corresponds to the anatomical location of the ischemic cortex as indicated by the intrinsic co-registration of the SQ and TQ data sets when the images are reconstructed

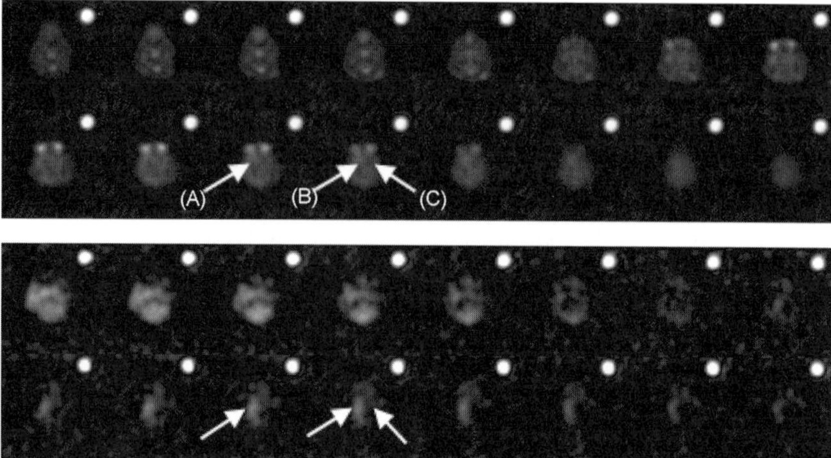

Figure 6 Three-dimensional single-quantum (top) and triple-quantum (bottom) sodium images from a monkey brain after 1.5 hours (bottom) and 4 hours of permanent right MCA occlusion. The circular structure at the superior aspect of each slice corresponds to a calibration standard with sodium in agarose gel. The arrows indicate ROIs where the time course of the TSC. Data acquisition times for these images were 5 (top) and 15 (bottom) minutes.

onto the same FOV. The SQ scan (top) also demonstrates a hyperintensity in the images (arrows labeled (A) and (B)), although the contrast is not as striking as in the TQ image. In both images, the circular structure on the anterior aspect of the animal's head corresponds to a calibration standard containing sodium chloride in agarose solution. The arrows show the location of the ischemic core (A) as well as two other regions of interest (ROI)— (C) for contralateral and (B) for another ischemic locus—in both data sets.

The increase in tissue sodium concentration in the ischemic cortex relative to the contralateral side is clearly demonstrated in Fig. 7 where the B_1-corrected, calibrated (i.e., converted to millimolar concentration of sodium) signal intensities at these regions of interest (A, B, and C labels in Fig. 6) are presented. The rise in TSC is clearly illustrated by this graph. Linear regression analysis of the data in these lines demonstrates significant differences between the slopes for the time course of TSC at the ischemic and contralateral hemispheres.

To investigate the existence of spatial variations in TSC accumulation across the ischemic brain hemisphere, the entire data set was subjected to a pixel-by-pixel linear regression analysis of the image intensity. Using this approach a spatial map of the TSC slope (i.e., the TSC as a function of time) can be obtained across the entire volume. Figure 8 presents the result of this analysis for the slices containing the ischemic and contralateral cortices

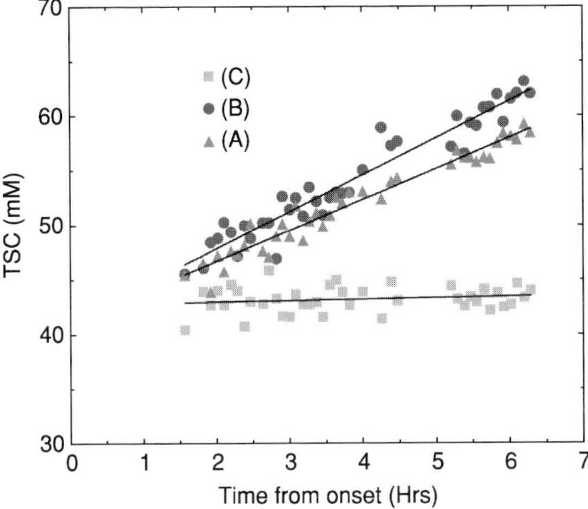

Figure 7   Time course of TSC for selected ROIs in the monkey brain after endovascular cerebral occlusion of the right MCA territory. Each data point represents a measurement made on a sodium image acquired in 5 minutes. The slopes of the curves demonstrate significant differences in TSC accumulation between the ischemic and nonischemic cortices. (See Color Insert.)

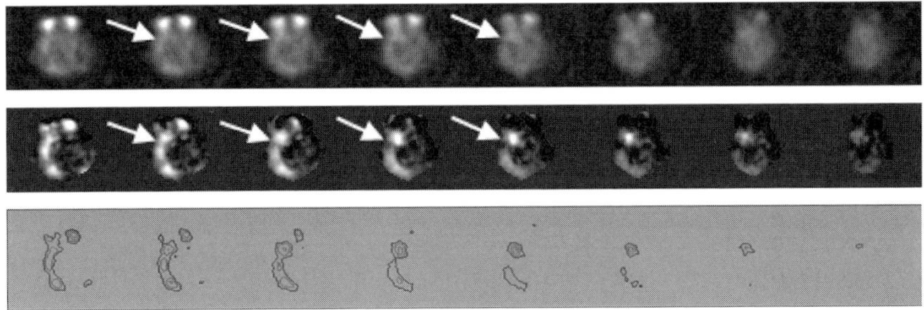

Figure 8   (Top) Selected partitions from a 3D sodium image from a monkey brain after 2 hours of permanent right MCA occlusion. The arrows indicate the positions of the ischemic cores. (Middle) Slope maps calculated for the partitions shown in the top image. This map is generated through pixel by pixel linear regression analysis of the signal intensity for the 40 images acquired during the study. This map clearly depicts a faster rise in TSC in the ischemic core (arrows). The contour plots in the bottom image clearly delineate the spatial heterogeneity of this map. In this contour plot each line corresponds to a change of 1.9 mM/hr in the slope of TSC. For the slope map only pixels with statistically significant slopes ($p < 0.001$) are shown. Only the pixels from the slope map with $r^2 > 0.8$ were used for the contour plot.

(bottom row of Fig. 6). This composite figure presents the partitions from the SQ image containing ischemic tissue (top), the corresponding TSC slope maps (middle), and the contour plots for the slope maps in the middle image (bottom). The slope maps only display pixels where statistically significant slopes ($p < 0.001$) are detected. The slope maps clearly demonstrate the presence of significant heterogeneity in the rate of TSC accumulation in the ischemic cortex. Note that a locus with non-zero slope is present in part of the volume corresponding to the left eye. This is an artifact that could be due to fluid accumulation during the imaging study. Measured slopes in this map range from 7.6 mM/hr at the center of the ischemic core (arrows) to about 1.9 mM/hr on the posterior aspect of the hemisphere.

The spatial variation in the TSC slope is best characterized by careful analysis of the contour plots in Fig. 8. In these plots continuous lines encompass areas above predetermined TSC accumulation rates. To make the analysis simpler, only pixels exhibiting a correlation coefficient larger than 0.8 are presented. For these pixels four intensity levels were chosen so that the spatial changes in the slope can be visualized in 1.9-mM/hr intervals (i.e., for the outermost line the volume within exhibits a slope higher than 1.9 mM/hr, whereas the volume of the innermost contour line exhibits slopes higher than 7.6 mM/hr). The presence of multiple contour levels within a particular volume is then indicative of large TSC accumulation rates.

A careful examination of the slope distribution for the slices is presented in Fig. 9 where an enlarged copy of one of the partitions in Figure 8 is presented alongside its corresponding contour plot (fourth slice from left to right). Note that in this contour plot the ischemic core (arrows) contains closely spaced lines, indicating that most of its underlying volume has TSC accumulation rates approaching 7.6 mM/hr. By contrast, most of the volume for the posterior part of the ischemic cortex has TSC accumulation rates

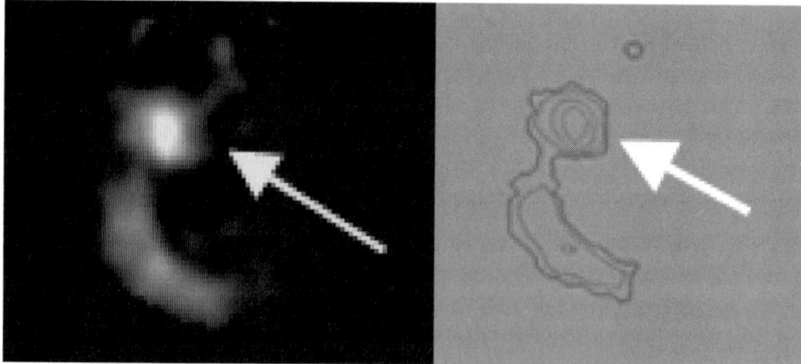

Figure 9 (Left) Enlarged map of the spatial distribution of the TSC slope for a partition in the sodium image of Figure 8 (fourth from left to right). The image clearly demonstrates the spatial heterogeneity of the slope. (Right) Contour map for the image on the right. Four contour levels spaced by 1.9 mM/hr are shown in the plot. The close proximity of the contour lines in the ischemic core (arrows) indicates a large rate of TSC accumulation. The presence of essentially two contour lines in the posterior aspect of the cerebral hemisphere demonstrates a much lower rate of TSC accumulation in this region.

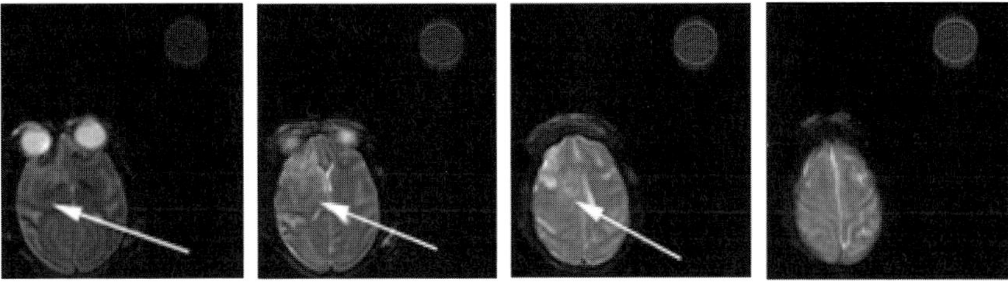

Figure 10   Selected T2 proton images from the animal in Figure 8. The images were acquired 7 hours after onset of ischemia. The arrows illustrate the location of the infarct. Considerable edema has formed in the ischemic hemisphere, leading to a mass effect and concomitant midline shift as demonstrated by the reduced ventricular space (TR = 2000 ms, TE = 120 ms, 256 × 128 matrix size).

below 3.8 mM/hr. Because the size of this latter volume is significantly larger than that of the ischemic cortex, the lower slope in this volume cannot be attributed to partial voluming effects and is therefore representative of an intrinsic heterogeneity in TSC accumulation over the ischemic hemisphere.

The presence of the infarction in the ischemic hemisphere is clearly demonstrated in the T2-weighted proton images of Fig. 10. These images were obtained at the end of the scanning session (6 hours after onset of ischemia) prior to removing the animal from the scanner. The location and distribution of the cerebral edema in these images corresponds well with the location of the ischemic core and penumbral tissues in the sodium images and slope maps shown above.

MR imaging of the temporary ischemia model is demonstrated in Fig. 11 where selected partitions from a time-average of four three-dimensional (3D) sodium images are presented. The images were collected between 1.6 and 2 hours after ischemia onset after an initial MR angiogram (MRA) (Fig. 12, left) was performed to confirm the occlusion of the MCA and PCA territories. The asymmetry in the sodium content of the brain is already apparent in these images. The circular structures at the inferior aspect of the animal's brain correspond to calibration standards with different concentrations of sodium (40, 80, and 120 mM from right to left) in 10% agarose.

Reperfusion of the MCA territory in this animal model is clearly demonstrated in Figure 12 where MRAs of the animal's head are presented immediately before and immediately after reperfusion. The MRAs clearly demonstrate the lack of flow in the PCA before and after balloon deflation. Changes in the rate of TSC accumulation across the right hemisphere were also observed in this model and are clearly depicted in the slope maps presented in Fig. 13. this figure presents slope maps generated using 3D sodium images before and after reperfusion. Notably, most of the changes in TSC accumulation took place in the MCA territory where a reduction in

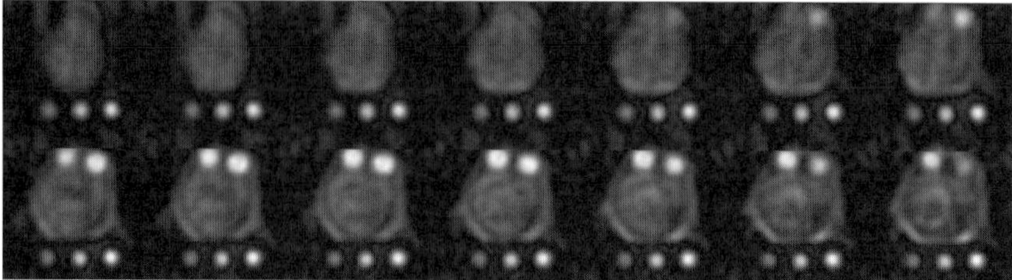

Figure 11 Selected partitions from a 3D time-averaged sodium image from a monkey brain. This image was obtained upon reconstruction of the time-average of four consecutive scans obtained between 1.6 and 2 hours after onset of ischemia. The voxel size is 0.02 cc, and each individual image was collected in 5.5 minutes using a TPI readout (816 scans, TE/TR = 0.5/100 ms, number of excitations (NEX) = 4) optimized for the relaxation times observed in the *in vivo* brain (biexponential relaxation with 3/15 ms for the fast/slow component, respectively).

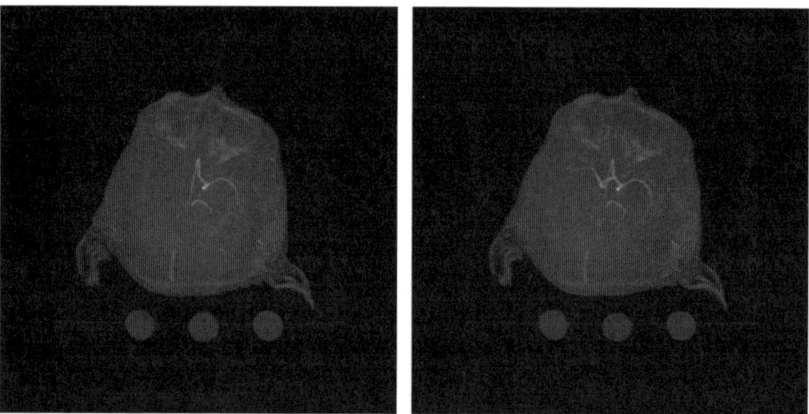

Figure 12 MR angiograms of the animal's neurovasculature before (left) and after (right) reperfusion of the MCA territory. The reperfusion was accomplished through deflation and removal of the balloon from the animal. Reperfusion took placc at 3.5 hours after onset of ischemia and was performed without moving the animal from the cradle or the patient table. The MR angiograms were collected in 3.5 minutes of imaging time using a 3D time of flight sequence optimized for operation at 3.0 T.

TSC accumulation was observed. These findings are consistent with the lack of 2,3,5-triphenyltetrazolium chloride (TTC) staining in the MCA territory on the "rough" histology sections (Fig. 14) and seem to indicate that reperfusion of the MCA territory has been of benefit in this model after 3.5 hours of ischemia. These findings are consistent with those observed in the permanent occlusion model described previously.

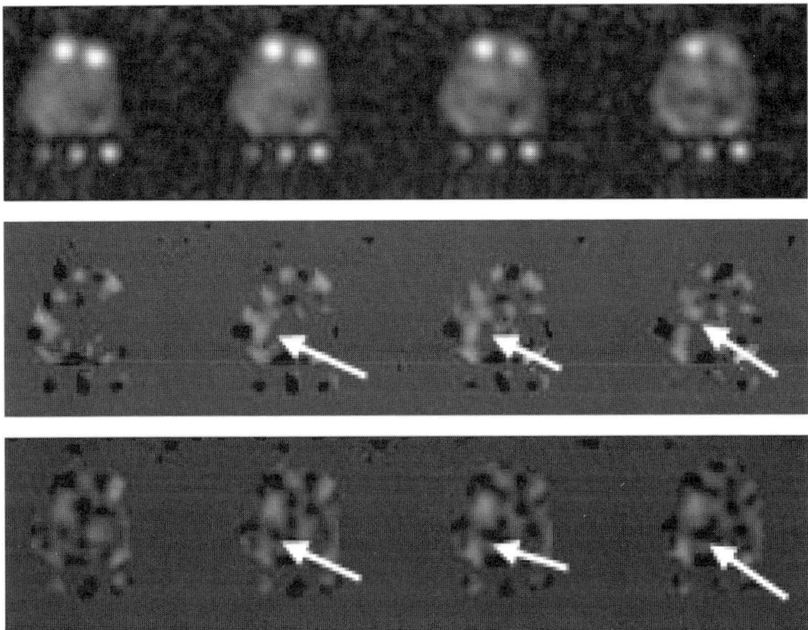

Figure 13   (Top) Selected partition of a sodium image and its corresponding slope map before (middle row) and after (bottom row) reperfusion of the MCA terrtitory. Changes in the rate of TSC accumulation are clearly observed in the region corresponding to the MCA territory (arrows). Only pixels with statistically slopes (P < 0.005) are presented in these maps.

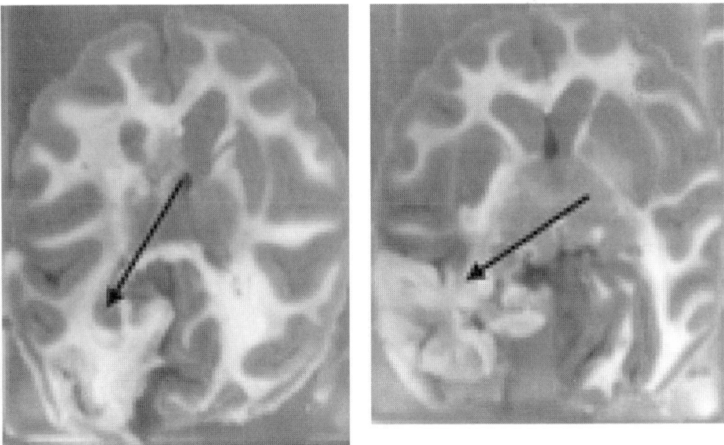

Figure 14   The TTC-stained sections from the brain of the animal in Fig. 13 (temporary ischemia model). The images demonstrate ischemic damage to the PCA territory which is the only vascular territory that remained occluded upon removal of the balloon catheter. (See Color Insert.)

V. Discussion

The existence of well-known ionic gradients across the cell-membrane in the brain and the high energy demands that they impose on the cell's normal metabolism (Erecinska, 1989; Erecinska *et al.*, 1991) have made monitoring of such ion fluxes an ideal means to understand the pathophysiology of stroke. Indeed, the stroke literature has many reports devoted to the study of tissue ion accumulation as a function of time in animal models of ischemia (Betz *et al.*, 1994; Gotoh *et al.*, 1985b; Hatashita and Hoff, 1990; Ito *et al.*, 1979; Menzies *et al.*, 1993; Minematsu *et al.*, 1992; O'Brien *et al.*, 1974; Schuier and Hossmann, 1980; Siegel *et al.*, 1973; Wang *et al.*, 2000; Yang *et al.*, 1992; Young *et al.*, 1987). Most of those studies, however, rely on the use of destructive techniques that require a large number of animals at each time point, which introduces a larger degree of variance on the data. Noninvasive techniques for the interrogation of such tissue parameters are thus highly desirable. Because of its NMR sensitivity the sodium ion is an ideal candidate for the noninvasive monitoring of brain ischemia.

The results presented in this chapter demonstrate that the change in TSC in ischemic tissue is readily measurable using sodium MRI at widely available magnetic field strengths (3.0 T). The changes in TSC as measured by MRI are also in good agreement with values reported in other animal models (Wang *et al.*, 2000) using invasive means. The larger spatial extend of the ischemic area in the animal models presented here allowed the evaluation of the spatial heterogeneity of the TSC accumulation, which is important if TSC is to be used as a surrogate marker for cell viability (Wang *et al.*, 2000). For both models, there is substantial heterogeneity in the accumulation of TSC across the ischemic hemisphere. The observation that the TQ signal intensity is elevated in the initial hours of ischemia seems to indicate that a significant ion flux across the cell membranes takes place shortly after the initial ischemic insult. The slow rise in TSC, on the other hand, seems to support a diffusional ion shift, perhaps from the plasma space, as the mechanism responsible for this gradual increase. Because of the intrinsic heterogeneity in TSC accumulation, tissues with different accumulation rates might be able to remain viable for longer periods of time than those where the initial rise in TSC is very rapid. For example, the tissue volumes below the 3.8-mM/hr TSC accumulation rate should remain viable for twice as long as the ischemic core. For the stroke models presented in this study, the volume of tissue below this tissue accumulation rate is clearly larger than the ischemic core, which suggests that reperfusion of the ischemic hemisphere could be of benefit beyond the 3-hour "window of opportunity" that is the cornerstone of the tPA treatment approach. In fact, as shown previously, the ischemic tissue damage to the MCA territory demonstrated in Figure 10 for this animal was not paralleled in the animal that underwent MCA reperfusion after 3.5 hours of temporary ischemia. These findings suggest that TSC accumulation could have a predictive value in the determination of ischemic tissue damage in the brain.

VI. Conclusions

We have demonstrated the use of quantitative, 3D sodium MRI for the monitoring of TSC changes in permanent and temporary primate models of focal brain ischemia. The combined use of an efficient imaging sequence with ultra-short TEs, as well as the measurement of the B_1 distribution across the imaging volume, permits accurate and reliable measurement of TSC using MRI. Our results demonstrate that TSC increases slowly and linearly as a function of time after focal brain ischemia and that this increase can be reversed without a concomitant development of brain infarction if the period of ischemia is below 3 hours.

This work was supported in part by PHS grants R01 NS44818 and R01 EB00291.

References

Albers, G. W. (1999). Expanding the window for thrombolytic therapy in acute stroke. The potential role of acute MRI for patient selection. *Stroke* **30**, 2230–2237.

Baird, A. E., Benfield, A., Schlaug, G., Siewert, B., Lovblad, K. O., Edelman, R. R., and Warach, S. (1997). Enlargement of human cerebral ischemic lesion volumes measured by diffusion-weighted magnetic resonance imaging. *Ann. Neurol.* **41**, 581–589.

Bansal, N., and Seshan, V. (1995). Three-dimensional triple quantum filtered 23Na imaging of rabbit kidney with weighted signal averaging. *J. Magn. Reson. Imaging* **5**, 1–7.

Bansal, N., Germann, M. J., Lazar, I., Malloy, C. R., and Sherry, A. D. (1992). *In vivo* Na−23 MR imaging and spectroscopy of rat brain during TMDOTP5-infusion. *J Magn. Reson. Imaging* **2**, 385–391.

Bansal, N., Germann, M. J., Seshan, V., Shires, G. T. III, Malloy, C. R., and Sherry, A. D. (1993). Thulium 1,4,7,10-tetraazacyclododecane-1,4,7,10-tetrakis(methylene phosphate as a 23Na shift reagent for the *in vivo* rat liver. *Biochemistry* **32**, 5638–5643.

Barber, P. A., Davis, S. M., Darby, D. G., Desmond, P. M., Gerraty, R. P., Yang, Q., Jolley, D., Donnan, G. A., and Tress, B. M. (1999). Absent middle cerebral artery flow predicts the presence and evolution of the ischemic penumbra. *Neurology* **52**, 1125–1132.

Betz, A., Keep, R., Beer, M., and Ren, X. (1994). Blood-brain barrier permeability and brain concentration of sodium, potassium, and chloride during focal ischemia. *J. Cereb. Blood Flow Metab.* **14**, 29–37.

Boada, F., Kharlamov, A., Kim, D. K., Reid, E. D., Hancu, I., and Jones, S. C. (April 21–27, 2001). Quantitative sodium MRI of evolving focal brain ischemia in the rat. International Society for Magnetic Resonance in Medicine, Glasgow, Scotland.

Boada, F., Gillen, J., Noll, D., Shen, G., and Thulborn, K. (1997). Data acquisition and post-processing strategies for fast quantitative sodium imaging. *Int. J. Imaging Systems Technol.* **8**, 544–550.

Boada, F., Gillen, J., Shen, G., Chang, S., and Thulborn, K. (1997). Fast three dimensional sodium imaging. *Magn. Reson. Med.* **37**, 706–715.

Boada, F., Christensen, J., Huang-Hellinger, F., Reese, T., and Thulborn, K. (1994). Quantitative *in vivo* tissue sodium concentration maps: The effects of biexponential relaxation. *Magn. Reson. Med.* **32**, 219–223.

Christensen, J., Barrere, B., Boada, F., Vevea, J., and Thulborn, K. (1996). Quantitative tissue sodium concentration mapping of normal rat brain. *Magn. Reson. Med.* **36**, 83–89.

Edelstein, W. A., Hutchison, J. M. S., Johnson, G., and Redpath, T. (1980). Spin warp NMR imaging and applications to human whole-body imaging. *Phys. Med. Biol.* **25**, 751–756.

Eliav, U., Shinar, H., and Navon, G. (1992). The formation of a second-rank tensor in Na 23 double-quantum filtered NMR as an indicator for order in a biological tissue. *J. Magn. Res.* **98**, 223–229.

Erecinska, M., Dagani, F., Nelson, D., Deas, J., and Silver, I. (1991). Relations between intracellular ions and energy netabolism: A study with monensin in synaptosomes, neurons and C6 glioma cells. *J. Neurosci.* **11**, 2410–2421.

Erecinska, M. (1989). Stimulation of the Na+/K+ pump activity during electrogenic uptake of acidic amino acid transmitters by rat brain synaptosomes. *J. Neurochem.* **52**, 135–139.

Flacke, S., Keller, E., Hartmann, A., Murtz, P., Textor, J., Urbach, H., Folkers, P., Traber, F., Gieseke, J., Block, W., Scheef, L., Leutner, C., Pauleit, D., and Schild, H. H. (1998). Improved diagnosis of early cerebral infarct by the combined use of diffusion and perfusion [in German]. *Rofo.* **168**, 493–501.

Gelderen, P., Veen, J. W.v.d., Creyghton, J. H. N., Mehlkopf, A. F., and Bovee, W. M. M. J. (1990). Separation of intra- and extracellular sodium by diffusion. 9th Annual Scientific Meeting Society of Magnetic Resonance in Medicine, 1990, New York.

Gotoh, O., Asano, T., Koide, T., and Takakura, K. (1985a). Ischemic brain edema following occlusion of the middle cerebral artery in the rat II: Alteration of the eicosanoid syntheis profile of brain microvessels. *Stroke* **16**, 110–113.

Gotoh, O., Asano, T., Koide, T., and Takakura, K. (1985b). Ischemic brain edema following occlusion of the middle cerebral artery in the rat I: The time courses of the brain water, sodium and potassium contents and blood-brain barrier permeabiity to 125I-albumin. *Stroke* **16**, 101–109.

Group, T. (1995). Tissue plasminogen activator for acute ischemic stroke. The National Institute of Neurological Disorders and Stroke rt-PA Stroke Study Group. *N. Engl. J. Med.* **333**, 1581–1587.

Gupta, R. K., and Gupta, P. (1982). Direct observation of resolved resonances from intra- and extra-cellular sodium-23 ions in NMR studies of intact cells and tissues using dysprosium (III) tripolyphosphate as paramagnetic shift reagent. *J. Magn. Reson.* **47**, 344–350.

Hancu, I., Boada, F., and Shen, G. (1999). Three-dimensional triple-quantum-filtered 23Na imaging of *in vivo* human brain. *Magn. Reson. Med.* **42**, 1146–1154.

Hatashita, S., and Hoff, J. (1990). Brain edema and cerebrovascular permeability during cerebral ischemia in rats. *Stroke* **21**, 582–588.

Hilal, S., Maudsley, A., Ra, J., Simon, H., Roschmann, P., Wittekoek, S., Cho, Z., and Mun, S. (1985). *In vivo* NMR imaging of sodium-23 in the human head. *J. Comput. Assist. Tomog.* **9**, 1–7.

Hilal, S., Oh, C., Mun, I., and Silver, A. (1992). Sodium imaging. *In* "Magnetic Resonance Imaging" (D. Stark and W. Bradley, Eds.), pp. 1091–1110. Mosby, St. Louis.

Hillis, A. E., Barker, P. B., Beauchamp, N. J., Gordon, B., and Wityk, R. J. (2000). MR perfusion imaging reveals regions of hypoperfusion associated with aphasia and neglect. *Neurology* **55**, 782–788.

Horowitz, M., Kassam, A., Nemoto, E., Arimoto, J., and Jungreis, C. (2001). An endovascular primate model for the production of a middle cerebral artery ischemic infarction. *Interventential Neuroradiology* **7**, 223–228.

Hossmann, K., Sakai, S., and Zimmermann, V. (1977). Cation activities in reversible ischemia of the cat brain. *Stroke* **8**, 77–81.

Hutchinson, R., Huntley, J. J. A., Zhou, H. Z., Ciesla, D. J., and Shapiron, J. I. (1993). Changes in double quantum filtered sodium intensity during prolonged ischemia in the isolated perfused heart. *Magn. Reson. Med.* **29**, 391–395.

Ito, U., Ohno, K., Nakamura, R., Suganuma, F., and Inaba, Y. (1979). Brain ddema during ischemia and after restoration of blood flow. *Stroke* **10**, 542–547.

Jelicks, L. A., and Gupta, R. K. (1989). Double-quantum NMR of sodium ions in cells and tissues. Paramagnetic quenching of extracellular coherence. *J. Magn. Res.* **81**, 586–592.

Jungreis, C. A., Nemoto, E., Boada, F., and Horowitz, M. B. (2003). Model of reversible cerebral ischemia in a monkey model. *AJNR Am. J. Neuroradiol.* **24**, 1834–1836.

Kalyanapuram, R., Sesham, V., and Bansal, N. (1998). Three-dimensional triple-quantum-filtered 23Na imaging of the dog head *in vivo*. *J. Magn. Res. Imaging* **8**, 1182–1189.

Kemp-Harper, R., Styles, P., and Wimperis, S. (1995). Three-dimensional triple-quantum filtration 23Na NMR imaging. *J. Magn. Reson. B* **108**, 280–284.

Kidwell, C., Saver, J., Mattiello, J., Starkman, S., Vinuela, F., Duckwiler, G., Gobin, Y., Jahan, R., Vespa, P., Kalafut, M., and Alger, J. (2000). Thrombolytic reversal of acute human cerebral ischemic injury shown by diffusion/perfusion magnetic resonance imaging. *Ann. Neurol.* **47**, 462–469.

Knubovets, T., Shinar, H., and Navon, G. (1998). Quantification of the contribution of extracellular sodium to Na-23 multiple-quantum-filtered NMR spectra of suspensions of human red blood cells. *J. Magn. Reson.* **131**, 92–96.

Krueger, K., Kugel, H., Grond, M., Thiel, A., Maintz, D., and Lackner, K. (2000). Late resolution of diffusion-weighted MRI changes in a patient with prolonged reversible ischemic neurological deficit after thrombolytic therapy. *Stroke* **31**, 2715–2718.

Labadie, C., Lee, J.-H., Vetek, G., and Springer, C. S. (1994). Relaxographic imaging. *J. Magn. Reson. B* **105**, 99–112.

Lee, J.-H., Labadie, C., and Springer, C. S. (1990). Relaxographic analysis of 23Na resonances. 9th Annual Scientific Meeting Society of Magnetic Resonance in Medicine, 1990. New York.

Lian, J., Zha, L., and Lowe, I. (1990). A NMR technique for measuring the RF magnetic field distributions of NMR coils. 9th Annual Scientific Meeting Society of Magnetic Resonance in Medicine, 1990. New York.

Lyon, R., Pekar, J., Moonen, C., and Mclaughlin, A. (1991). Double-quantum surface-coil NMR studies of sodium and potassium in the rat brain. *Magn. Reson. Med.* **18**, 80–92.

Memezawa, H., Smith, M., and Siesjo, B. (1992). Penumbral tissues salvaged by reperfusion following middle cerebral artery occlusion in rats. *Stroke* **23**, 552–559.

Menzies, S., Betz, A., and Hoff, J. (1993). Contributions of ions and albumin to the formation and resolution of ischemic brain edema. *J. Neurosurg.* **78**, 257–266.

Meyer, C., Bland, P., and Pipe, J. (1995). Retrospective correction of intensity inhomogeneities in MRI. *IEEE Trans. Med. Imaging* **14**, 36–41.

Minematsu, K., Li, L., Sotak, C., Davis, M., and Fisher, M. (1992). Reversible focal ischemia injury demonstrated by diffusion-weighted magnetic resonance imaging in rats. *Stroke* **23**, 1304–1311.

Moyher, S., Vigneron, D., and Nelson, S. (1995). Surface coil MR imaging of the human brain with an analytic reception profile correction. *J. Magn. Reson. Imaging* **5**, 139–144.

Navon, G. (1993). Complete elimination of the extracellular 23Na NMR signal in triple quantum filtered spectra of rat hearts in the presence of shift reagents. *Magn. Res. Med.* **30,** 503–506.

O' Brien, M., Jordan, M., and Waltz, A. (1974). Ischemic cerebral edema and the blood-brain barrier. *Arch. Neurol.* **30,** 461–465.

Pekar, J., and Leigh, J. S. (1986). Detection of biexponential relaxation in sodium-23 facilitated by double-quantum filtering. *J. Magn. Reson.* **69,** 582–584.

Pekar, J., Renshaw, P. F., and Leigh, J. S., Jr. (1987). Selective detection of intracellular sodium by coherence-transfer NMR. *J. Magn. Res.* **72,** 159–161.

Ra, J., Hilal, S., and Cho, Z. (1986). A method for *in vivo* MR imaging of the short T2 component of sodium-23. *Magn. Reson. Med.* **3,** 296–302.

Reddy, R., Insko, E. K., and Leigh, J. S. (1997). Triple quantum sodium imaging of articular cartilage. *Magn. Reson. Med.* **38,** 279–284.

Reddy, R., Bolinger, L., Shinnar, M., Noyszewski, E., and Leigh, J. S. (1995). Detection of residual quadrupolar interaction in human skeletal muscle and brain *in vivo* via multiple quantum filtered sodium NMR spectra. *Magn. Reson Med.* **33,** 134–139.

Rooney, W. D., and Springer, C. S., Jr. (1991). A comprehensive approach to the analysis and interpretation of the resonances of spins 3/2 from living systems. *NMR Biomed.* **4,** 209–226.

Rooney, W. D., and Springer, C. S., Jr. (1991). The molecular environment of intracellular sodium: 23Na NMR relaxation. *NMR Biomed.* **4,** 227–245.

Schepkin, V. D., Choy, I. O., and Budinger, T. F. (1996). Sodium alteration in isolated rat heart during cardioplegic arrest. *J. Appl. Physiol.* **81,** 2696–2702.

Schuier, F., and Hossmann, K. (1980). Experimental brain infarcts in cats. II. Ischemic brain edema. *Stroke* **11,** 593–601.

Schuierer, G., Ladebeck, R., Barfub, H., Hentschel, D., and Huk, W. (1991). Sodium-23 imaging of supratentorial lesions at 4.0T. *Magn. Reson. Med.* **22,** 1–9.

Seshan, V., Sherry, A. D., and Bansal, N. (1997). Evaluation of triple quantum-filtered 23Na NMR spectroscopy in the *in situ* rat liver. *Magn. Res. Med.* **38,** 821–827.

Shimizu, T., Naritomi, H., and Sawada, T. (1993). Sequential changes on Na23 MRI after cerebral infarction. *Neuroradiology* **35,** 416–419.

Siegel, B., Studer, R., and Potchen, E. (1973). Brain 22Na uptake in experimental cerebral microembolism. *J. Neurosurg.* **38,** 739–742.

Stolberger, R., Wach, P., McKinnon, G., Justich, E., and Ebner, F. (1988). RF-field mapping *in vivo.* 7th Annual Scientific Meeting Society for Magnetic Resonance in Medicine, 1988.

Sorensen, A., Koroshetz, W., Buonano, F., Schwamm, L., Reese, T., Weisskoff, R., Copen, W., Look, R., Rosen, B., and Gonzalez, R. (1996). Diffusion/perfusion Mismatch in MRI of acute human stroke. 4th Annual Meeting ISMRM, 1996. New York.

Talagala, S., and Gillen, J. (1991). Experimental determination of three dimensional RF magnetic field distribution of NMR coils. *J. Magn. Reson.* **94,** 493–500.

Tauskela, J. S., Dizon, J. M., Whang, J., and Katz, J. (1997). Evaluation of multiple-quantum-filtered 23Na NMR in monitoring intracellular Na content in the isolated perfused rat heart in the absence of a chemical shift reagent. *Magn. Reson. Med.* **127,** 115–127.

Thulborn, K., and Ackerman, J. (1983). Absolute molar concentraions by NMR in inhomogeneous B1, a scheme for analysis of in vivo metabolites. *J. Magn. Reson.* **55,** 357–371.

Thulborn, K., Boada, F., Christensen, J., Huang-Hellinger, F., Reese, T., and Kosewski, J. (1993). B1 correction maps and apparent water density maps as tools for quantitative functional MRI. 12th Annual Scientific Meeting Society of Magnetic Resonance in Medicine, 1993. New York.

Thulborn, K., Gidin, T., Davis, D., and Erb, P. (1999). Comprehensive MRI protocol for stroke management: Tissue sodium concentration as a measure of tissue viability in a non-human primate model and clinical studies. *Radiology* **139,** 26–34.

Tofts, P., and Wray, S. (1988). Critical assessment of methods of measuring metabolite concentrations by NMR spectroscopy. *NMR Biomed.* **1,** 1–10.

Ugurbil, K., Garwood, M., Rath, A., and Bendall, M. (1988). Amplitude- and frequency/phase-modulated refocusing pulses that induce plane rotations even in the presence of inhomogeneous B1 fields. *J. Magn. Reson.* **78,** 472–497.

van der Veen, J. W. C.v.d., Gelderen, P.v., Creyghton, J. H. N., and Bovee, W. M. M. J. (1993). Diffusion in red blood cell suspensions: Separation of the intracellular and extracellular NMR sodium signal. *Magn. Reson. Med.* **29,** 571–574.

Wang, Y., Hu, W., Perez-Trepichio, A., Ng, T., Furlan, A., Majors, A., and Jones, S. (2000). Brain tissue is a ticking clock telling time after arterial occlusion in rat focal cerebral ischemia. *Stroke* **31,** 1386–1392.

Warach, S., Dashe, J., and Edelman, R. (1996). Clinical outcome in ischemic stroke predicted by early diffusion-weighted and perfusion magnetic resonance imaging: A preliminary analysis. J. *Cereb. Blood Flow Metab.* **16,** 53–59.

Warach, S., Wielopolsk, P., and Edelman, R. (1993). Identification and characterization of the ischemic penumbra of acute human stroke using echo planar diffusion and perfusion imaging. 12th Annual Scientific Meeting Society of Magnetic Resonance in Medicine, 1993. New York.

Wimperis, S., and Wood, B. (19912). Triple-quantum sodium imaging. *J. Magn. Res.* **95,** 428–436.

Yang, G., Chen, S., Kinouchi, H., Chan, P., and Weinstein, P. (1992). Edema, cation content, and ATPase activity after middle cerebral artery occlusion in rats. *Stroke* **23,** 1331–1336.

Young, W., Rappaport, Z., Chalif, D., and Flamm, E. (1987). Regional brain sodium, potassium, and water changes in the rat middle cerebral artery occlusion model of ischemia. *Stroke* **18,** 751–759.

5

Quantum Dot Surfaces for Use
In Vivo and *In Vitro*

Byron Ballou
Molecular Biosensor and Imaging Center and Department of Biological Sciences
Carnegie Mellon University Pittsburgh, Pennsylvania 15213

I. Introduction

A. General Background

Quantum dots, first introduced for biological labeling in 1998 (Bruchez *et al.*, 1998; Chan and Nie, 1998) have proved to be extraordinarily useful fluorescence reagents that have significant advantages over other types of fluorescent dyes. They combine very high brightness, due both to high absorbency and high quantum yields, with unprecedented resistance to photobleaching. Emission wavelengths, governed primarily by composition and secondarily by size, range from the near-ultraviolet to the infrared. For a given quantum dot composition, emission bandwidths depend on the size range of

Current Topics in Developmental Biology, Vol. 70
Copyright 2005, Elsevier Inc. All rights reserved.

0070-2153/05 $35.00
DOI: 10.1016/S0070-2153(05)70005-3

the quantum dots and can be very narrow; <35 nm width is now routine for production quantum dots. Core-shell quantum dots can have quantum yields greater than 90%, and quantum dots may be excited in a broad spectral window shorter than their emission wavelength, with absorbency rising as wavelength decreases. For example, molar absorbencies of 6×10^6 at 450 nm and quantum yields in excess of 60% are found for the 655 nm polyethylene glycol (PEG)-coated cadmium selenide–zinc sulfide (CdSe-ZnS) core-shell quantum dots used in our laboratory. Because all quantum dots show rising absorbency at wavelengths shorter than their emission wavelengths, many quantum dots can be excited using one illumination window. (Readers are referred to the Quantum Dot Corporation Website, http://www.qdots.com, for more details and for several tutorials on the properties of quantum dots.)

The combination of high brightness, photostability, and narrow emission bandwidths with the ability to excite many colors naturally leads to the possibility of using multicolor combinations of quantum dots to label ("bar-code") large numbers of different objects (e.g., different cell types in a mixed population) (Chan and Nie, 1998; Gao and Nie, 2003, 2004; Han *et al.*, 2001; Jaiswal *et al.*, 2003; Lagerholm *et al.*, 2004; Voura *et al.*, 2004). Quantum dots may be used for single-molecule imaging in living cells (Dahan *et al.*, 2003). Finally, quantum dots are well suited for two-photon microscopy (Larson *et al.*, 2003).

There are potential drawbacks to the use of quantum dots: their large size and high molecular weights may limit applications that require measurement of molecular mobility, and attached quantum dots might interfere with molecular interactions. Finally, since the current generation of quantum dots is composed of toxic heavy metals (CdSe and cadmium telluride [CdTe] cores, with ZnS shells), toxicity might be anticipated if the quantum dots degrade during use. Several recent reviews have summarized progress in biological applications of quantum dots (Jaiswal and Simon, 2004; Parak *et al.*, 2003; Smith *et al.*, 2004). In this chapter we briefly review recent advances in the use of quantum dots for biological imaging, then summarize our work on the effects of chemically varying quantum dot surface properties to improve cellular uptake and imaging *in vivo*. Applications of nucleic acid conjugates are not discussed.

B. Stabilizing the Surface

Core-shell quantum dots are stable and highly fluorescent in nonpolar organic solvents but are not very fluorescent in aqueous solution. Much current work involves making surface coatings that preserve high fluorescence, confer stability and solubility in aqueous solution, and allow ready conjugation of biological molecules. During the usual core-shell manufacturing process, freshly prepared quantum dots are coated with trioctylphosphine oxide

(TOPO) (Dabbousi *et al.*, 1997; Hines and Guyot-Sionnest, 1996). Thus the quantum dots start with a hydrophobic surface. Several approaches have been reported to work reasonably well for creating a stable primary coat around the core-shell that preserves fluorescence in aqueous solution:

1. Coating the quantum dot with an amphiphilic (amp) polymer (Gao *et al.*, 2004; Watson *et al.*, 2003; Wu *et al.*, 2003)
2. Replacing the TOPO using organosulfur or other organophosphorous ligands that compete for binding sites on the quantum dot shell. Examples include mercaptoacetic acid (Chan and Nie, 1998), mercaptopropionic acid (Mitchell *et al.*, 1999), dihydrolipoic acid (DHLA) (Mattoussi *et al.*, 2000), DL-cysteine (Sukhanova *et al.*, 2004), and an organic phosphene oligomer (Kim and Bawendi, 2003)
3. Forming a micelle around quantum dots using a mixture of phosphatidylcholine and PEG-substituted phosphatidylethanolamine (Dubertret *et al.*, 2002), or by coating with a cone-shaped amphiphile that self-assembles around the quantum dot (Osaki *et al.*, 2004)
4. Creating a silica layer around the dot (Bruchez *et al.*, 1998; Chen and Gerion, 2004; Gerion *et al.*, 2001)
5. Adsorbing albumin directly to quantum dots (Hanaki *et al.*, 2003). These methods may be combined; for example, a primary cysteine coat was overcoated using polyallylamine (Sukhanova *et al.*, 2004). All these primary coats allow conjugation of biomolecules, usually by amide or thioether formation.

C. Minimizing Uptake *In Vitro* and *In Vivo*

Minimizing nonspecific aggregation and binding requires further modification of quantum dot surfaces. Most authors have used PEG conjugation to minimize nonspecific binding, as PEG derivatives for conjugation are readily available and work effectively in many systems (see reviews by Chapman, 2002; Greenwald, 2001; Harris and Chess, 2003; Harrington *et al.*, 2002; Molineux, 2002). A potential drawback to PEG conjugation is that neither tissue-cultured cells nor live animals metabolize PEG (*Ibid.*). Sugar or polysaccharide derivatives offer another potential way to avoid nonspecific binding or to cause binding to appropriate receptors (Osaki *et al.*, 2004).

D. Conjugates that Confer Specificity

Conjugation of biomolecules to the primary coat is usually performed by conventional methods, using active esters, carbodiimides, or maleimides. An interesting approach is the use of cationic proteins or polyhistidine tagged chimeric proteins for self-assembly onto DHLA-coated quantum dots. Binding is mediated by electrostatic interactions between the negatively charged surface of the quantum dot and the cationic protein or tag (Goldman *et al.*, 2002a,b; Mattoussi *et al.*, 2000; Voura *et al.*, 2004). This primary coat allows fluorescence resonance energy transfer (FRET) between adsorbed surface components and quantum dots (Clapp *et al.*, 2004; Medintz *et al.*, 2003a,b, 2004).

E. Binding and Uptake Using Cultured Cells *In Vitro*

Binding to cell surfaces nonspecifically can promote uptake (Derfus *et al.*, 2004a). Quantum dots layered on a surface can be taken up by cells that traverse the surface (Parak *et al.*, 2002). Binding to, and in some cases uptake by, specific cell surface receptors has been demonstrated in many cases (transferrin receptor [Chan and Nie, 1998], serotonin receptors [Rosenthal *et al.*, 2002], glycine receptor [Dahan *et al.*, 2003], and epidermal growth factor receptor [Lidke *et al.*, 2004]). Membrane labeling and monitoring of membrane integrity was performed using concanavalin A-biotin bound to streptavidin quantum dots (Minet *et al.*, 2004). Cell labeling by quantum dots after microinjection (Dubertret *et al.*, 2002), electroporation (Chen and Gerion, 2004), and cationic lipid–mediated cell entry (Derfus *et al.*, 2004a; Voura *et al.*, 2004) have all been demonstrated. Although internalized quantum dots frequently localize in endosomes, organelle-specific localization to the cell nucleus and to mitochondria has been shown using appropriate peptide conjugates to DHLA-coated quantum dots (Derfus *et al.*, 2004a) and to the nucleus using peptide conjugates of silica-coated quantum dots (Chen and Gerion, 2004); thus once quantum dots are transported across the cell membrane, tagging of intracellular structures is possible.

Derfus *et al.* (2004a) compared four methods of internalizing quantum dots: a membrane transport peptide, cationic lipids, electroporation, and microinjection. All methods except microinjection caused aggregation of the mercaptoacetic acid–coated quantum dots used in their experiments; cationic lipids provided the highest delivery.

We have successfully incorporated large numbers of polymer-coated quantum dots by using polyarginine conjugates. Results from our laboratory (Lagerholm *et al.*, 2004) are discussed in the following text.

F. Cell Tracking *In Vivo*

Two groups have used quantum dot–labeled cells to follow circulation and extravasation of labeled tumor cell in living mice. Hoshino *et al.* (2004) used quantum dots coated by cross-linked sheep albumin to label mouse EL-4 lymphoma cells. Although high concentrations of these quantum dot conjugates proved toxic to cells, stable nontoxic labeling could be obtained at lower input levels. After injection into mice, persistence of labeled cells in circulation could be followed using a fluorescence-activated cell sorter (FACS) for at least 5 days, and uptake in various tissues (mainly spleen and lungs) could be monitored by microscopy. Voura *et al.* (2004) used cationic lipid–mediated internalization to label mouse B16F10 melanoma cells with DHLA-capped quantum dots. Cells were co-labeled with organic dyes to determine whether any loss of quantum dots occurred, as would be shown by a loss of coincidence of the two labels; on the time scale of these experiments (5 hours), no such loss was detectable. Again, by comparison with organic dye labels, quantum dots had no effects on the ability of the cells to survive in circulation, extravasate (emerge from the vascular compartment), or form metastatic tumors. Both standard fluorescence microscopy and two-photon microscopy could be used to detect the labeled cells. The authors also showed that two populations of cells, severally labeled using different quantum dot emission colors, could be followed. Thus in two tumor cell models, quantum dot labeling is feasible for cell tracking *in vivo*. Both articles noted the potential for using multiple quantum dot labels to follow mixed cell populations.

G. Conjugates for Use *In Vivo*

Two nonspecific surfaces have been used *in vivo*. Oligophosphene-coated quantum dots were used to study lymphatic flow and sentinel lymph node accumulation (Kim *et al.*, 2004). The authors showed that rapid and ready detection of node accumulation was possible using both mouse and pig models; we may anticipate that one of the earliest medical uses for quantum dots will be in sentinel lymph node marking.

Our group found that polymer-coated quantum dots (Wu *et al.*, 2003) could be be used either for lymph node mapping or as long-circulating vascular markers, depending on their surfaces (Ballou *et al.*, 2004, and see below.)

Peptides and antibody fragments are the only moieties used to date for selective targeting *in vivo*. Akerman *et al.* (2002) coupled three different thiolated peptides to mercaptoacetic acid–surfaced ZnS-CdSe quantum dots by exchanging with the mercaptoacetic acid surface. The three peptides bound respectively to lung epithelium, tumor blood vessels, and a lymphatic cell marker also expressed in some tumors. All three peptide-coated quantum dots showed the same binding specificity as the corresponding peptides. After injection, each peptide–quantum dot conjugate homed to mouse lung or tumors as expected. Both aggregation *in vitro* and nonspecific uptake *in vivo* by the reticuloendothelial system (RES) were inhibited by making quantum dots approximately half-substituted by thiolated PEG, half by the peptide. The authors noted the possibility of using mixed peptide substituents to give greater specificity and improved targeting.

For tumor location *in vivo*, Gao *et al.* (2004) coated quantum dots with a triblock polymer consisting of polybutylacrylate–polyethylacrylate–polymethylacrylate substituted with an octylamine side chain on the methyacrylate moieties (~1/3 substitution.) This structure self-assembled on quantum dots, yielding a thick coat that conferred remarkable stability to pH changes. Conjugation with PEG (to minimize aggregation and increase circulating lifetime) and with an engineered antibody fragment directed to a prostate-specific antigen yielded quantum dots that targeted specifically to implanted prostate tumors in nude mice.

II. Laboratory Studies

A. Quantum Dot Surfaces for Use with Tissue-Cultured Cells

We have examined systematically the effects of changing surface charge and PEG substitution on quantum dot uptake into tissue-cultured cells (Lagerholm *et al.*, 2004). We began our experiments using quantum dots coated by hexadecylthiol, which yielded quantum dots with good stability and quantum yield, but they were soluble only in organic solvents. However, hydrophobic complexes with lecithins and other long-chain alkyl detergents yielded water-soluble, nonaggregating materials (Ernst *et al.*, in preparation). These lecithin-substituted surfaces could be conjugated to biomolecules by using partial substitution with phosphatidyl ethanolamine (see Dubertret *et al.*, 2002). When mixed with dodecyltrimethyl ammonium bromide, quantum dots with cationic surfaces were formed. We also used quantum dots coated with an amphiphilic polymer ("amp," polyacrylic acid partly substituted with octylamine [Wu *et al.*, 2003]), which gave good preservation of quantum yield (40–70% yields in dilute buffer solution); these quantum dots were stable in aqueous solution and showed minimal aggregation at neutral pH or above. We found that macrophages and dendritic cells take up lecithin-coated and amp-coated quantum dots spontaneously, and that uptake could be improved by adding a cationic surface to the quantum dots or reduced by conjugating PEG. Nonphagocytic cells did not take up the quantum dots (Lagerholm *et al.*, 2004).

In collaboration with Danith Ly (Carnegie Mellon University), we coupled biotinylated polyarginine (polyarg) residues (Wender *et al.*, 2000) to streptavidin-coated quantum dots. Bio-arg$_9$ caused rapid binding to and uptake into many cell types. Uptake occurred rapidly, was saturable only at very high levels, and the quantum dots were concentrated into endosomes of all cell types examined. Quantum dot fluorescence was stable for many weeks in the endosomes. Multiplex labeling of cells was demonstrated. In further experiments we found that the positive charge on the surface of quantum dots was the most important factor in binding and uptake, rather than the specific cation conjugated (Lagerholm *et al.*, 2004). Polyarg-surface quantum dots have since been commercialized (Qtracker Cell Labeling Kits, Quantum Dot Corporation; Website: http://www.qdots.com).

B. Quantum Dot Surfaces for *In Vivo* Studies

In collaboration with Quantum Dot Corporation, we obtained quantum dots with a primary amp coat and further substituted with several different molecular weight methoxy-terminated PEGs (mPEGs). We also prepared PEG conjugates that had carboxyl or amine functions on the unattached ends of conjugated PEGs (Fig. 1). This variety of surfaces, all based on the stable amp coat, allowed us to investigate systematically the effects of the molecular weight of the PEG substituent and of surface charges on circulating lifetime and sites of uptake *in vivo*. Mice were injected in the tail vein, and circulating lifetime was followed by successive bleeds at intervals after injection, or by imaging the anesthetized mouse. Sites of deposition were assessed by external imaging and by necropsy. Results may be summarized as follows:

1. Quantum dots are easily visible through the nude mouse skin; quantum dots that emit in the near-infrared (700–850 nm) range are more visible, as would be expected from our results using cyanine dyes for localization *in vivo* (Ballou *et al.*, 1997, 1998). The illumination wavelength is important in determining how deeply fluorescent quantum dots may be seen. Figure 2 shows the striking difference in the visibility of internal structures seen using 850 nm–emitting quantum dots with illumination at 450 nm or at 780 nm. The choice of the excitation wavelength, as well as the emission wavelength, can be used to determine what is detectable; either the surface or internal anatomy may be highlighted, at least on the scale of a small animal.

2. Carboxyl-coated quantum dots (unsubstituted amp- or carboxy-PEG) are rapidly taken up (half-life in circulation ~8 minutes, varying from 4–12 minutes.)

3. Amino-terminal PEG surfaces had varying half-lives in circulation depending on the molecular weight of the PEG; amino-PEG-3400 had a longer half-life than amino-PEG-2000. Since these particles were labeled

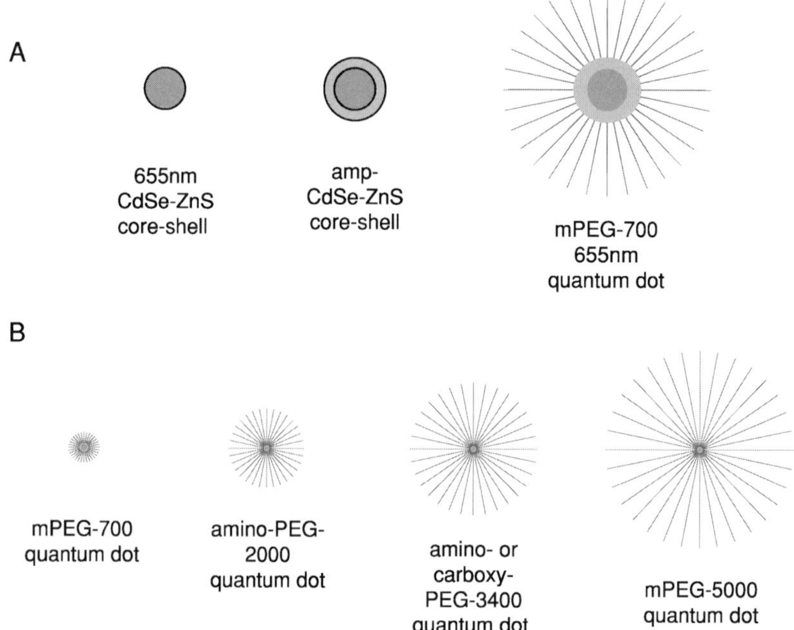

Figure 1 Quantum dot coatings used for this chapter. (A) 655 nm–emitting cadmium selenide core–zinc sulfide (CdSE-ZnS) shell quantum dot, showing approximate thickness of amphipathic coat (amp) and polyethylene glycol (PEG), average molecular weight 700 (PEG-700), conjugated to an amp-quantum dot. (B) Relative sizes of PEG conjugates used in this chapter. PEG polymer is drawn as if fully extended. (See Color Insert.)

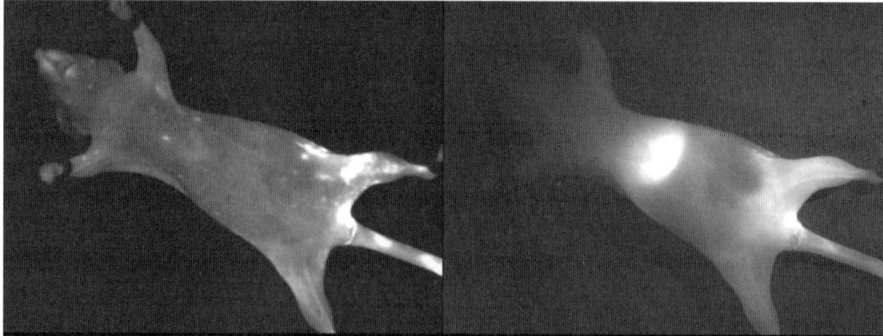

Figure 2 Mouse visualized 3 minutes after injection of amp-850 nm-emitting quantum dots. Left, 450 nm excitation light; right, 780 nm excitation. Identical emission filter (860 × 100 nm). Exposures made in rapid succession. Note that excitation at 450 nm emphasizes surface and subcutaneous features; at this early time point, the mouse liver is more visible using 780 nm excitation.

using carbodiimide coupling of diamino-PEGs, it is difficult to draw any firm conclusion because we do not at present know either the degree of substitution or the extent to which both amino-termini in any PEG are conjugated to the amp surface (hairpinning).

4. When using neutral mPEG, results varied depending on the length of the PEG and degree of substitution. In our published results (Ballou *et al.*, 2004), we contrasted the short lifetime of mPEG-700–coated quantum dots (12 minutes) with that of mPEG-5000–coated quantum dots (50–100 minutes.) Subsequent results using improved, more highly substituted PEG quantum dots have yielded half-lives in the 3- to 8-hour range for mPEG-5000. Increasing PEG chain length to 10K or 20K produced no further improvement in circulating half-life. In several early preparations there was a minority population of long-circulating quantum dots; this probably results from inhomogeneous surface modification.

Sites of deposition vary with the quantum dot surface. Amp-, amino-PEG, carboxy-PEG, and mPEG-700 quantum dots are deposited in the lymph nodes, liver, spleen, and, bone marrow (that is, the RES), although the sites vary somewhat with the surface. Figure 3 shows that sites of deposition of carboxy-PEG-3400 and amino-PEG-2000 are very similar. Deposition of uncharged PEG conjugates depends on the molecular size of the PEG, and probably on the density of substitution. As noted above, mPEG-750 quantum dots deposit in the RES, similar to charged quantum dots; on the other hand, mPEG-5000 quantum dots show very little deposition in the lymph nodes and spleen (Ballou *et al.*, 2004). Most of the injected dose of all types of quantum dots we used was excreted in the feces within 1–2 days.

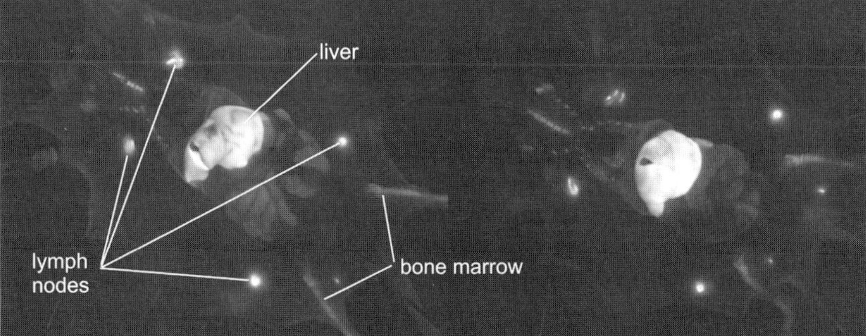

Figure 3 Sites of deposition of charged 655 nm–emitting quantum dots after tail vein injection. Left, carboxy-PEG-3400; right, amino-PEG-2000. Necropsy at 24 hours. Note fluorescence in liver, lymph nodes, and bone marrow. Excitation at 450 nm; emission at 655 nm.

C. Microscopic Detection

A great advantage of quantum dots is their relative resistance to standard conditions of processing for microscopy. We have used standard formaldehyde fixation, paraffin embedding, and hematoxylin staining to follow quantum dot deposition in tissues. By using quantum dots that emit at far-red or near-infrared wavelengths, fluorescence may be seen in stained sections with little background; Figure 4 shows one example. One day after injection, quantum dots deposited in the lymph nodes and spleen are found surrounding the germinal centers in large mononuclear cells; in liver, they are primarily localized in parenchymal cells.

D. Stability *In Vivo*

Many of the quantum dots taken up in tissues after injection remain fluorescent for many months after injection. Figure 5 shows one example; necropsy of a mouse imaged 8 months after injection clearly shows remaining fluorescence in the lymph nodes. At this point, most fluorescence had been cleared from the liver, spleen, and bone marrow. We have not determined whether nonfluorescent quantum dots are also retained for extended periods.

Although a fraction of the quantum dots can remain in animals for months, we also observed that some quantum dots may be extravasated very rapidly and deposited in the skin of injected mice. Figure 6 shows skin localization of amino-PEG-2000 at 39 minutes after tail vein injection.

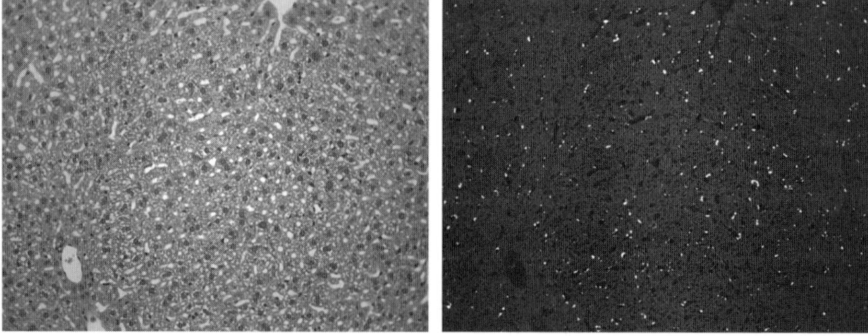

Figure 4 Liver section from mouse injected using 705 nm–emitting mPEG-5000 quantum dots. The mouse was injected using 300 pmol quantum dots, then necropsied 24 hours later. Tissues were fixed, paraffin-embedded, sectioned, and stained using standard methods. Left, hematoxylin-eosin stain; right, fluorescence. Excitation at 450 nm; emission at 705 nm. (See Color Insert.)

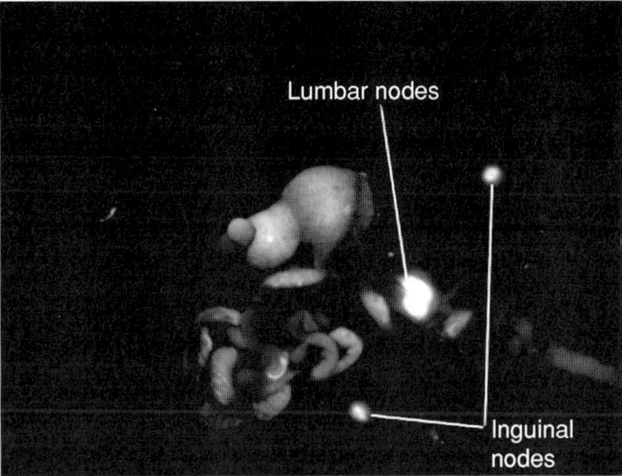

Figure 5 Nude mouse necropsied 8 months after injection of 360 pmol 655 nm–emitting quantum dots. Fluorescent lumbar and inguinal lymph nodes are indicated. Note that the background in the liver and digestive tract is actually low; the optical transfer function for the image was adjusted to display the mouse body. Excitation at 450 nm; emission at 655 nm.

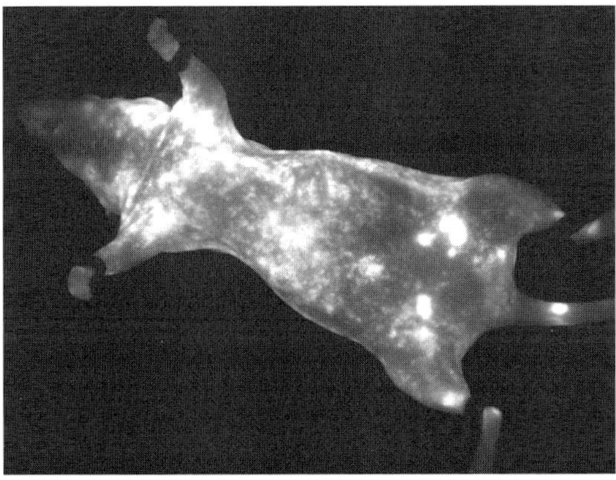

Figure 6 Skin fluorescence 39 minutes after injection of amino-PEG-2000 705 nm–emitting quantum dots. Excitation at 450 nm, emission at 705 nm.

E. Spectral Windows

We have used 605, 630, 645, 705, 755, 800, and 850 nm–emitting quantum dots. As expected, *in vivo* visualization is significantly improved closer to the infrared range. After allowing for quantum efficiency and camera sensitivity, the longer wavelength emitters are better for visualization of internal structures. Residual background fluorescence is essentially gone by 700 nm. However, for deep penetration, the illumination wavelength is critical. As noted, Fig. 1 shows that visualization at moderate depth (>2 mm) is much better when using 780 nm illumination than when using 450 nm despite the much lower absorbency of 850 nm quantum dots at 780 nm than at 450 nm. Visualization in the near-infrared range through the inhomogeneous scattering and absorbing medium of living tissues has been reviewed recently (Frangioni, 2003). Both theoretical modeling and empirical data suggest that there are spectral windows in the near-infrared range that offer acceptable compromises between light scattering by tissues (best minimized by using long wavelengths) and light absorption, due to hemoglobin at wavelengths below 650 nm and to both water and biomolecules, which have high absorbency in the near-infrared range (Lim *et al.*, 2003). Water absorption becomes prohibitive at wavelengths beyond ∼1350 nm. However, silicon-based cameras have sharply reduced efficiency beyond 900 nm, even when adequately cooled. Of the quantum dots we have used, 850 nm emitters seem best suited for visualization in whole animals using readily available cameras.

F. Imaging Lymphatics

Deliberate labeling of the lymphatic vessels is possible by subcutaneous injection, avoiding major blood vessels. We have used injection into the mouse tail to explore transport into the lymph nodes of nude mice. We monitored flow after injection into the flesh of nude mouse tails, using both 655 nm– and 800 nm–emitting quantum dots. Flow through lymphatics was pulsatile. Within 30 minutes, inguinal, axillary, and brachial lymph nodes could be seen. A surprise finding was that flow through the subsurface lymphatics between the inguinal and axillary nodes could also be seen noninvasively (Fig. 7).

We explored sentinel lymph node mapping using two mouse tumor systems. Quantum dots were injected into mouse tumors, and the tumors then were allowed to drain into the surrounding nodes. Considerable difficulty was encountered in imaging small lymph nodes adjacent to tumors because of the brightness of the tumor mass; however, these nodes could be visualized by masking the tumor, and their identity was confirmed histologically. Figure 8 shows one example; drainage to lymph nodes was rapid and readily visible, but masking of the tumor is helpful in displaying the result.

G. Toxicity

To date, our polymer-surfaced quantum dots have shown no toxicity in tissue-cultured cells. In a blind study, tissues from uninjected mice and mice injected with ∼300 pmol quantum dots, either amp only or PEG-amp, were submitted for examination by a veterinary pathologist at intervals up to 1

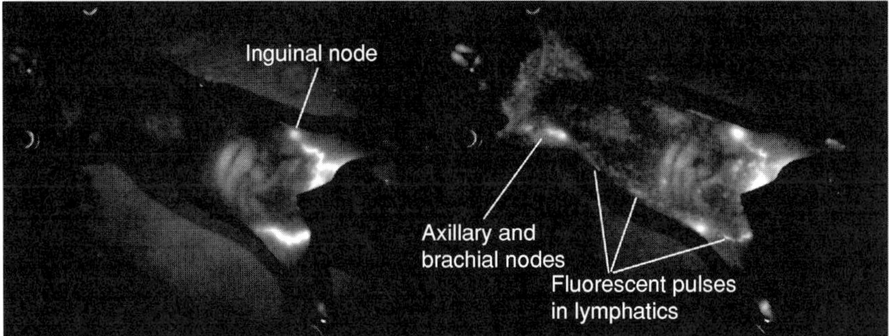

Figure 7 Flow into lymphatic vessels monitored using 800 nm quantum dots. A nude mouse was injected into the flesh of the tail, and successive images were taken at 1-minute intervals. In the left image (3 minutes after injection), note that the lymphatics are clearly labeled, while only the uppermost inguinal node has accumulated any significant fluorescence. In the right image (25 minutes post-injection), both inguinal nodes are labeled, as are the brachial and axillary nodes. Excitation at 450 nm confines visualization to immediate subcutaneous structures. Note the absence of fluorescence in the general vasculature and liver. The tail is masked.

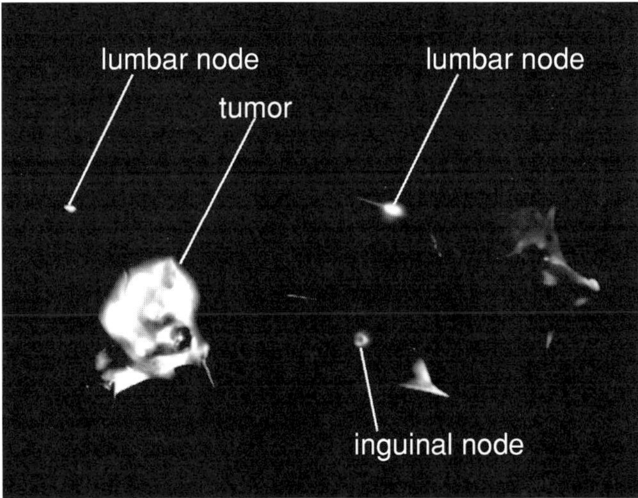

Figure 8 Sentinel lymph node detection. MH-15 teratocarcinoma was implanted into the right thigh of a mouse. When the tumor had grown to ~0.5 cm diameter, the tumor was injected using ~40 pmol carboxy-PEG quantum dots, 655 nm emission. Twenty minutes after injection, the animal was euthanatized and necropsied. Left, tumor is visualized directly; right, tumor is masked. Lumbar and inguinal lymph nodes are indicated. Excitation at 450 nm, emission at 655 nm.

year after injection. No significant toxicity was seen even in organs in which there had been significant accumulation of quantum dots (liver, spleen, and lymph nodes). Other authors have observed limited toxicity when using quantum dots of other types. Toxicity to tissue-cultured cells was demonstrated using core-only CdSe quantum dots coated by mercaptoacetic acid; preexposure to oxidizing conditions or long exposure to near-ultraviolet light rendered these quantum dots toxic. On the other hand, CdSe-ZnS core-shell quantum dots with various coatings proved nontoxic (Derfus *et al.*, 2004b). As noted previously, Hoshino *et al.* (2004) found toxicity at high levels of labeling when using albumin-coated CdSe-ZnS quantum dots. We conclude that properly coated quantum dots are necessary for long-term labeling in living cells.

III. Conclusion and Future Possibilities

We anticipate that quantum dots emitting in the range of 700–1300 nm will become widely available. New secondary coats will provide long circulating lifetimes and allow ready conjugation to biomolecules. If these coatings permit nonradiative (Foerster) energy transfer and are stable *in vivo*, new fluorescent sensors (Medintz *et al.*, 2004) may be developed for use in living animals. Combined modes of imaging integrated into one nanodevice should allow deep imaging by magnetic resonance imaging, gamma ray imaging, or positron emission tomography to supplement the limitations of fluorescence for visualization through tissue, while allowing the precision and sensitivity of fluorescence once the target is revealed during surgery (Josephson *et al.*, 2002). Improved tracking of mammalian and microbial cells, observing biodistribution of macromolecules and viruses in real time, and integrated physiological sensing in whole animals should be possible.

Currently fluorescence imaging is performed primarily in small animal models of human disease. Soon fluorescence imaging with quantum dots and other near-infrared probes will be used clinically. The immediate prospects are rapid identification of sentinel lymph nodes and of tumor cells within them; other potential uses are detecting tumors or other lesions that are superficial or accessible to endoscopy. The most exciting possibility is to extend combined-mode probes from small animal models to humans; fluorescence imaging during surgery could then be used to define precisely a lesion's extent and margins.

Note Added in Proof: This chapter was written in October 2004. Since then, many papers on biological applications of quantum dot have been published. The reader is especially directed to the reviews by Michalet *et al.* (*Science* **307**, 538–544, 2005), Medintz *et al.* (*Nat. Mater.* **4**, 435–446, 2005.), and Bruchez (*Curr. Opin. Chem. Biol.* **9**, 533–537, 2005.)

Acknowledgments

I am grateful to Lauren A. Ernst, Gregory W. Fisher, Berndt Christopher Lagerholm,* and Alan S. Waggoner of the Molecular Biosensor and Imaging Center: To Johathan W. Jarvik of the Department of Biology, Carnegie Mellon University; and to Marcel Bruchez, Theresa Harper, and Mare Schrier of Quantum Dot Corporation for collaborating on experiments and for many helpful discussions. This work was supported by National Institutes of Health grant No. R01 EB00364 and by funds of the Molecular Biosensor and Imaging Center, Carnegie Mellon University.

References

Akerman, M. E., Chan, W. C., Laakkonen, P., Bhatia, S. N., and Ruoslahti, E. (2002). Nanocrystal targeting *in vivo*. *Proc. Natl. Acad. Sci. USA* **99,** 12617–12621.

Ballou, B., Fisher, G. W., Deng, J. S., Hakala, T. R., Srivastava, M., and Farkas, D. L. (1998). Cyanine fluorochrome-labeled antibodies *in vivo*: Assessment of tumor imaging using Cy3, Cy5, Cy5.5, and Cy7. *Cancer Detect. Prev.* **22,** 251–257.

Ballou, B., Fisher, G. W., Hakala, T. R., and Farkas, D. L. (1997). Tumor detection and visualization using cyanine fluorochrome-labeled antibodies. *Biotechnol. Prog.* **13,** 649–658.

Ballou, B., Lagerholm, B. C., Ernst, L. A., Bruchez, M. P., and Waggoner, A. S. (2004). Noninvasive imaging of quantum dots in mice. *Bioconjug. Chem.* **15,** 79–86.

Bruchez, M. (2005). Turning all the lights on: Quantum dots in cellular assays. *Curr. Opin. Chem. Biol.* **9,** 533–537.

Bruchez, M. Jr., Moronne, M., Gin, P., Weiss, S., and Alivisatos, A. P. (1998). Semiconductor nanocrystals as fluorescent biological labels. *Science* **281,** 2013–2016.

Chan, W. C., and Nie, S. (1998). Quantum dot bioconjugates for ultrasensitive nonisotopic detection. *Science* **281,** 2016–2018.

Chapman, A. P. (2002). PEGylated antibodies and antibody fragments for improved therapy: A review. *Adv. Drug Deliv. Rev.* **54,** 531–545.

Chen, F., and Gerion, D. (2004). Fluorescent CdSe/ZnS nanocrystal-peptide conjugates for long-term, nontoxic imaging and nuclear targeting in living cells. *Nano Lett.* **4,** 1827–1832.

Clapp, A. R., Medintz, I. L., Mauro, J. M., Fisher, B. R., Bawendi, M. G., and Mattoussi, H. (2004). Fluorescence resonance energy transfer between quantum dot donors and dye-labeled protein acceptors. *J. Am. Chem. Soc. 126,* **30,** 301–310.

Dabbousi, B. O., Rodriguez-Viejo, J., Mikulec, F. V., Heine, J. R., Mattoussi, H., Ober, R., Jensen, K. F., and Bawendi, M. G. (1997). (CdSe)ZnS core-shell quantum dots: Synthesis and optical and structural characterization of a size series of highly luminescent materials. *J. Phys. Chem. B* **101,** 9463–9475.

Dahan, M., Levi, S., Luccardini, C., Rostaing, P., Riveau, B., and Triller, A. (2003). Diffusion dynamics of glycine receptors revealed by single-quantum dot tracking. *Science* **302,** 442–445.

Derfus, A. M., Chan, W. C. W., and Bhatia, S. N. (2004a). Intracellular delivery of quantum dots for live cell labeling and organelle tracking. *Adv. Mater.* **16,** 961–966.

Derfus, A. M., Chan, W. C. W., and Bhatia, S. N. (2004b). Probing the cytotoxicity of semiconductor quantum dots. *Nano Lett.* **4,** 11–18.

*Current affiliation: Department of Cell and Developmental Biology, University of North Carolina, Chapel Hill, North Carolina.

Dubertret, B., Skourides, P., Norris, D. J., Noireaux, V., Brivanlou, A. H., and Libchaber, A. (2002). *In vivo* imaging of quantum dots encapsulated in phospholipid micelles. *Science* **298,** 1759–1762.

Frangioni, J. V. (2003). *In vivo* near-infrared fluorescence imaging. *Curr. Opin. Chem. Biol.* **7,** 626–634.

Gao, X., Cui, Y., Levenson, R. M., Chung, L. W., and Nie, S. (2004). *In vivo* cancer targeting and imaging with semiconductor quantum dots. *Nat. Biotechnol.* **22,** 969–976.

Gao, X., and Nie, S. (2003). Molecular profiling of single cells and tissue specimens with quantum dots. *Trends Biotechnol.* **21,** 371–373.

Gao, X., and Nie, S. (2004). Quantum dot-encoded mesoporous beads with high brightness and uniformity: Rapid readout using flow cytometry. *Anal. Chem.* **76,** 2406–2410.

Gerion, D., Pinaud, F., Williams, S. C., Parak, W. J., Zanchet, D., Weiss, S., and Alivisatos, A. P. (2001). Synthesis and properties of biocompatible water-soluble silica-coated CdSe/ZnS semiconductor quantum dots. *J. Phys. Chem. B* **105,** 8861–8871.

Goldman, E. R., Anderson, G. P., Tran, P. T., Mattoussi, H., Charles, P. T., and Mauro, J. M. (2002a). Conjugation of luminescent quantum dots with antibodies using an engineered adaptor protein to provide new reagents for fluoroimmunoassays. *Anal. Chem.* **74,** 841–847.

Goldman, E. R., Balighian, E. D., Mattoussi, H., Kuno, M. K., Mauro, J. M., Tran, P. T., and Anderson, G. P. (2002b). Avidin: A natural bridge for quantum dot-antibody conjugates. *J. Am. Chem. Soc.* **124,** 6378–6382.

Greenwald, R. B. (2001). PEG drugs: An overview. *J. Control. Release* **74,** 159–171.

Han, M., Gao, X., Su, J. Z., and Nie, S. (2001). Quantum-dot-tagged microbeads for multiplexed optical coding of biomolecules. *Nat. Biotechnol.* **19,** 631–635.

Hanaki, K., Momo, A., Oku, T., Komoto, A., Maenosono, S., Yamaguchi, Y., and Yamamoto, K. (2003). Semiconductor quantum dot/albumin complex is a long-life and highly photostable endosome marker. *Biochem. Biophys. Res. Comm.* **302,** 496–501.

Harrington, K. J., Mubashar, M., and Peters, A. M. (2002). Polyethylene glycol in the design of tumor-targetting radiolabelled macromolecules—lessons from liposomes and monoclonal antibodies. *Q. J. Nucl. Med.* **46,** 171–180.

Harris, J. M., and Chess, R. B. (2003). Effect of pegylation on pharmaceuticals. *Nat. Rev. Drug Discov.* **2,** 214–221.

Hines, M. A., and Guyot-Sionnest, P. (1996). Synthesis and characterization of strongly luminescing ZnS-capped CdSe nanocrystals. *J. Phys. Chem.* **100,** 468–471.

Hoshino, A., Hanaki, K., Suzuki, K., and Yamamoto, K. (2004). Applications of T-lymphoma labeled with fluorescent quantum dots to cell tracing markers in mouse body. *Biochem. Biophys. Res. Comm.* **314,** 46–53.

Jaiswal, J. K., Mattoussi, H., Mauro, J. M., and Simon, S. M. (2003). Long-term multiple color imaging of live cells using quantum dot bioconjugates. *Nat. Biotechnol.* **21,** 47–51.

Jaiswal, J. K., and Simon, S. M. (2004). Potentials and pitfalls of fluorescent quantum dots for biological imaging. *Trends Cell. Biol.* **14,** 497–504.

Josephson, L., Kircher, M. F., Mahmood, U., Tang, Y., and Weissleder, R. (2002). Near-infrared fluorescent nanoparticles as combined MR/optical imaging probes. *Bioconjug. Chem.* **13,** 554–560.

Kim, S., and Bawendi, M. G. (2003). Oligomeric ligands for luminescent and stable nanocrystal quantum dots. *J. Am. Chem. Soc.* **125,** 14652–14653.

Kim, S., Lim, Y. T., Soltesz, E. G., De Grand, A. M., Lee, J., Nakayama, A., Parker, J. A., Mihaljevic, T., Laurence, R. G., Dor, D. M., Cohn, L. H., Bawendi, M. G., and Frangioni, J. V. (2004). Near-infrared fluorescent type II quantum dots for sentinel lymph node mapping. *Nat. Biotechnol.* **22,** 93–97.

Lagerholm, B., Christoffer, Wang, Miaomiao, Ernst, Lauren, A. Ly, Danith, H. Liu, Hongjian Bruchez, M. P., and Waggoner, A. S. (2004). Multicolor coding of cells with cationic peptide coated quantum dots. *Nano Lett.* **4**, 2019–2022.

Larson, D. R., Zipfel, W. R., Williams, R. M., Clark, S. W., Bruchez, M. P., Wise, F. W., and Webb, W. W. (2003). Water-soluble quantum dots for multiphoton fluorescence imaging in vivo. *Science* **300**, 1434–1436.

Lidke, D. S., Nagy, P., Heintzmann, R., Arndt-Jovin, D. J., Post, J. N., Grecco, H. E., Jares-Erijman, E. A., and Jovin, T. M. (2004). Quantum dot ligands provide new insights into erbB/HER receptor-mediated signal transduction. *Nat. Biotechnol.* **22**, 198–203.

Lim, Y. T., Kim, S., Nakayama, A., Stott, N. E., Bawendi, M. G., and Frangioni, J. V. (2003). Selection of quantum dot wavelengths for biomedical assays and imaging. *Mol. Imaging* **2**, 50–64.

Mattoussi, H., Mauro, J. M., Goldman, E. R., Anderson, G. P., Sundar, V. C., Mikulec, F. V., and Bawendi, M. G. (2000). Self-assembly of CdSe-ZnS quantum dot bioconjugates using an engineered recombinant protein. *J. Am. Chem. Soc.* **122**, 12142–12150.

Medintz, I. L., Clapp, A. R., Mattoussi, H., Goldman, E. R., Fisher, B., and Mauro, J. M. (2003a). Self-assembled nanoscale biosensors based on quantum dot FRET donors. *Nat. Mater.* **2**, 630–638.

Medintz, I. L., Goldman, E. R., Lassman, M. E., and Mauro, J. M. (2003b). A fluorescence resonance energy transfer sensor based on maltose binding protein. *Bioconjug. Chem.* **14**, 909–918.

Medintz, I. L., Trammell, S. A., Mattoussi, H., and Mauro, J. M. (2004). Reversible modulation of quantum dot photoluminescence using a protein-bound photochromic fluorescence resonance energy transfer acceptor. *J. Am. Chem. Soc.* **126**, 30–31.

Medintz, I. L., Uyeda, H. T., Goldman, E. R., and Mattoussi, H. (2005). Quantum dot bioconjugates for imaging, labeling and sensing. *Nat. Mater.* **4**, 435–446.

Michalet, X., Pinaud, F. F., Bentolila, L. A., Tsay, J. M., Doose, S., Li, J. J., Sundaresan, G., Wu, A. M., Gambhir, S. S., and Weiss, S. (2005). Quantum dots for live cells, *in vivo* imaging, and diagnostics. *Science* **307**, 538–544.

Minet, O., Dressler, C., and Beuthan, J. (2004). Heat stress induced redistribution of fluorescent quantum dots in breast tumor cells. *J. Fluor.* **14**, 241–247.

Mitchell, G. P., Mirkin, C. A., and Letsinger, R. L. (1999). Programmed assembly of DNA functionalized quantum dots. *J. Am. Chem. Soc.* **121**, 8122–8123.

Molineux, G. (2002). Pegylation: Engineering improved pharmaceuticals for enhanced therapy. *Cancer Treat. Rev.* **28** (Suppl. A), 13–16.

Osaki, F., Kanamori, T., Sando, S., Sera, T., and Aoyama, Y. (2004). A quantum dot conjugated sugar ball and its cellular uptake. On the size effects of endocytosis in the subviral region. *J. Am. Chem. Soc.* **126**, 6520–6521.

Parak, W. J., Boudreau, R., Le Gros, M., Gerion, D., Zanchet, D., Micheel, C. M., Williams, S. C., Alivisatos, A. P., and Larabell, C. (2002). Cell motility and metastatic potential studies based on quantum dot imaging of phagokinetic tracks. *Adv. Mater. (Weinheim, Germany)* **14**, 882–885.

Parak, W. J., Gerion, D., Pellegrino, T., Zanchet, D., Micheel, C., Williams, S. C., Boudreau, R., Le Gros, M. A., Larabell, C. A., and Alivisatos, A. P. (2003). Biological applications of colloidal nanocrystals. *Nanotechnology* **14**, R15–R27.

Rosenthal, S. J., Tomlinson, I., Adkins, E. M., Schroeter, S., Adams, S., Swafford, L., McBride, J., Wang, Y., De Felice, L. J., and Blakely, R. D. (2002). Targeting cell surface receptors with ligand-conjugated nanocrystals. *J. Am. Chem. Soc.* **124**, 4586–4594.

Smith, A. M., Gao, X., and Nie, S. (2004). Quantum-dot nanocrystals for *in-vivo* molecular and cellular imaging. *Photochem. Photobiol.* **80**, 377–385.

                                                    Byron Ballou

Sukhanova, A., Devy, J., Venteo, L., Kaplan, H., Artemyev, M., Oleinikov, V., Klinov, D., Pluot, M., Cohen, J. H. M., and Nabiev, I. (2004). Biocompatible fluorescent nanocrystals for immunolabeling of membrane proteins and cells. *Anal. Biochem.* **324,** 60–67.

Voura, E. B., Jaiswal, J. K., Mattoussi, H., and Simon, S. M. (2004). Tracking metastatic tumor cell extravasation with quantum dot nanocrystals and fluorescence emission-scanning microscopy. *Nat. Med.* **10,** 993–998.

Watson, A., Wu, X., and Bruchez, M. (2003). Lighting up cells with quantum dots. *Biotechniques* **34,** 296–303.

Wender, P. A., Mitchell, D. J., Pattabiraman, K., Pelkey, E. T., Steinman, L., and Rothbard, J. B. (2000). The design, synthesis, and evaluation of molecules that enable or enhance cellular uptake: Peptoid molecular transporters. *Proc. Natl. Acad. Sci. USA* **97,** 13003–13008.

Wu, X., Liu, H., Liu, J., Haley, K. N., Treadway, J. A., Larson, J. P., Ge, N., Peale, F., and Bruchez, M. P. (2003). Immunofluorescent labeling of cancer marker Her2 and other cellular targets with semiconductor quantum dots. *Nat. Biotechnol.* **21,** 41–46.

6

In Vivo Cell Biology of Cancer Cells Visualized with Fluorescent Proteins

Robert M. Hoffman
AntiCancer, Inc.
San Diego, California 92111

This chapter describes a new cell biology where the behavior of individual cells can be visualized in the living animal. Previously it has been demonstrated that fluorescent proteins can be used for whole-body imaging of metastatic tumor growth, bacterial infection, and gene expression. An example of the new cell biology is dual-color fluorescence imaging using red fluorescent protein (RFP)-expressing tumors transplanted in green fluorescent protein (GFP)-expressing transgenic mice. These models show with great clarity the details of tumor–stroma interactions and especially tumor-induced angiogenesis, tumor-infiltrating lymphocytes, stromal fibroblasts, and macrophages. Another example is the color coding of cells with RFP or GFP such that both cell types can be simultaneously visualized *in vivo*. Stem cells can also be visualized and tracked *in vivo*. Mice in which the regulatory elements of the stem cell marker nestin drive GFP expression enable nascent vasculature to be visualized interacting with transplanted RFP-expressing cancer cells. Nestin-driven GFP expression can also be used to visualize hair follicle stem cells. Dual-color cells expressing GFP in the nucleus and RFP in the cytoplasm enable real-time visualization of nuclear–cytoplasm dynamics including cell cycle events and apoptosis. Highly elongated cancer

Current Topics in Developmental Biology, Vol. 70

0070-2153/05 $35.00
DOI: 10.1016/S0070-2153(05)70006-5

cells in capillaries in living mice were observed within skin flaps. The migration velocities of the cancer cells in the capillaries were measured by capturing images of the dual-color fluorescent cells over time. The cells in the capillaries elongated to fit the width of these vessels. The use of the dual-color cancer cells differentially labeled in the cytoplasm and nucleus and associated fluorescent imaging provide a powerful tool to understand the mechanism of cancer cell migration and deformation in small vessels. © 2005, Elsevier Inc.

I. Introduction

Animal models of cancer that use the stable expression of green fluorescent protein (GFP) and red fluorescent protein (RFP) have made it possible to directly observe cell behavior in live animals. The interaction of cells with each other and entry of tumor cells into the circulation are important factors in metastasis. These processes can be observed using fluorescent proteins in real time (Yamauchi *et al.*, 2005). Before the introduction of GFP, intravital imaging was limited to the study of cells that were transiently labeled with vital dyes.

Initial studies of tumor biology that used stable GFP expression focused on static images and examination of metastases. The first use of stable GFP expression to characterize cancer cells *in vivo* was by Chishima *et al.* (1997a). Cell motility, shape changes, and migration of carcinoma cells expressing GFP in live animals *in vivo* was first described by Farina *et al.* (1998). Huang *et al.* (2002) demonstrated that GFP-transduced H1299 lung cancer cells allowed the imaging of local invasion at the single-cell level. Tumor cells in blood vessels were readily imaged. Wyckoff *et al.* (2000) have measured tumor cell density in the blood using intravital GFP imaging. Wyckoff *et al.* (2000) have used GFP imaging to view cells in time-lapse images within a single optical section using a confocal microscope. Naumov *et al.* (1999) visualized CHO-K1 cells that stably express GFP. Fine cellular details such as pseudopodial projections, even after extended periods of *in vivo* growth, were visualized by GFP expression. Mook *et al.* (2003) generated a GFP-expressing rat adenocarcinoma cell line (CC531s) that forms metastases in rat liver after administration to the portal vein. Initial arrest of colon cancer cells in sinusoids of the liver was due to size restriction. Tumor cells divided exclusively intravascularly during the first 4 days. We visualized the trafficking of metastatic cells targeting the liver via the portal vein using GFP-expressing cancer cells. Within 72 hours after transplantation on the ascending colon in nude mice, metastasis was visualized *ex vivo* on a single-cell basis around the portal vein by GFP imaging (Wang *et al.*, 2004). Al-Mehdi *et al.* (2000) observed the steps in early hematogenous metastasis of tumor cells expressing GFP in subpleural microvessels in intact, perfused mouse and rat lungs. Metastatic tumor cells attached to the endothelia of pulmonary precapillary arterioles and capillaries. Extravasation of tumor cells was rare. Early tumor colonies were observed entirely within the blood vessels. Brown *et al.* (2001) showed that multiphoton laser-scanning microscopy could provide high-resolution three-dimensional images of angiogenesis gene expression and that this technique could be used to investigate deeper regions of GFP-expressing tumors in dorsal skinfold chambers. To monitor the activity of the vascular endothelial growth factor (VEGF) promoter, Fukumura *et al.* (1997, 1998) made transgenic mice that express GFP under control of the VEGF promoter. Multiphoton laser scanning microscopy showed that the tumor was able to induce activity of the VEGF promoter; GFP-positive stromal cells were seen at least 200 μm into the tumor with this technique.

Yamamoto *et al.* (2004) reported the genetic engineering of dual-color fluorescent cells with one color in the nucleus and the other in the cytoplasm that allows real-time nuclear-cytoplasmic dynamics to be visualized in living cells *in vivo* as well as *in vitro*. To obtain the dual-color cells, RFP was expressed in the cytoplasm of HT1080 human fibrosarcoma cells, and GFP linked to histone H2B was expressed in the nucleus. Nuclear GFP expression enabled visualization of nuclear dynamics, whereas simultaneous cytoplasmic RFP expression enabled visualization of nuclear-cytoplasmic ratios as well as simultaneous cell and nuclear shape changes. Thus, total cellular dynamics can be visualized in the dual-color cells in real time (Fig. 1). Common carotid artery injection of dual-color cells and a reversible skin flap enabled the external visualization of the dual-color cells in microvessels in the mouse brain where extreme elongation of the cell body as well as the nucleus occurred.

II. Whole-Body Imaging of Tumor Growth and Metastasis

The use of fluorescent proteins for both whole-body and intravital cellular imaging is reviewed in the following text. External whole-body imaging of mice with primary and metastatic tumors that are genetically labeled with the fluorescent proteins GFP and RFP is a simple but powerful tool for investigating tumor development. The technology is based on the bright intrinsic fluorescence of GFP and RFP, which is partly caused by the high quantum yield of these fluorophores (Heim *et al.*, 1995; Matz *et al.*, 1999). For tumor cells to be visualized with this technique, they must be transduced with GFP or RFP genes so that they become brightly fluorescent. This can be accomplished by *in vitro* (Chishima *et al.*, 1997b; Yang *et al.*, 2000) and *in vivo* (Hasegawa *et al.*, 2000) selection of such fluorescent tumor cells.

To produce metastasis in mice, the genetically fluorescent tumors should be transplanted orthotopically (Bouvet *et al.*, 2000, 2002; Chishima *et al.*,

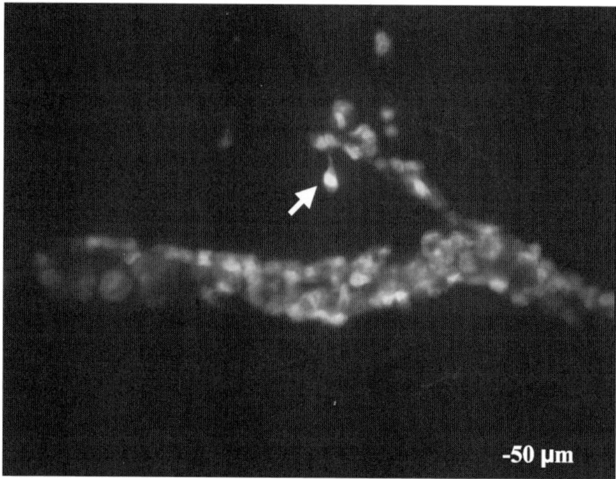

Figure 1 Extravasation of tumor cells from vessels containing numerous dual-color HT1080 fibrosarcoma cells in a living mouse. HT1080 human fibrosarcoma cells were labeled with histone-H2B-GFP in the nucleus and with retroviral RFP in the cytoplasm. Cells were injected in the heart and visualized in skin vessels by whole-body imaging. From K. Yamauchi and R. M. Hoffman, unpublished data. (See Color Insert.)

1997b,c; Hastings *et al.*, 2001; Maeda *et al.*, 2000; Rashidi *et al.*, 2000; Yang *et al.*, 1998, 1999a,b, 2001).

Once the GFP-expressing tumors have developed and metastases have formed, individual tumor cells can be detected in the live mouse by use of whole-body imaging with fairly simple equipment. A fluorescence light box with fiberoptic lighting at ~490 nm and appropriate filters, placed on top of the light box, can be used to image large tumors and can be viewed with the naked eye (Yang *et al.*, 2000). Alternatively, the light box can be linked to a camera with an appropriate filter to enable images to be displayed on a monitor and digitally stored (Yang *et al.*, 2000). To visualize smaller tumors and metastases, the animal can be put on a fluorescence dissecting microscope that incorporates a light source and filters for excitation at ~490 nm. Fluorescence emission can be observed through a 520 nm long-pass filter (Yang *et al.*, 2000). The animals can be irradiated at 490 nm for long periods without harming them or bleaching the GFP or RFP fluorescence. Images can be processed with standard software and the imaging procedures can be repeated as often as necessary without harming the animal. Therefore, with these techniques real-time tracking of tumor growth and metastasis is feasible.

Reversible skin flaps can also be introduced onto different parts of the animal to examine single tumor cells or small colonies on internal organs (Yang *et al.*, 2002). The skin flaps are rendered reversible by simple suturing.

External imaging can then be done through the relatively transparent body walls of the mouse, which include the skull, by use of a fluorescence dissecting microscope. Blood vessels growing on tumors can also be observed using skin flaps because they contrast with the fluorescence of the tumors (Yang *et al.*, 2002). Examples of findings from studies of these techniques are described below.

Transplanted mice with metastatic lesions of GFP-expressing tumors in the colon, brain, liver, lymph nodes, and bone have been used to produce images of metastasis. These images are used for real-time quantitative measurement of primary and metastatic tumor growth for each of these organs. GFP-expressing cells emit a bright fluorescence signal compared with background fluorescence from other tissue. The signal is so strong and selective that external images of GFP-expressing tumors and their metastases can be obtained in freely moving animals (Yang *et al.*, 2000).

In another study, an RFP-expressing human pancreatic tumor cell line was introduced as tissue fragments into the pancreases of nude mice by surgical orthotopic implantation. As the tumors were growing, whole-body optical imaging was used to track, in real time, the growth of the primary tumor and the formation of metastatic lesions that developed in the spleen, bowel, portal lymph nodes, omentum, and liver. The images were used for quantification of tumor growth in each of these organs (Katz *et al.*, 2003a). Whole-body imaging with this model was used to compare standard and experimental agents for pancreatic cancer (Katz *et al.*, 2003b, 2004).

Peyruchaud *et al.* (2001) established a GFP-expressing bone-metastasis subclone of MDA-MB-231 (B02/GFP.2) by repeated *in vivo* passages in bone, by use of the heart injection model. When injected into the tail vein of mice, the selected cells grew preferentially in bone. Whole-body fluorescence imaging of the live mice showed that bone metastases could be detected about 1 week before radiologically distinctive osteolytic lesions developed. Furthermore, when the tumor-bearing mice were treated with a bisphosphonate, progression of established osteolytic lesions, and the expansion of breast cancer cells within bone, were inhibited.

The MDA-MB-435 human breast carcinoma cell line produced widespread osteolytic skeletal metastases following injection into the left ventricle of the heart. Osseous metastases localized predominantly to trabecular regions, especially proximal and distal femur, proximal tibia, proximal humerus, and lumbar vertebrae. GFP permitted detection of single cells and microscopic metastases in bone at early time points (Harms and Welch, 2003).

Using a different approach, human ovarian tumor cells (SKOV3.ip1) were made to express GFP by infection with a replication-deficient adenoviral (Ad) vector encoding GFP (Chaudhuri *et al.*, 2001). The infected cells showed high GFP fluorescence, and when implanted into mice, intraperitoneal tumors as small as 0.2 mm in diameter could be detected by whole-body imaging within

24 hours (Chaudhuri *et al.*, 2003). In another study, however, GFP-expressing tumors could not be detected by whole-body imaging until 7 days after subcutaneous tumor cell inoculation (Choy *et al.*, 2003). These results strongly contrast with the results described previously. This discrepancy shows the need to use appropriate instrumentation and techniques for whole-body imaging.

III. Whole-Body and Intravital Imaging of Angiogenesis and Individual Tumor Cells

Tumor angiogenesis can also be visualized by use of GFP techniques. The footpads of mice are quite transparent with few resident blood vessels and are therefore ideal for quantitative imaging of tumor angiogenesis in intact animals. Vessels can be seen because of their striking contrast to the GFP fluorescence of the tumor tissue (Yang *et al.*, 2001). These researchers injected GFP-expressing Lewis lung carcinoma cells subcutaneously into the footpad of nude mice and, using whole-body imaging. They found that capillary density increased linearly over 10 days. Similarly, when GFP-expressing MDA-MB-435 cells were orthotopically transplanted to the mouse mammary fat pad, whole-body optical imaging showed that blood vessel density increased linearly over about 20 weeks.

Reversible skin flaps can be used to visualize angiogenesis on GFP-expressing tumors transplanted onto internal organs, in addition to examining the tumor itself (Yang *et al.*, 2002). These angiogenesis mouse models can be used for real-time *in vivo* evaluation of agents inhibiting or promoting tumor angiogenesis (Yang *et al.*, 2001, 2002). Opening a reversible skin-flap in the light path greatly reduces attenuation of the fluorescent signal, thereby increasing the sensitivity of tumor detection by many times. This procedure also greatly increases the depth at which tumor cells can be observed. Single GFP-expressing tumor cells can thus be seen on numerous internal organs. GFP glioma cells seeded on the brain can be visualized through a scalp skin-flap. GFP lung tumor microfoci, which represent a few malignant cells, can be viewed through a skin flap over the chest wall, and contralateral micrometastases can also be examined by use of a corresponding skin flap. Peritoneal wall skin flaps can be used to examine GFP-pancreatic tumors and their angiogenic microvessels. For the liver, a skin flap allowed imaging of physiologically relevant micrometastases that had originated in an orthotopically implanted GFP tumor. Single tumor cells on the liver, which had been introduced through an intraportal injection, were also detectable (Yang *et al.*, 2002).

Ilyin *et al.* (2001) visualized glioma cells in rats by inserting a fiberoptic endoscope through a preimplanted guide cannula. Tumor monitoring was coupled to confocal microscopy so that visualization of the fluorescent signals from the C-6 glioma-GFP cells that had been preimplanted in the brain was very sensitive. Funovics *et al.* (2003) described the design and construction of a miniaturized multichannel near-infrared endoscopic imaging system developed for high-resolution imaging of mice. This endoscope was used to visualize tumor cells transplanted orthotopically in mice. This device should be useful for *in vivo* imaging using fluorescent proteins (Kelly *et al.*, 2004).

IV. Dual-Color Imaging

GFP- and RFP-labeled HT1080 human fibrosarcoma cells allow the determination of clonality by fluorescence visualization of metastatic colonies after mixed implantation of the red and green fluorescent cells. Resulting pure red or pure green colonies were scored as clonal, whereas mixed yellow colonies were scored as nonclonal. In a spontaneous metastasis model originating from footpad injection in severe combined immunodeficient (SCID) mice, 95% of the resulting lung colonies were either pure green or pure red, indicating monoclonal origin, whereas 5% were of mixed color, indicating polyclonal origin. In an experimental lung metastasis model established by tail vein injection in SCID mice, clonality of lung metastasis was dependent on cell number. With a minimum number of cells was injected, almost all (96%) colonies were pure red or green and therefore monoclonal. When a large number of cells were injected, almost all (87%) colonies were mixed color and therefore heteroclonal. Spontaneous metastases may be clonal because they are rare events, thereby supporting the rare-cell clonal-origin-of-metastasis hypothesis. The clonality of the experimental metastasis model depended on the number of input cells (Yamamoto *et al.*, 2003a).

Dual-color lung metastases were visualized by external fluorescence imaging in live animals through skin flap windows over the chest wall. Lung metastases were observed on the lung surface of all mice. SCID mice tolerated multiple surgical procedures well for direct-view imaging via skin flap windows. Real-time metastatic growth of the two different colored clones in the same lung was externally imaged with resolution and quantification of green, red, or yellow colonies in live animals. The simultaneous real-time dual-color imaging of metastatic colonies gives rise to the possibility of color-coded imaging of clones of cancer cells carrying various forms of gene of interest (Yamamoto *et al.*, 2003b).

V. Dual-Color Tumor–Host Models

A transgenic GFP nude mouse with ubiquitous GFP expression has been developed. The GFP nude mouse was obtained by crossing nontransgenic nude mice with the transgenic C57/B6 mouse in which the β-actin promoter drives GFP expression in essentially all tissues (Okabe *et al.*, 1997). In crosses between *nu/nu* GFP male mice and *nu/+* GFP female mice, the embryos fluoresced green. Approximately 50% of the offspring of these mice were GFP nude mice. Newborn mice and adult mice fluoresced very bright green and could be detected with a simple blue light–emitting diode flashlight with a central peak of 470 nm and a bypass emission filter. In the adult mice, the organs all brightly expressed GFP, including the heart, lungs, spleen, pancreas, esophagus, stomach, and duodenum. The following systems were dissected out and shown to have brilliant GFP fluorescence: the entire digestive system from tongue to anus; the male and female reproductive systems; the brain and spinal cord; and the circulatory system, including the heart and major arteries and veins. The skinned skeleton highly expressed GFP. Pancreatic islets showed GFP fluorescence. The spleen cells were also GFP positive. RFP-expressing human cancer cell lines, including PC-3-RFP prostate cancer, HCT-116-RFP colon cancer, MDA-MB-435-RFP breast cancer, and HT1080-RFP fibrosarcoma were transplanted to the transgenic GFP nude mice. All of these human tumors grew extensively in the transgenic GFP nude mouse (Yang *et al.*, 2004).

This model shows with great clarity the details of the tumor–stroma interaction, especially tumor-induced angiogenesis and tumor-infiltrating lymphocytes. The GFP-expressing tumor vasculature, both nascent and mature, of the GFP host mouse could be readily distinguished interacting with the RFP-expressing tumor cells. GFP-expressing dendritic cells were observed contacting RFP-expressing tumor cells with their dendrites. GFP-expressing macrophages were observed engulfing RFP-expressing cancer cells. GFP lymphocytes were seen surrounding cells of the RFP tumor, which eventually regressed (Yang *et al.*, 2003).

Duda *et al.* (2004) noted that at the time of transplantation tumor fragments contain "passenger" cells: endothelial cells and other stromal cells from the original host. They investigated the fate of GFP-labeled endothelial and nonendothelial stromal cells after transplantation in syngeneic mice. Angiogenic stroma associated with tumor or adipose tissue persists when transplanted, remains functional, and governed the initial neovascularization of grafted tissue fragments for more than 4 weeks after implantation. The passenger endothelial cells survive longer than other stromal cells, which are replaced by host-activated fibroblasts after 3 weeks. The transplantability of tumor stroma suggests that the angiogenic potential of a tumor xenograft, which determines its viability, depends on the presence of passenger endothelial cells and other stromal cells within the xenograft.

VI. Intravital Imaging of GFP-Expressing Cells

Intravital videomicroscopy can be used to visualize sequential steps in metastasis by use of CHO-K1 cells that stably express GFP. In mouse liver, the stages from the initial arrest of cells in the microvasculature up to the growth and angiogenesis of metastases have been recorded (Naumov *et al.*, 1999). Individual nondividing cells, as well as micrometastases and macrometastases, were visualized and quantified; additional cellular details such as pseudopodial projections were also detected. Micrometastases were found to preferentially grow and survive near the liver surface. A small population of single cells persisted over the 11-day observation period and the investigators believe these cells may represent dormant tumor cells (Naumov *et al.*, 1999).

GFP techniques have also shown that CT-26 (mouse colon carcinoma) cells inhibit the development of liver metastasis in BALB/c mice that receive intraportal injections of GFP-expressing tumor cells (Guba *et al.*, 2001). Intravital microscopy of the livers of these mice showed that growth of primary tumors promoted dormancy of single tumor cells for up to 7 days. Immunohistological staining for Ki-67 confirmed that these solitary cells were indeed dormant. By contrast, in the absence of a primary tumor, GFP-expressing tumor cells quickly developed into micrometastases. Thus, primary CT-26 tumor implants seem to inhibit tumor metastasis by promotion of a state of single-cell dormancy.

The *ras* oncogene promotes growth of micrometastases into macroscopic metastases. Two types of cells with constituitively active *ras,* NIH 3T3 and T24 H-*ras*–transformed (PAP2) fibroblasts, both of which were expressing GFP, were injected into mouse liver. Subsequent examination by GFP intravital imaging established that only the micrometastases formed by *ras*–transformed cells went on to produce macroscopic metastases; most NIH 3T3 micrometastases just disappeared. Furthermore, PAP2 metastases had a significantly higher proportion of proliferating cells than apoptotic cells, whereas NIH 3T3 metastases had low proliferation and a high proportion of apoptotic cells (Varghese *et al.*, 2002).

VII. Imaging GFP Tumor Cells in Blood Vessels

Following injection of tumor cells stably expressing GFP into the tail vein of mice, it was possible to visualize single tumor cells in blood vessels (Chishima *et al.*, 1997b). With intravital microscopy, Naumov *et al.* (1999) visualized GFP tumor cells colonizing various organs after extravasation. Huang (Huang *et al.*, 1999), Li (Li *et al.*, 2000), and their respective co-workers visualized GFP tumor cell–vessel interaction by use of skin window chambers in rodents and observed angiogenic effects very early in tumor colony formation. In an orthotopic mammary-pad injection model, Wyckoff *et al.* (2000) also visualized tumor–vessel interaction using the fluorescence of the tumor cells. Al-Mehdi *et al.* (2000) saw what they claimed to be intravascular tumor colony formation by tracking GFP expression in a lung perfusion study. In addition, Moore *et al.* (1998) have also visualized vessels in a GFP-expressing rodent cell line.

Rat tongue carcinoma cell lines expressing GFP have been used to investigate the formation of micrometastasis. The cells were injected into the portal vein and then tracked by use of intravital videomicroscopy (Ito *et al.*, 2001). The two cell types—LM-GFP metastatic and E2-GFP nonmetastatic tongue carcinoma cells—immediately got stuck in the sinusoidal vessels near terminal portal venules. The E2-GFP cells disappeared from the liver sinusoid within 3 days, whereas a substantial number of LM-GFP cells remained in the liver, possibly because these cells formed stable attachments to the sinusoidal wall. Upon examination of the process with a confocal laser scanning microscope, only LM-GFP cells were shown to grow in the liver.

Mook *et al.* (2003) noted that initial arrest of colon cancer cells in sinusoids of the liver was due to size restriction after injection of the CC531S-GFP rat tumor cell line. Adhesion of cancer cells to endothelial cells was never found. Instead, endothelial cells retracted rapidly and interactions were observed only between cancer cells and hepatocytes. Tumors developed exclusively intravascularly during the first 4 days. In conclusion, initial steps in the classic metastatic cascade such as adhesion to endothelium and extravasation are not essential for colon cancer metastasis in liver.

Wong *et al.* (2001) showed that death of transformed, metastatic, rat embryo cells, which were expressing GFP, occurred via apoptosis in the lungs 24–48 hours after injection into the circulation. The researchers established that BCL-2 overexpression was causing inhibition of apoptosis in culture, and this mechanism also conferred resistance *in vivo* for 24–48 hours after injection. This inhibition of apoptosis led to a greater number of macroscopic metastases.

Large detectable metastases did not form after athymic mice were given an intravenous injection of chromosome 6–transduced tumor cells expressing GFP (Goldberg *et al.*, 1999). However, fluorescence microscopy revealed micrometastases (single cells or clusters of fewer than 10 cells) in the lungs, suggesting that these cells managed to lodge themselves in the lungs but failed to proliferate. Cells isolated from mouse lungs up to 60 days after the injection were able to grow in culture and formed tumors when injected into skin; therefore, the cells were still viable, but dormant. This result implies that the gene(s) on chromosome 6 interfere specifically with growth-regulatory response in the lung, but not in the skin.

An isogenic pair of metastatic (M4A4) and nonmetastatic (NM2C5), GFP-labeled human breast cancer cell lines derived from the same patient and inoculated into the mammary glands of nude mice was used to visualize the dissemination patterns and fate of cells that escaped spontaneously from the resulting tumors. After tumors appeared, fluorescing single tumor cells were regularly seen in the lungs, even in animals inoculated with NM2C5, which fails to form secondary tumors in other organs. The sensitivity of the technique confirmed the continuing presence of scattered NM2C5 cells after primary tumor resection, although they formed no metastases by 6 months. These self-disseminated human tumor cells were retrievable from the tissues and were still viable and malignant, manifested by indefinite proliferation *in vitro* and green fluorescence and local tumorigenicity *in vivo*. Therefore, these scattered tumor cells were still alive but rendered indefinitely quiescent by the microenvironmental conditions in the lung tissue. Although many of the cells disseminating from M4A4 tumors grew into fluorescing metastases in the lungs, others remained solitary and quiescent. Therefore, even in a clonally derived cell population with metastatic properties, many cells do not, or cannot, mobilize the organ-specific growth properties needed to generate metastases (Goodison *et al.*, 2003).

Chang *et al.* (2000) used GFP-labeled tumor cells and CD31 and CD105 to identify endothelial cells; they showed that colon carcinoma xenografts had mosaic vessels with focal regions where no CD31/CD105 immunoreactivity was detected and tumor cells appeared to contact the vessel lumen.

Farina *et al.* (1998) developed a model that directly examines the motility of metastatic primary tumor cells *in situ*. A metastatic rat breast cancer cell line was established that constitutively expresses GFP. Animations of meta-static tumor cells moving in the mammary fat pad of live rats were generated by intravital imaging of the primary tumor *in situ* on a laser scanning confocal microscope.

Wyckoff *et al.* (2000) used metastatic (MTLn3) and nonmetastatic (MTC) cell lines derived from the rat mammary adenocarcinoma 13762 NF, express-sing GFP to measure tumor cell density in the blood, individual tumor cells in the lungs, and lung metastases. Metastatic cells showed greater orienta-tion toward blood vessels, whereas nonmetastatic cells fragment when inter-acting with vessels. These results demonstrate that a major difference in intravasation between metastatic and nonmetastatic cells is detected in the primary tumor.

We have recently shown that the neural stem cell marker nestin is expressed in hair follicle stem cells and the blood vessel network intercon-necting hair follicles in the skin of transgenic mice with nestin regulatory-element–driven GFP (ND-GFP) (Amoh *et al.*, 2004; Li *et al.*, 2003). The hair follicles were shown to give rise to the nestin-expressing blood vessels in the skin. We also visualized tumor angiogenesis by dual-color fluorescence imaging in ND-GFP transgenic mice after transplantation of the murine melanoma cell line B16F10 expressing RFP. ND-GFP was highly expressed in proliferating endothelial cells and nascent blood vessels in the growing tumor. Results of immunohistochemical staining showed that the blood vessel–specific antigen CD31 was expressed in ND-GFP–expressing nascent blood vessels. ND-GFP expression was diminished in the vessels with increased blood flow. Progressive angiogenesis during tumor growth was readily visualized during tumor growth by GFP expression. RFP tumor cells were visualized inside ND-GFP blood vessels. Doxorubicin inhibited the nascent tumor angiogenesis as well as tumor growth in the ND-GFP mice transplanted with B16F10-GFP (Amoh *et al.*, 2005).

Yamamoto *et al.* (2004) have genetically engineered dual-color fluorescent cells with one color in the nucleus and the other in the cytoplasm that enables real-time nuclear-cytoplasmic dynamics to be visualized in living cells *in vivo* as well as *in vitro*. Nuclear-GFP expression enabled visualization of nuclear dynamics, whereas simultaneous cytoplasmic RFP expression enabled visualization of nuclear-cytoplasmic ratios as well as simultaneous cell and nuclear shape changes. Thus, total cellular dynamics can be visualized in the living dual-color cells in real time. The cell cycle position of individual living cells was readily visualized by the nuclear-cytoplasmic ratio and nuclear morphology. Real-time induction of apoptosis was observed by nuclear size changes and progressive nuclear fragmentation. Mitotic cells were visualized by whole-body imaging after injection in the mouse ear. Common carotid artery injection of dual-color cells and a reversible skin flap enabled the external visualization of the dual-color cells in microvessels in the mouse brain where extreme elongation of the cell body as well as the nucleus occurred. Dual-color cells in various positions of the cell cycle were visualized in excised mouse lungs after tail vein injection of the dual-color cells. In the lung, the dual-color cells were observed frequently juxtaposing their nuclei, suggesting a potential novel form of cell–cell communication.

After the dual-color HT1080 cells were injected in the hearts of nude mice, a skin-flap on the abdomen was made and spread on a flat stand. Highly elongated cancer cells were observed in capillaries in the skin-flap in living mice with a color charge-coupled device (CCD) camera. The migration velocities of the cancer cells in the capillaries were measured by capturing images of the dual-color fluorescent cells over time. The cells in the capillaries elongated to fit the width of these vessels. The average length of the major axis of the cancer cells in the capillaries increased to 3.97 times their normal length. The nuclei increased their length 1.64 times in the capillaries. Cancer cells that were arrested in capillaries more than 8 μm in diameter could migrate up to 48.3 μm/hr. The data suggest that the minimum diameter of capillaries where cancer cells are able to migrate is approximately 8 μm. Extravasation was found to be infrequent in the HT1080 cells, which were visualized by dual-color imaging to be frequently undergoing apoptosis. Mouse mammary tumor cells, in contrast, frequently extravasated beginning approximately 18 hours after cell injection. The cytoplasm of the

MMT cells exited the vessels first and subsequently the whole cancer cell extravasated. The use of the dual-color cancer cells differentially labeled in the cytoplasm and nucleus and associated fluorescence imaging provide a powerful tool to understand the mechanism of cancer cell migration and deformation in small vessels and extravasation (Yamauchi *et al.*, 2005; Yamauchi, K. and Hoffman, R. M., unpublished data).

VIII. Clinically Applicable Models of GFP Tumor Imaging

Several studies have focused on delivering the GFP gene selectively to tumors in order to provide a marker for the development of new metastases. Hasegawa *et al.* (2000) administered the GFP gene to nude mice with human stomach tumors growing intraperitoneally in order to visualize future regional and distant metastases. GFP retroviral supernatants were injected intraperitoneally from day 4 to day 10 following implantation of the cancer cells. A laparotomy was done every other week so that tumor growth and metastasis formation could be visualized by GFP expression. No normal tissues were transduced by the GFP retrovirus. Two weeks after retroviral GFP delivery, GFP-expressing tumor cells were observed in the gonadal fat, greater omentum, and intestine, indicating that the tumors were efficiently transduced by the GFP gene and could be detected by its expression. Second and third laparotomies were done at weeks 4 and 6, respectively. GFP-expressing tumor cells were found spreading to lymph nodes in the mesentery. The fourth laparotomy, at 8 weeks, showed widespread GFP-expressing tumor growth including metastasis to the liver. Thus, reporter gene transduction of the primary tumor enabled detection of its subsequent metastasis. This reporter gene therapy model could be applied to primary tumors before resection or other treatment and thus provide an early detection system for metastasis and recurrence (Hasegawa *et al.*, 2000).

Human MKN45-GFP stomach cancer cells were injected into the peritoneal cavity of BALB/c nude mice to determine whether the carcinoembryonic antigen promoter could direct the GFP gene to the tumor (Kaneko *et al.*, 2001). A carcinoembryonic antigen (CEA) enhanced fluorescent protein (EGFP) plasmid was thus introduced in the peritoneal cavity using liposomes. GFP-fluorescent tumor nodules were subsequently detected by fluorescence stereomicroscopy.

In another approach, a GFP gene was conjugated to transferrin to target disseminated tumors *in vivo* (Sato *et al.*, 2000). When GFP gene conjugates were systemically administered through the tail vein to nude mice that had been subcutaneously inoculated with tumor, GFP expression was detected only in the tumor.

Varda-Bloom *et al.* (2001) developed a tissue-specific gene therapy to the angiogenic blood vessels of tumor metastasis using an adeno-based vector containing the murine preproendothelin-1 (PPE-1) promoter driving GFP. High specific activity of PPE-1 was achieved by systemic administration of the adenoviral vector to mice bearing Lewis lung carcinoma tumors. This effect was detected by GFP expression in the new vasculature of primary tumors and lung metastasis. The highest area of expression was in the angiogenic endothelial cells of the metastasis.

GFP on a herpes simplex virus type 1–Epstein-Barr virus (HSV-1/EBV) vector has been administered to tumor-bearing animals. However, persistent GFP expression was not achieved in this study (Qi *et al.*, 2000).

Umeoka *et al.* (2004) described a new approach to visualizing tumors whose fluorescence can be detected using tumor-specific replication-competent adenovirus in combination with Ad-GFP, a replication-deficient adenovirus expressing GFP. An adenovirus 5 vector (OBP-301) was constructed in which the human telomerase reverse transcriptase promoter element drives expression of E1A and E1B genes linked with an internal ribosome entry site. When human lung and colon cancer cell lines were infected with Ad-GFP at a low multiplicity of infection, GFP expression could not be detected. In the presence of OBP-301, however, Ad-GFP replicated in these tumor cells and showed strong green signals. In contrast, co-infection with OBP-301 and Ad-GFP did not show any signals in normal cells such as fibroblasts and vascular endothelial cells. Subcutaneous tumors could be visualized after intratumoral injection of OBP-301 and Ad-GFP in nude mice. Within 3 days of treatment, the fluorescence of the expressed GFP became visible. Intrathoracic administration of Ad-GFP and OBP-301 could visualize disseminated A549 tumor nodules in mice after intrathoracic implantation.

IX. Fluorescent Reporter Gene for Human T Cells

Normal, human, peripheral blood T lymphocytes were transduced with a retroviral herpes simplex virus–thymidine kinase (HSV-TK-GFP; vGFPTKfus) and nucleus-restricted green fluorescence was observed. Sorting allowed for selection of GFP-expressing T lymphocytes. The ability to target GFP-expressing T lymphocytes to tumors could have many clinical uses (Paquin *et al.*, 2001).

Singbartl *et al.* (2001) described the development of a CD2-enhanced GFP-transgenic mouse to characterize lymphocyte trafficking. A CD2-GFP plasmid including the CD2 promoter, the GFP transgene, and the CD2 locus control region was injected into B6CBA/F1 pronuclei to construct the transgenic mice. GFP CD8$^+$ T cells were imaged in cremaster muscle venules treated with both tumor necrosis factor-α and interferon-γ.

Panoskaltsis-Mortari *et al.* (2004) developed a mouse system in which to track the migration and homing of cells in a setting of bone marrow transplantation–induced graft-versus-host disease (GVHD). After systemic infusion using GFP cells, a fluorescence stereomicroscope outfitted with a color CCD camera was used for imaging. Whole-body images of anesthetized mice taken at various time points after cell infusion revealed the early migration of allogeneic cells to peripheral lymphoid organs, with later infiltration of GVHD target organs. Localization of GFP cells could be seen through the skin of shaved mice, and internal organs were easily discernible. After allogeneic or syngeneic GFP cell infusion, representative mice were dissected to better visualize deeper internal organs and tissues. Infusion of different cell populations revealed distinct homing patterns. This method thus provides a simple way to identify the critical time points for expansion of the transplanted cells in various organs.

X. Bone Marrow Protection by Transfer of Drug-Resistance Genes Coupled to GFP

A retroviral vector expressing human O^6-methylguanine-DNA methyltransferase (MGMT) and GFP was developed for stem cell protection in a murine transplant model. Mice transplanted with transduced cells showed significant resistance to the myelosuppressive effects of temozolomide, a DNA-methylating drug, and O^6-benzylguanine, a drug that inhibits MGMT. Following drug treatment, increases in GFP-positive peripheral blood cells were seen. Secondary transplant experiments proved that selection had occurred at the stem cell level. Such an approach could be used clinically in the future to protect bone marrow against chemotherapy (Sawai *et al.*, 2001).

XI. Molecular Imaging

p53 tumor-suppressor functions were visualized *in vivo* by GFP imaging. By use of the antiapoptotic gene *BCL-2* or a dominant-negative caspase 9 *(C9D*N), GFP whole-body imaging established that disruption of the apoptosis pathway downstream of p53 leads to *Eμ-myc*-GFP lymphoma expansion that phenocopies p53 loss in *Eμ-myc* transgenic mice. This finding shows that GFP whole-body imaging can be used to identify individual genes that affect tumor growth. Such information could be used to identify genes predicting aggressive tumor growth (Schmitt *et al.*, 2002a).

XII. Chemotherapy Effects of a Senescence Program Controlled by p53 and p16^{INK4a}

The GFP primary lymphomas derived from *Eμ-myc* transgenic mice respond to chemotherapy by undergoing apoptosis and engaging a premature senescence program controlled by p53 and p16^{INK4a}. Therefore, tumors respond poorly to cyclophosphamide therapy if their p53 or *INK4a/ARF* genes are disrupted; this can be seen with GFP whole-body imaging. It has also been shown that mice bearing tumors capable of drug-induced senescence have a much better prognosis after chemotherapy than those harboring tumors with senescence defects. These findings indicated that premature senescence can contribute to treatment outcome *in vivo* and provide new insights into the molecular genetics of drug resistance, which can be applied clinically (Schmitt *et al.*, 2002b).

XIII. Conclusions and Future Directions

Tumor cells stably expressing GFP *in vivo* are a powerful new tool for cancer research. Stability of expression has been studied by Naumov *et al.* (1999), who noted that all the CHO-K1-GFP cells used in their study were stably fluorescent (measured by flow cytometry) even after 24 days of growing in medium where they were deprived of selective pressure. This finding implies that GFP can be stably expressed in cells *in vivo*. This feature has proved true for all cells studied so far and is exemplified by the generation of extensive GFP-expressing metastases.

The use of GFP-expressing tumor cells in transplanted mice, fresh tissue, or live animals (Chishima *et al.*, 1997b; Naumov *et al.*, 1999) has provided new insights into the real-time growth and metastatic behavior of cancer. Several independent studies, which include an extensive comparison between metastases of GFP-transduced rat mammary carcinoma cell and the parental cell line (Chishima *et al.*, 1997b,d; Farina *et al.*, 1998) have shown that GFP transduction and expression does not affect metastatic behavior. GFP can be transfected into any cell type of interest and used as a cytoplasmic marker to show the general outlines of cells *in vivo* as well as fine morphological details such as long slender pseudopodial projections (Naumov *et al.*, 1999).

The development of tumor cells that stably express GFP at high levels has enabled investigation of tumor and metastastic growth in a completely noninvasive manner by use of whole-body imaging (Yang *et al.*, 2000). For the first time, tumor growth and metastatic studies, including drug evaluations, can be done and quantified in real time in nonperturbed individual animals. The potential of this technology is very high.

A further advantage of GFP-expressing cells is the increased contrast between brightly fluorescent tumor tissue and blood vessels within it. The ability to visualize and quantify blood-vessel development in metastases *in vivo* will greatly facilitate studies of angiogenesis and the testing of effects of antiangiogenic agents on metastatic development (Naumov *et al.*, 1999; Yang *et al.*, 2001).

The GFP approach has several important advantages over other optical approaches to imaging. In comparison with the luciferase reporter, GFP has a much stronger signal and therefore can be used to image unrestrained animals. The fluorescence intensity of GFP is very strong since the quantum yield is approximately 0.9 (Cormack *et al.*, 1996; Crameri *et al.*, 1996; Delagrave *et al.*, 1995; Heim *et al.*, 1995; Morin and Hastings, 1971). The protein sequence of GFP has also been "humanized," which enables it to be highly expressed in mammalian cells (Zolotukhin *et al.*, 1996). In addition, GFP fluorescence is fairly unaffected by the external environment since the chromaphore is protected by the three-dimensional structure of the protein (Cody *et al.*, 1993). *In vivo*, GFP fluorescence is mainly limited by light scattering which, as noted previously, can be overcome by skin flaps (Yang *et al.*, 2002) and endoscopes (Ilyin *et al.*, 2001) so that single cells can be imaged externally. Longer-wavelength fluorescence proteins, such as RFP, can also be used to reduce scatter.

An improved method of whole-body GFP imaging made use of a laser excitation source and band-pass filters matched specifically to GFP and constitutive tissue fluorescence emission bands. Processing of the primary GFP fluorescence images acquired by a CCD camera subtracted background tissue autofluorescence. This approach achieved 100% sensitivity and specificity for *in vivo* detection of a 10%-transfected BxPC3 pancreatic tumor after subcutaneous grafting or orthotopic implantation in the pancreas of nude mice (Wack *et al.*, 2003).

The luciferase reporter technique requires that animals be anesthetized and restrained so that sufficient photons to construct an image can be collected. Furthermore, this process must be carried out in a virtually light-free environment and animals must be injected with the luciferin substrate, which must reach every tumor cell in order to be useful. This limitation precludes studies that would be perturbed by anesthesia, restraint, or substrate injection and also makes high-throughput screening unfeasible. Expression of firefly luciferase (Luc) can be used to visualize tumor cell growth and regression in response to various therapies in mice. However, detection of Luc-labeled cells *in vivo* was limited to 1000 human tumor cells (Contag *et al.*, 2000; Sweeney *et al.*, 1999). The clearance of the luciferin results in an unstable luciferase signal (Burgos *et al.*, 2003). The high-intensity signal produced by GFP, however, allows unrestrained animals to be imaged without any perturbation or substrate—irradiation with nondamaging blue light is the only step needed. Images can be captured with fairly simple apparatus and there is no need for darkness.

A fusion vector containing the luciferase gene, the monomeric RFP gene, and the herpes simplex thymidine kinase gene was tested *in vivo*. A highly sensitive cooled CCD camera compatible with both luciferase and fluorescence imaging compared these two signals from the fused reporter gene using a lentivirus vector in 293T cells implanted in nude mice. The signal from RFP was approximately 1000 times stronger than luciferase (Ray *et al.*, 2004).

Near-infrared probes activated by the action of proteases (Bremer *et al.*, 2001; Weissleder *et al.*, 1999) can also be used for optical imaging of tumors. This approach requires substrate injection and the tumor must contain a specific protease that cleaves the substrate. Tumors on normal tissues (e.g., the liver) that contain these proteases cannot be visualized because background signals are too high; there is not sufficient distinction between tumor and normal tissue to obtain useful external images (Hirsch *et al.*, 2001).

For clinical application of GFP developments, future studies may make use of the approach of Hasegawa *et al.* (2000), who used retroviral GFP vectors that were selectively targeted to tumors for the purpose of identifying future metastasis. GFP labeling of tumor-infiltrating lymphocytes (Paquin *et al.*, 2001) and bone marrow transduced with GFP-linked genes that confer chemoresistance (Sawai *et al.*, 2001) also have intriguing clinical potential. Umeoka *et al.* (2004) have shown that intratumoral or intrathoracic administration of Ad-GFP in combination with another vector containing a telomerase promoter driving the E1A and E1B genes could possibly be used clinically in the future to label tumors with GFP.

The applications of *in vivo* cellular imaging with fluorescent proteins should markedly expand with the development of proteins with new colors. Tsien *et al.* (2004) have taken the *Discoma* RFP and converted it through multiple amino acid substitutions into a monomer. With further genetic modification this group has created from the *Discoma* RFP monomer a series of modified proteins with multiple new colors from yellow-orange to red-orange. These new colored proteins include mBanana, mOrange, dTomato, tdTomato, mTangerine, mStrawberry, and mCherry with increasingly longer emission maxima. It is expected that many additional colored proteins will be isolated from various organisms and modified to produce even more colors. The availability of large number of different colored proteins will enable simultaneous imaging of multiple cellular events *in vivo*.

Acknowledgments

These studies were supported in part by National Cancer Institute grants CA99258, CA101600, and CA103563.

References

Al-Mehdi, A. B., Tozawa, K., Fisher, A. B., Shientag, L., Lee, A., and Muschel, R. J. (2000). Intravascular origin of metastasis from the proliferation of engothelium-attached tumor cells: A new model for metastasis. *Nat. Med.* **6**, 100–102.

Amoh, Y., Li, L., Yang, M., Moossa, A. R., Katsuoka, K., Penman, S., and Hoffman, R. M. (2004). Nascent blood vessels in the skin arise from nestin-expressing hair follicle cells. *Proc. Natl. Acad. Sci. USA* **101**, 13291–13295.

Amoh, Y., Li, L., Yang, M., Jiang, P., Moossa, A. R., Katsuoka, K., and Hoffman, R. M. (2005). Hair-follicle-derived blood vessels vascularize tumors in skin and are inhibited by doxorubicin. *Cancer Res.* **65**, 2337–2343.

Bouvet, M., Wang, J. W., Nardin, S. R., Nassirpour, R., Yang, M., Baranov, E., Jiang, P., Moossa, A. R., and Horrman, R. M. (2002). Real-time optical imaging of primary tumor growth and multiple metastatic events in a pancreatic cancer orthotopic model. *Cancer Res.* **62**, 1534–1540.

Bouvet, M., Yang, M., Nardin, S., Wang, X., Jiang, P., Baranov, E., Moossa, A. R., and Hoffman, R. M. (2000). Chronologically-specific metastatic targeting of human pancreatic tumors in orthotopic models. *Clin. Exp. Metastasis* **18**, 213–218.

Bremer, C., Tung, C. H., and Weissleder, R. (2001). *In vivo* molecular target assessment of matrix metalloproteinase inhibition. *Nat. Med.* **7**, 743–748.

Brown, E. B., Campbell, R. B., Tsuzuki, Y., Xu, L., Carmeliet, P., Fukumura, D., and Jain, R. K. (2001). *In vivo* measurement of gene expression, angiogenesis and physicological function in tumors using multiphoton laser scanning microscopy. *Nat. Med.* **7**, 864–868.

Burgos, J. S., Rosol, M., Moats, R. A., Khankaldyyan, V., Kohn, D. B., Nelson, M. D., Jr., and Laug, W. E. (2003). Time course of bioluminescent signal in orthotopic and heterotopic brain tumors in nude mice. *Biotechniques* **34**, 1184–1188.

Chang, Y. S., di Tomaso, E., McDonald, D. M., Jones, R., Jain, R. K., and Munn, L. L. (2000). Mosaic blood vessels in tumors: Frequency of cancer cells in contact with flowing blood. *Proc. Natl. Acad. Sci. USA* **97**, 14608–14613.

Chaudhuri, T. R., Mountz, J. M., Rogers, B. E., Partridge, E. E., and Zinn, K. R. (2001). Light-based imaging of green fluorescent protein-positive ovarian cancer xenografts during therapy. *Gynecol. Oncol.* **82**, 581–589.

Chaudhuri, T. R., Cao, Z., Krasnykh, V. N., Stargel, A. V., Belousova, N., Partridge, E. E., and Zinn, K. R. (2003). Blood-based screening and light based imaging for the early detection and monitoring of ovarian cancer xenografts. *Technol. Cancer Res. Treat.* **2**, 171–180.

Chishima, T., Miyagi, Y., Wang, X., Yamaoka, H., Shimada, H., Moossa, A. R., and Hoffman, R. M. (1997b). Cancer invasion and micrometastasis visualized in live tissue by green fluorescent protein expression. *Cancer Res.* **57**, 2042–2047.

Chishima, T., Miyagi, Y., Wang, X., Baranov, E., Tan, Y., Shimada, H., Moossa, and A. R., and Hoffman, R. M. (1997c). Metastatic patterns of lung cancer visualized live and in process by green fluorescent protein expression. *Clin. Exp. Metastasis* **15**, 547–552.

Chishima, T., Miyagi, Y., Wang, X., Tan, Y., Shimada, H., Moossa, A., and Hoffman, R. M. (1997d). Visualization of the metastatic process by green fluorescent protein expression. *Anticancer Res.* **17**, 2377–2384.

Chishima, T., Yang, M., Miyagi, Y., Li, L., Tan, Y., Baranov, E., Shimada, H., Moossa, A. R., Penman, S., and Hoffman, R. M. (1997). Governing step of metastasis visualized *in vitro*. *Proc. Natl. Acad. Sci. USA* **94**, 11573–11576.

Choy, G., O' Connor, S., Diehn, F. E., Costouros, N., Alexander, H. R., Choyke, P., and Libutti, S. K. (2003). Comparison of noninvasive fluorescent and bioluminescent small animal optical imaging. *Biotechniques* **35**, 1022–1026, 1028–1030.

Cody, C. W., Prasher, D. C., Westler, V. M., Prendergast, F. G., and Ward, W. W. (1993). Chemical structure of the hexapeptide chromophore of the *Aequorea* green fluorescent protein. *Biochemistry* **32**, 1212–1218.

Contag, C. H., Jenkins, D., Contag, P. R., and Negrin, R. S. (2000). Use of reporter genes for optical measurements of neoplastic disease *in vivo*. *Neoplasia* **2**, 41–52.

Cormack, B., Valdivia, R., and Falkow, S. (1996). FACS-optimized mutants of the green fluorescent protein (GFP). *Gene* **173**, 33–38.

Crameri, A., Whitehorn, E. A., Tate, E., and Stemmer, W. P. C. (1996). Improved green fluorescent protein by molecular evolution using DNA shuffling. *Nat. Biotechnol.* **14**, 315–319.

Delagrave, S., Hawtin, R. E., Silva, C. M., Yang, M. M., and Youvan, D. C. (1995). Red-shifted excitation mutants of the green fluorescent protein. *Biotechnology* **13**, 151–154.

Duda, D. G., Fukumura, D., Munn, L. L., Booth, M. F., Brown, E. B., Huang, P., Seed, B., and Jain, R. K. (2004). Differential transplantability of tumor-associated stromal cells. *Cancer Res.* **64**, 5920–5924.

Farina, K. L., Wyckoff, J. B., Rivera, J., Lee, H., Segall, J. E., Condeelis, J. S., and Jones, J. G. (1998). Cell motility of tumor cells visualized in living intact primary tumors using green fluorescent protein. *Cancer Res.* **58**, 2528–2532.

Fukumura, D., Yuan, F., Monsky, W. L., Chen, Y., and Jain, R. K. (1997). Effect of host microenvironment on the microcirculation of human colon adenocacinoma. *Am. J. Pathol.* **151**, 679–988.

Fukumura, D., Xavier, R., Sugiura, T., Chen, Y., Park, E. C., Lu, N., Selig, M., Nielsen, G., Taksir, T., Jain, R. K., and Seed, B. (1998). Tumor induction of VEGF promoter activity in stromal cells. *Cell* **94**, 715–725.

Funovics, M. A., Alencar, H., Su, H. S., Khazaie, K., Weissleder, R., and Mahmood, U. (2003). Miniaturized multichannel near infrared endoscope for mouse imaging. *Mol. Imaging* **2**, 350–357.

Goldberg, S. F., Harms, J. F., Quon, K., and Welch, D. R. (1999). Metastasis-suppressed C8161 melanoma cells arrest in lung but fail to proliferate. *Clin. Exp. Metastasis* **17**, 601–607.

Goodison, S., Kawai, K., Hihara, J., Jiang, P., Yang, M., Urquidi, V., Hoffman, R. M., and Tarin, D. (2003). Prolonged dormancy and site-specific growth potential of cancer cells spontaneously disseminated from nonmetastatic breast tumors as revealed by labeling with green fluorescent protein. *Clin. Cancer Res.* **9**, 3808–3814.

Guba, M., Cernaianu, G., Koehi, G., Geissler, E. K., Jauch, K. W., Anthuber, M., Falk, W., and Steinbauer, M. (2001). A primary tumor promotes dormancy of solitary tumor cells before inhibiting angiogenesis. *Cancer Res.* **61**, 5575–5579.

Harms, J. F., and Welch, D. R. (2003). MDA-MB-435 human breast carcinoma metastasis to bone. *Clin. Exp. Metastasis* **20**, 327–334.

Hasegawa, S., Yang, M., Chishima, T., Miyagi, Y., Shimada, H., Moossa, A. R., and Hoffman, R. M. (2000). *In vivo* tumor delivery of the green fluorescent protein gene to report future occurrence of metastasis. *Cancer Gene Ther.* **7**, 1336–1340.

Hastings, R. H., Burton, D. W., Quintana, R. A., Biederman, E., Gujral, A., and Deftos, L. J. (2001). Parathyroid hormone-related protein regulates the growth of orthotopic human lung tumors in athymic mice. *Cancer* **92**, 1402–1410.

Heim, R., Cubitt, A. B., and Tsien, R. Y. (1995). Improved green fluorescence. *Nature* **373**, 663–664.

Hirsch, F. R., Prindiville, S. A., Miller, Y. E., Franklin, W. A., Dempsey, E. C., Murphy, J. R., Bunn, P. A., Jr., and Kennedy, T. C. (2001). Fluorescence versus white-light bronchoscopy for detection of preneoplastic lesions: A randomized study. *J. Natl. Cancer Inst.* **93**, 1385–1391.

Huang, M. S., Wang, T. J., Liang, C. L., Huang, H. M., Yang, I. C., Yi-Jan, H., and Hsiao, M. (2002). Establishment of fluorescent lung carcinoma metastasis model and its real-time microscopic detection in SCID mice. *Clin. Exp. Metastasis* **19**, 359–368.

Huang, Q., Shan, S., Braun, R. D., Lanzen, J., Anyrhambatla, G., Kong, G., Borelli, M., Corry, P., Dewhirst, M. W., and Li, C. Y. (1999). Noninvasive visualization of tumors in rodent dorsal skin window chambers. *Nat. Biotechnol.* **17**, 1033–1035.

Ilyin, S. E., Flynn, M. C., and Plata-Salaman, C. R. (2001). Fiber-optic monitoring coupled with confocal microscopy for imaging gene expression *in vitro* and *in vivo*. *J. NeuroSci. Methods* **108**, 91–96.

Ito, S., Nakanishi, H., Ikehara, Y., Kato, T., Kasai, Y., Ito, K., Akiyama, S., Nakao, A., and Tatematsu, M. (2001). Real-time observation of micrometastasis formation in the living mouse liver using a green fluorescent protein gene-tagged rat tongue carcinoma cell line. *Int. J. Cancer* **93**, 212–217.

Kaneko, K., Yano, M., Yamano, T., Tsujinaka, T., Miki, H., Akiyama, Y., Taniguchi, M., Fujiwara, Y., Doki, Y., Inoue, M., Shiozaki, H., Kaneda, Y., and Monden, M. (2001). Detection of peritoneal micrometastases of gastric carcinoma with green fluorescent protein and carcinoembryonic antigen promoter. *Cancer Res.* **61**, 5570–5574.

Katz, M. H., Bouvet, M., Takimoto, S., Spivack, D., Moossa, A. R., and Hoffman, R. M. (2003b). Selective antimetastatic activity of cytosine analog CS-682 in a red fluorescent protein orthotopic model of pancreatic cancer. *Cancer Res.* **63**, 5521–5525.

Katz, M. H., Bouvet, M., Takimoto, S., Spivack, D., Moossa, A. R., and Hoffman, R. M. (2004). Survival efficacy of adjuvant cytosine-analogue CS-682 in a fluorescent orthotopic model of human pancreatic cancer. *Cancer Res.* **64**, 1828–1833.

Katz, M. H., Takimoto, S., Spivack, D., Moossa, A. R., Hoffman, R. M., and Bouvet, M. (2003). A novel red fluorescent protein orthotopic pancreatic cancer model for the preclinical evaluation of chemotherapeutics. *J. Surg. Res.* **113**, 151–160.

Kelly, K., Alencar, H., Funovics, M., Mahmood, U., and Weissleder, R. (2004). Detection of invasive colon cancer using a novel, targeted, library-derived fluorescent peptide. *Cancer Res.* **64**, 6247–6251.

Li, C. Y., Shan, S., Huang, Q., Braun, R. D., Lanzen, J., Hu, K., Lin, P., and Dewhirst, M. W. (2000). Initial stages of tumor cell-induced angiogenesis: Evaluation via skin window chambers in rodent models. *J. Natl. Cancer Inst.* **92**, 143–147.

Li, L., Mignone, J., Yang, M., Matic, M., Penman, S., Enikolopov, G., and Hoffman, R. M. (2003). Nestin expression in hair follicle sheath progenitor cells. *Proc. Natl. Acad. Sci. USA* **100**, 9958–9961.

Maeda., H., Segawa, T., Kamoto, T., Yoshida, H., Kakizuka, A., Ogawa, O., and Kakehi, Y. (2000). Rapid detection of candidate metastatic foci in the orthotopic inoculation model of androgen-sensitive prostate cancer cells introduced with green fluorescent protein. *Prostate* **45,** 335–340.

Matz, M. V., Fradkov, A. F., Labas, Y. A., Savitsky, A. P., Zaraisky, A. G., Markelov, M. L., and Lukyanov, S. A. (1999). Fluorescent proteins from nonbioluminescent *Anthozoa* species. *Nat. Biotechnol.* **17,** 969–973.

Mook, O. R., Van Marle, J., Vreeling-Sindelarova, H., Jonges, R., Frederiks, W. M., and Van Noorden, C. J. (2003). Visualization of early events in tumor formation of eGFP-transfected rat colon cancer cells in liver. *Hepatology* **38,** 295–304.

Moore, A., Marecos, E., Simonova, M., Weissleder, R., and Bogdanov, A., Jr. (1998). Novel gliosarcoma cell line expressing green fluorescent protein: A model for quantitative assessment of angiogenesis. *Microvasc Res.* **56,** 145–153.

Morin, J., and Hastings, J. (1971). Energy transfer in a bioluminescent system. *J. Cell Physiol.* **77,** 313–318.

Naumov, G. N., Wilson, S. M., MacDonald, I. C., Schmidt, E. E., Morris, V. L., Groom, A. C., Hoffman, R. M., and Chambers, A. F. (1999). Cellular expression of green fluorescent protein, coupled with high-resolution *in vivo* videomicroscopy, to monitor steps in tumor metastasis. *J. Cell Sci.* **112,** 1835–1842.

Okabe, M., Ikawa, M., Kominami, K., Nakanishi, T., and Nishimune, Y. (1997). 'Green mice' as a source of ubiquitous green cells. *FEBS Lett.* **407,** 313–319.

Panoskaltsis-Mortari, A., Price, A., Hermanson, J. R., Taras, E., Lees, C., Serody, J. S., and Blazar, B. R. (2004). *In vivo* imaging of graft-versus-host-disease in mice. *Blood* **103,** 3590–3598.

Paquin, A., Jaalouk, D. E., and Galipeau, J. (2001). Retrovector encoding a green fluorescent protein-herpes simplex virus thymidine kinase fusion protein serves as a versatile suicide/reporter for cell and gene therapy applications. *Hum. Gene Ther.* **12,** 13–23.

Peyruchaud, O., Winding, B., Pecheur, I., Serre, C. M., Delmas, P., and Clezardin, P. (2001). Early detection of bone metastases in a murine model using fluorescent human breast cancer cells: Application to the use of the bisphosphonate zoledronic acid in the treatment of osteolytic lesions. *J. Bone Miner. Res.* **16,** 2027–2034.

Qi, J., Link, C. J., Jr., and Wang, S. (2000). Direct observation of GFP gene expression transduced with HSV-1/EBV amplicon vector in unfixed tumor tissue. *Biotechniques* **28,** 206–208.

Rashidi, B., Yang, M., Jiang, P., Baranov, E., An, Z., Wang, X., Moossa, A. R., and Hoffman, R. M. (2000). A highly metastatic Lewis lung carcinoma orthotopic green fluorescent protein model. *Clin. Exp. Metastasis* **18,** 57–60.

Ray, P., De, A., Min, J-J, Tsien, R. Y., and Gambhir, S. S. (2004). Imaging tri-fusion multimodality reporter gene expression in living subjects. *Cancer Res.* **64,** 1323–1330.

Sato, Y., Yamauchi, N., Takahashi, M., Sasaki, K., Fukaura, J., Neda, H., Fujii, S., Hirayama, M., Itoh, Y., Koshita, Y., Kogawa, K., Kato, J., Sakamaki, S., and Niitsu, Y. (2000). *In vivo* gene delivery to tumor cells by transferrin-streptavidin-DNA conjugate. *FASEB J.* **14,** 2108–2118.

Sawai, N., Zhou, S., Vanin, E. F., Houghton, P., Brent, T. P., and Sorrentino, B. P. (2001). Protection and *in vivo* selection of hematopoietic stem cells using temozolomide, O6-benzylguanine, and an alkyltransferase-expressing retroviral vector. *Mol. Ther.* **3,** 78–87.

Schmitt, C. A., Fridman, J. S., Yang, M., Baranov, E., Hoffman, R. M., and Lowe, S. W. (2002a). Dissecting p53 tumor suppressor functions *in vivo*. *Cancer Cell* **1,** 289–298.

Schmitt, C. A., Fridman, J. S., Yang, M., Lee, S., Baranov, E., Hoffman, R. M., and Lowe, S. W. (2002b). Senescence program controlled by p53 and p16 INK4a contributes to the outcome of cancer therapy. *Cell* **109,** 335–346.

Singbartl, K., Thatte, J., Smith, M. L., Wethmar, K., Day, K., and Ley, K. (2001). A CD2-green fluorescence protein-transgenic mouse reveals very late antigen-4-dependent CD8+ lymphocyte rolling in inflamed venules. *J. Immunol.* **166**, 7520–7526.

Sweeney, T. J., Mailander, V., Tucker, A. A., Olomu, A. B., Zhang, W., Cao, Y., Negrin, R. S., and Contag, C. H. (1999). Visualizing the kinetics of tumor-cell clearance in living animals. *Proc. Natl. Acad. Sci. USA* **96**, 12044–12049.

Umeoka, T., Kawashima, T., Kagawa, S., Teraishi, F., Taki, M., Nishizaki, M., Kyo, S., Nagai, K., Urata, Y., Tanaka, N., and Fujiwara, T. (2004). Visualization of intrathoracically disseminated solid tumors in mice with optical imaging by telomerase-specific amplification of a transferred green fluorescent protein gene. *Cancer Res.* **64**, 6259–6265.

Varda-Bloom, N., Shaish, A., Gonen, A., Levanon, K., Greenbereger, S., Ferber, S., Levkovitz, H., Castel, D., Goldberg, I., Afek, A., Kopolovitc, Y., and Harats, D. (2001). Tissue-specific gene therapy directed to tumor angiogenesis. *Gene Ther.* **8**, 819–827.

Varghese, H. J., Davidson, M. T., Mac Donald, I. C., Wilson, S. M., Nadkarni, K. V., Groom, A. C., and Chambers, A. F. (2002). Activated ras regulates the proliferation/apoptosis balance and early survival of developing micrometastases. *Cancer Res.* **62**, 887–891.

Wack, S., Hajri, A., Heisel, F., Sowinska, M., Berger, C., Whelan, M., Marescaux, J., and Aprahamian, M. (2003). Feasibility, sensitivity, and reliability of laser-induced fluorescence imaging of green fluorescent protein-expressing tumors *in vivo*. *Mo. Ther.* **7**, 765–773.

Wang, J., Yang, M., and Hoffman, R. M. (2004). Visualizing portal vein metastatic trafficking to the liver with green fluorescent protein-expressing tumor cells. *Anticancer Res.* **24**, 3699–3702.

Weissleder, R., Tung, C. H., Mahmood, U., and Bogdanov, A., Jr. (1999). *In vivo* imaging of tumors with protease-activated near-infrared fluorescent probes. *Nat. Biotechnol.* **17**, 375–378.

Wong, C. W., Lee, A., Shientag, L., Yu, J., Dong, Y., Kao, G., Al-Mehdi, A. B., Bernhard, E. J., and Muschel, R. J. (2001). Apoptosis: An early event in metastatic inefficiency. *Cancer Res.* **61**, 333–338.

Wyckoff, J. B., Jones, J. G., Condeelis, J. S., and Segall, J. E. (2000). A critical step in metastasis: *In vivo* analysis of intravasation at the primary tumor. *Cancer Res.* **60**, 2504–2511.

Yamamoto, N., Jiang, P., Yang, M., Xu, M., Yamauchi, K., Tsuchiya, H., Tomita, K., Wahl, G. M., Moossa, A. R., and Hoffman, R. M. (2004). Cellular dynamics visualized in live cells *in vitro* and *in vivo* by differential dual-color nuclear-cytoplasmic fluorescent-protein expression. *Cancer Res.* **64**, 4251–4256.

Yamamoto, N., Yang, M., Jiang, P., Xu, M., Tsuchiya, H., Tomita, K., Moossa, A. R., and Hoffman, R. M. (2003a). Determination of clonality of metastasis by cell-specific color-coded fluorescent-protein imaging. *Cancer Res.* **63**, 7785–7790.

Yamamoto, N., Yang, M., Jiang, P., Xu, M., Tsuchiya, H., Tomita, K., Moossa, A. R., and Hoffman, R. M. (2003b). Real-time imaging of individual fluorescent-protein color-coded metastatic colonies *in vivo*. *Clin. Exp. Metastasis* **20**, 633–638.

Yamauchi, K., Yang, M., Jiang, P., Yamamoto, N., Xu, M., Amoh, Y., Tsuji, K., Bouvet, M., Tsuchiya, H., Tomita, K., Moossa, A. R., and Hoffman, R. M. (2005). Real-time *in vivo* dual-color imaging of intracapillary cancer cell and nucleus deformation and migration. *Cancer Res.* **65**, 4246–4252.

Yang, M., Baranov, E., Jiang, P., Sun, F. X., Li, X. M., Li, L., Hasegawa, S., Bouvet, M., Al-Tuwaijri, M., Chishima, T., Shimada, H., Moossa, A. R., Penman, S., and Hoffman, R. M. (2000). Whole-body optical imaging of green fluorescent protein-expressing tumors and metastases. *Proc. Natl. Acad. Sci. USA* **97**, 1206–1211.

Yang, M., Baranov, E., Li, X. M., Wang, J. W., Jiang, P., Li, L., Moossa, A. R., Penman, S., and Hoffman, R. M. (2001). Whole-body and intravital optical imaging of angiogenesis in orthotopically implanted tumors. *Proc. Natl. Acad. Sci. USA* **98**, 2616–2621.

Yang, M., Baranov, E., Wang, J-W., Jiang, P., Wang, X., Sun, F. X., Bouvet, M., Moossa, A. R., Penman, S., and Hoffman, R. M. (2002). Direct external imaging of nascent cancer, tumor progression, angiogenesis, and metastasis on internal organs in the fluorescent orthotopic model. *Proc. Natl. Acad. Sci. USA* **99**, 3824–3829.

Yang, M., Chishima, T., Wang, X., Baranov, E., Shimada, H., Moossa, A. R., and Hoffman, R. M. (1999b). Multi-organ metastatic capability of Chinese ovary cells revealed by green fluorescent protein (GFP) expression. *Clin. Exp. Metastasis* **17**, 417–422.

Yang, M., Hasegawa, S., Jiang, P., Wang, X., Tan, Y., Chishima, T., Shimada, H., Moossa, A. R., and Hoffman, R. M. (1998). Widespread skeletal metastatic potential of human lung cancer revealed by green fluorescent protein expression. *Cancer Res.* **58**, 4217–4221.

Yang, M., Jiang, P., Sun, F. X., Hasegawa, S., Baranov, E., Chishima, T., Shimada, H., Moossa, A. R., and Hoffman, R. M. (1999a). A fluorescent orthotopic bone metastasis model of human prostate cancer. *Cancer Res.* **59**, 781–786.

Yang, M., Li, L., Jiang, P., Moossa, A. R., Penman, S., and Hoffman, R. M. (2003). Dual-color fluorescence imaging distinguishes tumor cells from induced host angiogenic vessels and stromal cells. *Proc. Natl. Acad. Sci. USA* **100**, 14259–14262.

Yang, M., Reynoso, J., Jiang, P., Li, L., Moossa, A. R., and Hoffman, R. M. (2004). Transgenic nude mouse with ubiquitous green fluorescent protein expression as a host for human tumors. *Cancer Res.* **64**, 8651–8656.

Zolotukhin, S., Potter, M., Hauswirth, W. W., Guy, J., and Muzyczka, N. (1996). A 'humanized' green fluorescent protein cDNA adapted for high-level expression in mammalian cells. *J. Virol.* **70**, 4646–4654.

Further Reading

Hoffman., R. M. (2002). Green fluorescent protein imaging of tumour growth, metastasis, and angiogenesis in mouse models. *Lancet Oncology* **3**, 546–556.

MacDonald, T. J., Tabrizi, P., Shimada, H., Zlokovic, B. V., and Laug, W. E. (1998). Detection of brain tumor invasion and micrometastasis *in vivo* by expression of enhanced green fluorescent protein. *Neurosurgery* **43**, 1437–1442.

Shaner, N. C., Campbell, R. E., Steinbach, P. A., Giepmans, B. N., Palmer, A. E., and Tsien, R. Y. (2004). Improved monomeric red, orange, and yellow fluorescent proteins derived from Discosoma sp. red fluorescent protein. *Nat. Biotechnol.* **22**, 1567–1572.

7

Modulation of Tracer Accumulation in Malignant Tumors: Gene Expression, Gene Transfer, and Phage Display

Uwe Haberkorn
Department of Nuclear Medicine, University of Heidelberg
Clinical Cooperation Unit Nuclear Medicine
German Cancer Research Center, D-69120 Heidelberg, Germany

Assessment of gene function following the completion of human genome sequencing may be done using radionuclide imaging procedures. These procedures are needed for the evaluation of genetically manipulated animals or new designed biomolecules which requires a thorough understanding of physiology, biochemistry and pharmacology. The experimental approaches will involve many new technologies including *in vivo* imaging with SPECT and PET. Nuclear medicine procedures may be applied for the determination of gene function and regulation using established and new tracers or using *in vivo* reporter genes such as genes encoding enzymes, receptors, antigens or transporters. Visualization of *in vivo* reporter gene expression can be done using radiolabeled substrates, antibodies or ligands. Combinations of specific promoters and *in vivo* reporter genes may deliver information about the regulation of the corresponding genes. Furthermore, protein-protein interactions and activation of signal transduction pathways may be visualized non-invasively. The role of radiolabeled antisense molecules for the analysis of mRNA content has to be investigated. However, possible applications are therapeutic intervention using triplex oligonucleotides with therapeutic isotopes which can be brought near to specific DNA sequences to induce DNA strand breaks at selected loci. Imaging of labeled siRNA's makes sense if these are used for therapeutic purposes in order to assess the delivery of these new drugs to their target tissue. Finally, new biomolecules will be developed by

Current Topics in Developmental Biology, Vol. 70
0070-2153/05 $35.00

DOI: 10.1016/S0070-2153(05)70007-7

bioengineering methods which may be used for isotope-based diagnosis and treatment of disease. © 2005, Elsevier Inc.

I. Introduction

Specific accumulation of radioactive probes can be used for both diagnostic and therapeutic purposes. During recent years considerable research has focused on the molecular characterization of carcinogenesis, tumor progression, and the tumor-specific immune response of patients, which offers alternative concepts of cancer diagnosis and cancer treatment. Furthermore, after the identification of new genes by high-throughput methods, functional information is required to investigate the role of these genes in living organisms. This can be done by analysis of gene expression, protein–protein interaction, or biodistribution of new molecules and may result in new diagnostic and therapeutic procedures, which include visualization of and interference with gene transcription and the development of new biomolecules to be used for diagnosis and treatment.

II. Gene Expression: Noninasive Visualization Using Antisense Oligonucleotides

Antisense RNA and DNA techniques were originally developed to modulate the expression of specific genes. These techniques originated from studies in bacteria demonstrating that these organisms are able to regulate gene replication and expression by the production of small complementary RNA molecules in an opposite (antisense) direction. Base pairing between the oligonucleotide and the corresponding target messenger RNA (mRNA) leads to highly specific binding and specific interaction with protein synthesis. Several laboratories showed that synthetic oligonucleotides complementary to mRNA sequences downregulated the translation of various oncogenes (Mukhopadhyay *et al.*, 1991; Zamecnik and Stepehnson, 1978).

Silencing of genes is also possible by a mechanism based on double-stranded RNA (dsRNA). dsRNA is cleaved by a ribonuclease (dicer) to yield short RNAs of 21–25 nucleotide length (siRNA). After interaction of these siRNAs with a complex of cellular proteins to form an RNA-induced silencing complex (RISC), the RISC binds to the complementary RNA and inhibits its translation into a protein. This is known as RNA interference (RNAi) and can be used for treatment either by application of synthetic oligonucleotides or after introduction of DNA-bearing vectors that produce RNA hairpins in the cell, which are cleaved to the corresponding siRNAs (Hannon, 2002; Moss, 2003; Sui *et al.*, 2002; Zeng *et al.*, 2002).

In addition to their use as therapeutics for specific interaction with RNA processing, oligonucleotides have been proposed for diagnostic imaging and treatment of tumors. Assuming a total human gene number between 24,000 and 30,000, calculations that take into account alternative polyadenylation and alternative splicing result in an mRNA number between 46,000 and 85,000 (Claverie, 2001). It is expected that an oligonucleotide with more than 12 nucleobases represents a unique sequence in the whole genome (Woolf *et al.*, 1992). Since these short oligonucleotides can easily be produced, antisense imaging using radiolabeled oligonucleotides offer a high number of new tracers with high specificity. Prerequisites for the use of radiolalebeled antisense oligonucleotides are ease of synthesis, stability *in vivo*, uptake into the cell, accumulation of the oligonucleotide inside the cell, interaction with the target structure, and minimal nonspecific interaction with other macromolecules. For the stability of radiolabeled antisense molecules, nuclease resistance of the oligonucleotide, stability of the oligo-linker complex, and a stable binding of the radionuclide to the complex are required. In this respect, modifications of the phosphodiester backbone such as phosphorothioates, methylphosphonates, peptide nucleic acids, or gapmers (mixed backbone oligonucleotides) result in at least a partial loss in cleavage by RNAses.

Evidence indicates that receptor-coupled endocytosis is the low-capacity mechanism by which oligonucleotides enter cells (Iversen *et al.*, 1992; Loke *et al.*, 1989). Subcellular fractionation experiments showed a sequestration of the oligonuleotides in the nuclei and the mitochondria of cervix carcinoma (HeLa) cells (Loke *et al.*, 1989). This phenomenon of fractionation, problems with *in vivo* stability of the oligonucleotides, and the low stability of the hybrid oligo-RNA structures may prevent successful imaging of gene expression. Furthermore, binding to other polyanions such as heparin-based on-charge interaction result in unspecific signals.

However, successful antisense imaging has been reported in several studies. Accumulation of ^{111}In-labeled c-*myc* antisense probes with a phosphorothioate-backbone occurred in mice bearing c-*myc*-overexpressing mammary tumors (Dewanjee *et al.* 1994). Imaging was also possible with a transforming growth factor-α antisense oligonucleotide, an antisense phosphorothioate oligodeoxynucleotide for the mRNA of glial fibrillary acidic protein, and a ^{125}I-labeled antisense peptide nucleic acid targeted to the initiation codon of the luciferase mRNA in rat glioma cells permanently transfected with the luciferase gene (Cammilleri *et al.*, 1996; Kobori *et al.*, 1999; Shi *et al.*, 2000; Urbain *et al.*, 1995). Furthermore, positron emission tomography (PET) was used for the assessment of the biodistribution and kinetics of ^{18}F-labeled oligonucleotides (Tavitian *et al.*, 1998). In addition, ^{90}Y-labeled phosphorothioate antisense oligonucleotides have been proposed as targeted radionuclide therapeutic agents for malignant tumors (Watanabe *et al.*, 1999).

However, data obtained from mRNA profiling do not faithfully represent the proteome because the mRNA content seems to be a poor indicator of the corresponding protein levels (Anderson and Seilhamer, 1977; Futcher *et al.*, 1999; Gygi *et al.*, 1999). Direct comparison of mRNA and protein levels in mammalian cells—either for several genes in one tissue or for one gene product in many cell types—revealed only poor correlations with up to 30-fold variations. Furthermore, mRNA is labile, leading to spontaneous chemical degradation and degradation by enzymes that may be dependent on the specific sequence and result in nonuniform degradation of RNA. This phenomenon introduces quantitative biases that are dependent on the time after the onset of tissue stress or death. Proteins are generally more stable and exhibit slower turnover rates in most tissues. In addition, a substantial fraction of interesting intracellular events is located at the protein level, for example, operating primarily through phosphorylation/dephosphorylation and the migration of proteins. Also of note, proteolytic modifications of membrane-bound precursors appear to regulate the release of a large series of extracellular signals such as angiotensin or tumor necrosis factor.

Because protein levels often do not reflect mRNA levels, antisense imaging may be not a generally applicable approach for a clinically useful description of biological properties of tissues. Therefore, antisense imaging for the determination of transcription by hybridization of the labeled antisense probe to the target mRNA makes sense in cases where RNA and protein content are highly correlated. Consequently, successful imaging was possible in cases where the expression of the protein was proven or the gene of interest was introduced by an expression vector (Cammilleri *et al.*, 1996; Dewanjee *et al.* 1994; Kobori *et al.*, 1999; Si *et al.*, 2000; Urbain *et al.*, 1995). In the absence of such a correlation between mRNA and protein content, the diagnostic use of antisense imaging seems questionable. Therapeutic applications may use triplex oligonucleotides with therapeutic isotopes such as Auger electron emitters, which can be brought near to specific DNA sequences to induce DNA strand breaks at selected loci. Imaging of labeled siRNAs makes sense if these are used for therapeutic purposes to assess the delivery of these new drugs to their target tissue.

III. Gene Transfer

A. *In Vivo* Reporter Genes

With the increasing availability of intrinsically fluorescent proteins that can be fused to virtually any protein of interest, their application as fluorescent biosensors has extended to dynamic imaging studies of cellular biochemistry, even at the level of organelles or compartments participating in specific processes (Fred *et al.*, 2001). Fluorescence imaging allows the determination of cell to cell variation, the extent of variation in cellular responses, the mapping of processes in multicellular tissues, and visualization of intracellular gradients in enzymatic activities, such as phosphorylation and GTPase activity.

However, for the examination of whole organisms and organ systems in deeper parts of the body, *in vivo* reporter systems are promising. Biological systems are more complex than cell cultures because external stimuli may affect and trigger cells. Therefore, noninvasive dynamic *in vivo* measurements are needed to study gene regulation in the physiological context of complex organisms. These *in vivo* reporters also may be used for the characterization of promoter regulation involved in signal transduction, gene regulation during changes of the physiological environment, and gene regulation during pharmacological intervention. This may be done by combining specific promoter elements with an *in vivo* reporter gene.

In vivo reporter genes can be visualized noninvasively using radionuclide-based methods, magentic resonance imaging (MRI), or methods based on the detection of fluorescence or luminescence. In this respect, genes encoding for enzymes, receptors, antigens, and transporters have been used. Enzyme activity can be assessed by the accumulation of the metabolites of radiolabeled specific substrates, receptors by the binding and internalization of ligands, or antigens by binding of antibodies and transporters by the uptake of their substrates. Since expression of the herpes simplex vsirus–thymidine kinase (HSV-tk) gene leads to phosphorylation of specific substrates and to the accumulation of the resulting negatively charged metabolite (Fig. 1), this gene can be used as an *in vivo* reporter gene (Alauddin *et al.*, 1999; deVries *et al.*, 2000; Gambhir *et al.*, 1999; Haberkorn *et al.*, 1997, 1998; Haubner *et al.*, 2000; Hospers *et al.*, 2000; Hustinx *et al.*, 2001; Monclus *et al.*, 1995; Saito *et al.*, 1982; Tjuvajev *et al.*, 1995; Wiebe *et al.*, 1999). A general problem with this gene is the fact that the affinity of these specific substrates for the nucleoside transport systems as well as for the enzyme is rather low; this may be a limiting factor for cellular accumulation (Haberkorn and Altmann, 2001). Therefore, at present the ideal tracer for HSV-tk imaging has not been identified and more efforts are needed to synthesize radiolabeled compounds with improved biochemical properties. Another option is the use of mutant genes. In order to improve the detection of low levels of PET reporter gene expression, a mutant herpes simplex virus type 1-thymidine kinase (HSV1-sr39tk) has been used as an *in vivo* reporter gene for PET (Gambhir *et al.*, 2000). Successful transfer of this mutant gene resulted in enhanced uptake of the specific substrates [8-^{3}H]penciclovir, and 8-[^{18}F]fluoropenciclovir in C6 rat glioma cells with a twofold increase in tracer accumulation as compared to the value obtained in tumor cells transduced with the wild-type HSV-tk gene.

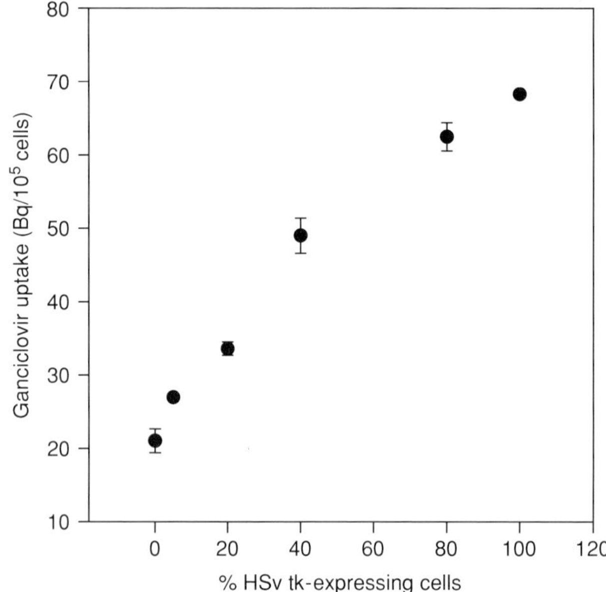

Figure 1 Uptake of the specific substrate ganciclovir in different mixtures of control cells and HSV-tk-expressing rat prostate adenocarcinoma cells after 4 hours of incubation. Mean values and standard deviation (n = 3).

Two receptor genes have been used to visualize genetically modified tumor cells. The dopamine D_2 receptor gene is an endogenous gene that is not likely to result in an immune response. Furthermore, the corresponding tracer 3-(2'-[^{18}F]-fluoroethyl)spiperone (FESP), which is in clinical use, rapidly crosses the blood–brain barrier and can be produced at high specific activity. After gene transfer with an adenoviral-directed hepatic gene delivery system, the tracer uptake in nude mice and in transplanted stable tumor cells was proportional to *in vitro* data of hepatic FESP accumulation, dopamine receptor ligand binding, and the D_2 receptor mRNA (MacLaren *et al.*, 1999). In addition, tumors modified to express the D_2 receptor accumulated significantly more FESP than wild-type tumors. In modified non-small cell lung cell lines expressing the human type 2 somatostatin receptor and transplanted in nude mice, a 5- to 10-fold higher accumulation of a ^{99m}Tc- or ^{188}Re-labeled somatostatin receptor binding peptide was obtained (Zinn *et al.*, 2000).

The low expression of tumor-associated antigens on target cells for radio-immunotherapy may be encountered by the transfer of the specific gene. Therefore, the gene for the human carcinoembryonic antigen (CEA) has been transfected in a human glioma cell line, resulting in high levels of CEA expression (Raben *et al.*, 1996). In these modified tumor cells high binding of an ^{131}I-labeled CEA antibody was observed in cell culture experiments and by scintigraphic imaging.

The gene of the human sodium iodide symporter (hNIS) has been tranferred in a variety of tumor models (Boland *et al.*, 2002; Carlin *et al.*, 2000; Cho *et al.*, 2000; Haberkorn *et al.*, 2001b, 2003; LaPerle *et al.*, 2002; Mandell *et al.*, 1999; Nakamoto *et al.*, 2000; Shimura *et al.*, 1997; Sieger *et al.*, 2003; Smit *et al.*, 2000, 2002; Spitzweg *et al.*, 2000, 2001). The corresponding protein seems promising because in addition to iodine it accepts pertechnetate, which is commonly available. Transfer of the hNIS gene into tumor cells caused a significant increase in iodide uptake by a factor of 84 to 235 (Fig. 2). Animal studies with wild-type and hNIS-expressing tumors in rats showed a maximum uptake after 1 hour and a continuous disappearance of the radioactivity from the body and the hNIS-expressing tumors (Haberkorn *et al.*, 2001b, 2003; Sieger *et al.*, 2003). Although the NIS activity is asymmetrical (favoring iodide influx), there is obviously an efflux activity with the consequence that in cells that do not organify iodide by coupling to tyrosine residues, the concentration of intracellular iodide will drop proportionally to the external iodide concentration. However, the hNIS gene may be used together with [121]I, [124]I, or even with ^{99m}Tc-pertechnetate as a simple reporter system for the visualization of other genes in bicistronic vectors, which allow co-expression of two different genes.

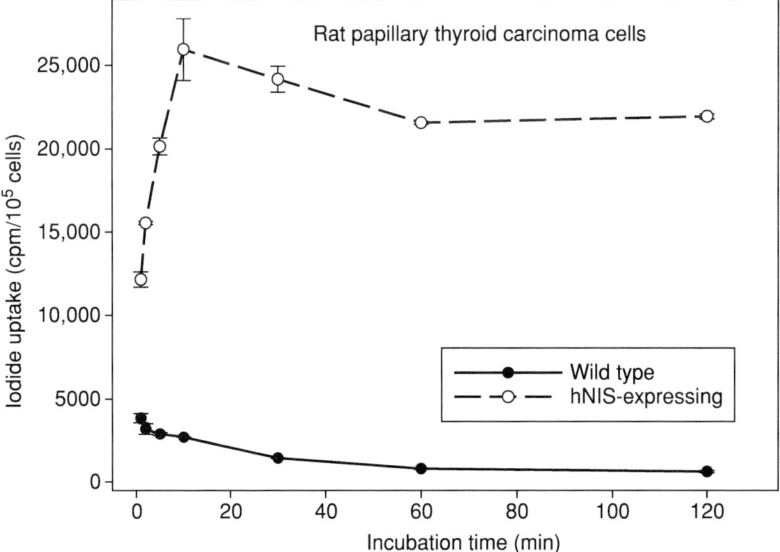

Figure 2 Time dependence of [125]I⁻ uptake in wild-type rat thyroid carcinoma cells and in an hNIS-expressing cell line. Mean values and standard deviation (n = 3).

In comparison to the other reporter genes described, the NIS gene has the advantage that it is not likely to interact with underlying cell biochemistry and that its substrate pertechnetate is widely available. Iodide is not metabolized in most tissues and, although sodium influx may be a concern, no adverse effects have been observed to date (Haberkorn and Altmann, 2002). The HSV-tk gene may alter the cellular behavior toward apoptosis by changes in the deoxynucleotide (dNTP) pool (Oliver *et al.*, 1997); antigens may cause immunoreactivity and receptors may result in second messenger activation, such as triggering signal transduction pathways. However, these possible interactions require detailed study in future experiments. For the D_2 receptor system a mutant gene has been applied that shows uncoupling of signal transduction (Liang *et al.*, 2001).

B. Gene Therapy with Suicide Genes

Recent treatment modalities include the transfer of foreign genes into normal or tumor tissue. These therapeutics are based on several different principles as follow:

1. Protection of normal tissues such as the bone marrow, which are normally targets for cytotoxic drugs; this may be achieved, for example, by the transfer of the gene for the drug efflux pump glycoprotein p.

2. Improvement of the host antitumor response by increasing the antitumor activity of immune competent cells that infiltrate the tumor or by modifying the tumor cells to enhance their immunogeneity. This approach relies on the introduction of genes that are responsible for the production of foreign surface antigens or that result in local production and secretion of cytokines.

3. Reversion of the malignant phenotype, either by suppression of oncogene expression or by introduction of normal tumor suppressor genes. The inactivation of oncoproteins may be performed by introduction of genes for intracellular antibodies (intrabodies) against these oncogenes or by use of antisense oligonucleotides and ribozymes.

4. Direct killing of tumor cells by the transfer of cytotoxic or prodrug-activating genes (Altmann and Haberkorn, 2003; Haberkorn and Altmann, 2001).

Noninvasive tools are needed to evaluate the efficiency of gene transfer in terms of gene transcription for the clinical application of these new treatment modalities. In that respect, nuclear medicine procedures offer a high sensitivity in the picomolar range. Labeling of substrates with radioactive isotopes and administration of very low amounts of these tracers allows the assessment of biochemical or physiological processes without any interference to the phenomena to be studied.

Expression of HSV-tk has been studied in several tumor models after viral as well as nonviral transfer of the gene. The principle of *in vivo* HSV-tk imaging was first demonstrated by Saito *et al.* (1982) for the visualization of HSV encephalitis. In all studies enhanced uptake of specific substrates such as 5-iodo-2'-fluoro-2'deoxy-1-b-D-arabinofuranosyluracil (FIAU), fluoro-deoxycytidine (FCdR), 5-fluoro-1-(2'-deoxy-fluoro-β-D-ribofuranosyl)uracil (FFUdR), ganciclovir (GCV), 8-[^{18}F]fluoroganciclovir (PGCV), 9-(4-[^{18}F]-fluoro-3-hydroxymethylbutyl)-guanine ([^{18}F]FHBG), and 9-[(3-^{18}F-fluoro-1-hydroxy-2-propoxy)methyl]-guanine ([^{18}F]-FHPG) was seen in genetically modified tumor cells *in vitro* and *in vivo* (Alauddin *et al.*, 1999; deVries *et al.*, 2000; Gambhir *et al.*, 1999; Haberkorn *et al.* 1997, 1998; Haubner *et al.*, 2000; Hospers *et al.*, 2000; Hustinx *et al.*, 2001; Jacobs *et al.*, 2001; Monclus *et al.*, 1995; Saito *et al.*, 1982; Tjuvajev *et al.*, 1995; Wiebe *et al.*, 1999). Furthermore, GCV, FFUdR, and FIAU uptake was highly correlated with the percentage of HSV-tk-expressing cells and the growth inhibition as measured in bystander experiments (Germann *et al.*, 1998; Tjuvajev *et al.*, 1995, 1998). In rats infected with adenovirus particles there was a significant positive correlation between the percent injected dose of FGCV retained per gram of liver and the levels of hepatic HSV-tk expression (Gambhir *et al.*, 1999).

To elucidate the transport mechanism of the specific HSV-tk substrate, GCV inhibition/competition experiments were performed in rat hepatoma and human mammary carcinoma cells. The nucleoside transport in mammalian cells is known to be heterogeneous with two classes of nucleoside transporters—the equilibrative, facilitated diffusion systems and the concentrative, sodium-dependent systems. In these experiments competition for all concentrative nucleoside transport systems and inhibition of the GCV transport by the equilibrative transport systems was observed, whereas the pyrimidine nucleo-base system showed no contribution to the GCV uptake (Haberkorn *et al.*, 1997, 1998). In human erythrocytes acyclovir has been shown to be transported mainly by the purine nucleobase carrier (Mahony *et al.*, 1988). Due to a hydroxymethyl group on its side chain, GCV has a stronger similarity to nucleosides and, therefore, may also be transported by a nucleoside transporter. Moreover, the 3'-hydroxyl moiety of nucleosides was shown to be important for their interaction with the nucleoside transporter (Gati *et al.*, 1984).

The GCV uptake was much lower than the thymidine uptake in rat hepatoma cells and human mammary carcinoma cells (Haberkorn *et al.*, 1997, 1998). Therefore, in addition to low infection efficiency of the current viral delivery systems, slow transport of the substrate and its slow conversion into the phosphorylated metabolite are the limiting factors for the therapeutic success of the HSV-tk/GCV system. Co-transfection with nucleoside transporters or the use of other substrates for HSV-tk with higher affinities for nucleoside transport and phosphorylation by HSV-tk may improve therapy outcome.

Retroviral transfer of the gene for the *Drosophila melanogaster* multisubstrate deoxyribonucleoside kinase (Dm-dNK) was done to evaluate this gene as a new potential suicide and *in vivo* reporter gene (Altmann *et al.*, 2004). Thereafter, uptake measurements were performed in wild-type and HSV-tk-expressing and Dm-dNK-expressing cell lines using different radiolabeled potential substrates: thymidine, fluorodeoxyuridine, iododeoxyuridine, bromodeoxyuridine, fluorodeoxycytidine, chlorodeoxyadenosine, FIAU, GCV, BVDU, iododeoxycytidine, and gemcitabine. Dm-dNK-expressing cells showed an enhanced uptake of different radiolabeld nucleoside analogs with a different pattern compared with HSV-tk. Furthermore, the enzyme confers enhanced uptake of gemcitabine as well as enhanced sensitivity against the drug.

The effects of cytosine deaminase (CD) gene transfer were evaluated in human glioblastoma cells. When exposed to ^{3}H-fluorocytosine (5-FC) these cells produced ^{3}H-5 fluorouracil (FU; ^{3}H-5-FU), whereas in the control cells only ^{3}H-5-FC was detected (Haberkorn *et al.*, 1996). Moreover, significant amounts of 5-FU were found in the medium of cultured cells, which may account for the bystander effect observed in previous experiments. However, uptake studies revealed only a moderate and nonsaturable accumulation of radioactivity in the tumor cells and lack of inhibition by hypoxanthine or uracil, suggesting that 5-FC enters the cells only via diffusion. Although a significant difference in 5-FC uptake was seen between CD-positive cells and controls after 48 hours incubation, no difference was observed after 2 hours of incubation. Furthermore, a rapid efflux could be demonstrated. Therefore, 5-FC transport and 5-FU efflux may be limiting factors for this therapeutic procedure and quantitation with PET. To evaluate the 5-FC uptake *in vivo*, a rat prostate adenocarcinoma cell line was transfected with a retroviral vector bearing the *Escherichia coli* CD gene. The cells were found to be sensitive to 5-FC exposure, but there was a loss of sensitivity over time. This may be due to inactivation of the viral promoter (cytomegalovirus) used in this vector. *In vivo* studies with PET and ^{18}FC showed no preferential accumulation of the tracer in CD-expressing tumors, although high-performance liquid chromatography analysis revealed a production of 5-FU that was detectable in both tumor lysates and in the blood of the animals (Haberkorn, 1999).

A comparison of the functional properties of bacterial CD and yeast CD expressed in COS-1 cells revealed that both recombinant enzymes utilized cytosine with equal efficacy, but 5-FC was a poor substrate for the bacterial CD, with an apparent catalytic efficiency 280-fold lower than that observed for the yeast CD (Hamstra *et al.*, 1999). Furthermore, after retroviral infection of tumor cell lines with the different genes, the inhibitory concentration of 50% (IC_{50}) of 5-FC was 30-fold lower in yeast CD-infected cells than in cells with expression of the bacterial CD gene. In subcutaneous human colorectal carcinoma xenografts in nude mice *in vivo,* magnetic resonance spectroscopy was used to measure yeast CD transgene expression in genetically modified tumors by direct detection of CD-catalyzed conversion of 5-FC to 5-FU (Stegman *et al.,* 1999). A three-compartment model revealed first-order kinetics, suggesting that the yeast CD was not saturated *in vivo* in the presence of measured intratumoral 5-FC concentrations above the *in vitro* determined affinity (K_m) values.

C. Radionuclide Therapy in Genetically Modified Tumors

Currently used viral vectors for gene therapy of cancer show low infection efficiency leading to moderate or low therapy effects. This problem could be solved using an approach that leads to accumulation of radioactive isotopes with beta emission. Isotope-trapping centers in the tumor create a cross-firing of beta particles, thereby efficiently killing transduced as well as nontransduced tumor cells. Currently, transfer of genes for the NIS, the norepinephrine transporter, or the thyroid peroxidase has been evaluated in different tumor models.

The first step in the process of iodide trapping in the thyroid is the active transport of iodide together with sodium ions into the cell, which is mediated by the NIS. This process acts against an electrochemical gradient and requires energy; it is coupled to the action of Na+/K+-ATPase and is stimulated by thyroid-stimulating hormone (TSH) (Marcocci *et al.,* 1984; Nakamura *et al.,* 1988, 1990; Paire *et al.,* 1997; Weiss *et al.,* 1984). Since the cloning of the human and rat cDNA sequences, several experimental studies have been performed that investigated the recombinant expression of the hNIS gene in malignant tumors by viral transfer of the hNIS gene under the control of different promoter elements (Boland *et al.,* 2002; Carlin *et al.,* 2000; Cho *et al.,* 2000; Dai *et al.,* 1996; Haberkorn *et al.,* 2001, 2003; La Perle *et al.,* 2002; Mandell *et al.,* 1999; Nakamoto *et al.,* 2000; Shimura *et al.,* 1997; Sieger *et al.,* 2003; Smanik *et al.,* 1996; Smit *et al.,* 2000, 2002; Spitzweg *et al.,* 2000, 2001). Although all of them reported high initial iodide uptake in the genetically modified tumors (Fig. 3), differing results have been obtained with regard to the efficiency of radioiodine treatment based on NIS gene transfer with very high doses given to tumor-bearing mice.

In vitro, a rapid efflux of iodide occurred with 80% of the radioactivity released into the medium after 20 minutes (Haberkorn *et al.,* 2001b, 2003; Nakamoto *et al.,* 2000; Sieger *et al.,* 2003; Smit *et al.,* 2000, 2002; Huang *et al.,* 2001). Since the effectiveness of radioiodine therapy depends not only on the type and amount, but also on the biological half-life of the isotope in the tumor, a therapeutically useful absorbed dose seems unlikely for that type of experiment.

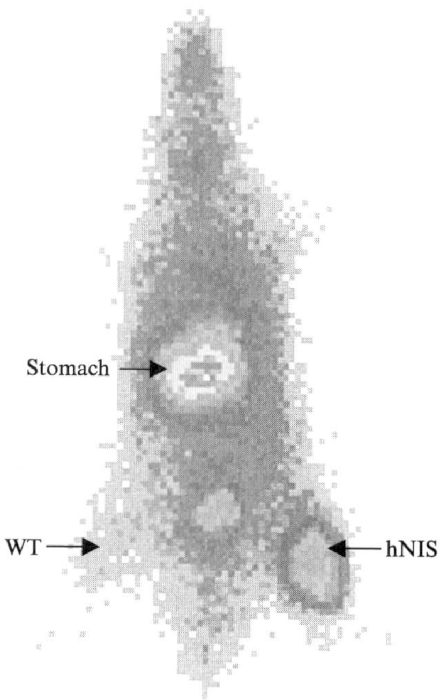

Figure 3 Scintigraphic image of a tumor-bearing male Copenhagen rat subcutaneously transplanted with hNIS-expressing (hNIS) or wild-type (WT) prostate adenocarcinoma cells (left thigh) at 2 hours after injection of $^{131}I^-$. (See Color Insert.)

A significant iodide efflux was also seen *in vivo* when doses were applied that are commonly administered to patients; only 0.4 ± 0.2 (1200 MBq/m^2) and 0.24 ± 0.02 % (2400 MBq/m^2) of the injected dose per gram in the hNIS-expressing tumors was observed at 24 hours after tracer administration (Haberkorn *et al.*, 2003). Nakamoto *et al.* found less than 1% of the injected radioactivity at 24 hours after ^{131}I administration in modified MCF7 mammary carcinomas, although initially a high uptake was seen (Nakamoto *et al.*, 2000). This corresponds to a very short half-life of ^{131}I (approximately 7.5 hours) in rat prostate carcinomas, which has also been described by Nakamoto *et al.* (2000) for human mammary carcinomas with a calculated biological half-life of 3.6 hours. In contrast, differentiated thyroid carcinoma showed a biological half-life of less than 10 days and normal thyroid of approximately 60 days (Berman *et al.*, 1968). However, *in vitro* clonogenic assays revealed selective killing of NIS-expressing cells in some studies (Boland *et al.*, 2002; Carlin *et al.*, 2000; Mandell *et al.*, 1999; Spitzweg *et al.*, 2000). In addition, bystander effects have been suggested in three-dimensional spheroid cultures (Carlin *et al.*, 2000). *In vivo* experiments in stably trans-fected human prostate carcinoma cells showed a long biological half-life of 45 hours (Spitzweg *et al.*, 2000). This resulted in a significant tumor reduc-tion (84 ± 12%) after a single intraperitoneal application of a very high ^{131}I dose of 111 MBq (Spitzweg *et al.*, 2000, 2001). The authors concluded that transfer of the NIS gene causes effective radioiodine doses in the tumor and might therefore represent a potentially curative therapy for prostate cancer. To improve therapy outcome, Smit *et al.* (2002) investigated the effects of low-iodide diets and thyroid ablation on iodide kinetics. The half-life in NIS-expressing human follicular thyroid carcinomas without thyroid ablation and under a regular diet was very short—3.8 hours. In thyroid-ablated mice maintained on a low-iodide diet, the half-life of radioiodide was increased to 26.3 hours, which may be due to diminished renal clearance of radioiodine and lack of iodide trapping by the thyroid. Subcutaneous injection of 74 MBq in thyroid-ablated nude mice maintained on a low-iodide diet postponed tumor development. However, 9 weeks after therapy tumors had developed in 4 of the 7 animals. The estimated tumor dose in these animals was 32.2 Gy (Smit *et al.*, 2002).

However, these studies used very high doses: in a mouse 74 MBq and 111 MBq correspond to administered doses of 11,100 MBq/m^2 and 16,650 MBq/m^2, respectively. This is far more than the doses used in patients. In rat prostate carcinomas treatment with amounts of ^{131}I corresponding to those given to patients (1200 MBq ^{131}I/m^2 and 2400 MBq ^{131}I/m^2) resulted in only 3 Gy absorbed dose in the genetically modified tumors (Haberkorn *et al.*, 2003). Since approximately 80 Gy has been described as necessary to achieve elimination of metastases in patients with thyroid cancer, this is not likely to induce a significant therapeutic effect in the tumors (Maxon *et al.*, 1983). Furthermore, the experiments were performed under ideal conditions with 100% NIS-expressing cells in the tumors. Given the low infection efficiency of currently viral vectors *in vivo,* the absorbed dose in a clinical study would be considerably lower.

There are also other differences in these studies, such as tracer administra-tion, time of treatment, animal models, and tumor models. Therefore, differences in the biodistribution of iodide and the biochemical properties of the tumor cells may lead to differences in iodide retention.

Some authors tried to simultaneously transfer the NIS and the thyroper-oxidase gene in order to prolong the iodine retention time in tumors (Noland *et al.*, 2002; Huang *et al.*, 2001). Boland *et al.* (2002) observed iodide organification in cells co-infected with both the NIS and the thyroid peroxi-dase (TPO) gene in the presence of exogenous hydrogen peroxide. However, the levels of iodide organification obtained were too low to significantly increase the iodide retention time. In a variety of different cell lines, includ-ing human anaplastic thyroid carcinoma and rat hepatoma cells, we were not able to measure TPO enzyme activity or enhanced accumulation of iodide irrespective of very high amounts of human TPO (hTPO) protein after retroviral transfer of the hTPO gene (Haberkorn *et al.*, 2001a). In contrast, Huang *et al.* (2001) observed an increased radioiodide uptake (by a factor of 2.5) and retention (by a factor of 3) and enhanced tumor cell apoptosis after transfection of non-small cell lung cancer cells with both human NIS and TPO genes (Huang *et al.*, 2001). However, a 72% efflux occurred *in vitro* during the first 30 minutes, indicating a very low hTPO activity in the genetically modified cells. Therefore, other modulations of iodide retention in tumor cells should be evaluated in future studies.

Lithium has been reported to reduce the release of iodine from the thyroid and has been used to enhance the efficacy of radioiodine treatment of differentiated thyroid cancer (Koong *et al.*, 1999). When the biological half-life was less than 3 days, lithium prolonged the effective half-life by more than 50% (Koong *et al.*, 1999). In Fischer rat thyroid low serum 5% (FRTL-5) cells and in primary cultures of porcine thyroid follicles, 2 mM lithium suppressed TSH-induced iodide uptake, iodide uptake stimulated by 8-bromo-cAMP, iodine organification, and *de novo* thyroid hormone formation (Lazarus *et al.*, 1998; Urabe *et al.*, 1991). Lithium is concentrated by the thyroid and inhibits thyroidal iodine uptake, iodotyrosine coupling, alters thyroglobulin structure, and inhibits thyroid hormone secretion (Gershengorn *et al.*, 1976; Lazarus *et al.*, 1998; Sedvall *et al.*, 1968; Temple *et al.*, 1972; Urabe *et al.*, 1991). Therefore, if enhanced iodide trapping in the thyroid by lithium relies on interaction with iodine coupling to tyrosine residues or inhibition of thyroid hormone secretion, an organification process is still needed to obtain a sufficient iodine accumulation in the tumor. However, experiments with hepatoma and thyroid carcinoma cells in our laboratory showed no significant effect of lithium in hNIS-expressing hepatoma cells (Haberkorn *et al.*, 2004; Sieger *et al.*, 2003). In rat papillary thyroid carcinomas the application of ^{131}I in clinically relevant amounts did not result in therapeutically useful absorbed doses in the NIS-expressing tumors (Fig. 4) *in vivo*, even under optimized conditions of thyroid ablation and treatment with lithium carbonate (Haberkorn *et al.*, 2004).

A further option to increase therapy outcome is the use of biologically more effective isotopes. Dadachova *et al.* (2002) compared ^{188}Re-perrhenate with ^{131}I for treatment of NIS-expressing mammary tumors. In a xenografted breast cancer model in nude mice, ^{188}Re-perrhenate exhibited NIS-dependent uptake into the mammary tumor. Dosimetry showed that ^{188}Re-perrhenate delivered a 4.5 times higher dose than ^{131}I and may provide enhanced therapeutic efficacy. Furthermore, the high linear energy transfer (LET)-emitter astatine-211 has been suggested as an isotope with high radiobiologic effectiveness (Nakamoto *et al.*, 2000; Petrich *et al.*, 2002). First experiments showed that the tracer uptake in NIS-expressing cell lines

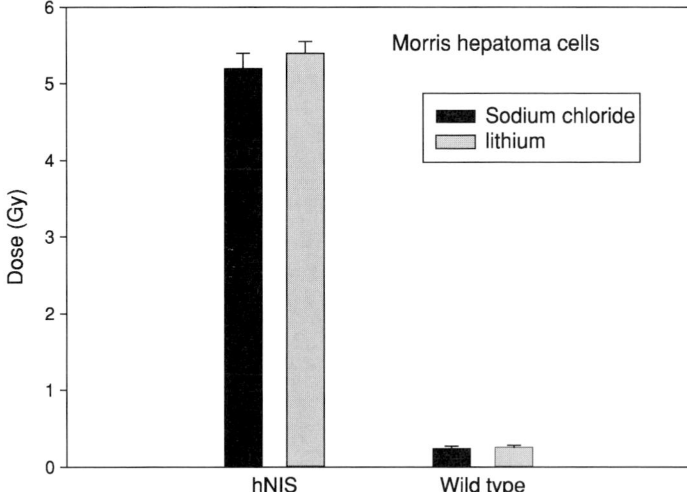

Figure 4 Results of dosimetry after administration of therapeutically relevant amounts of ^{131}I in wild-type rat papillary thyroid carcinoma cells and in hNIS-expressing cells. The rats were treated with lithium or with sodium chloride (controls). Mean values and standard deviation (n = 3).

increased up to 350-fold for ^{123}I, 340-fold for ^{99m}TcO$_4^-$, and 60-fold for ^{211}At. Although all radioisotopes showed a rapid efflux, higher absorbed doses in the tumor were found for for ^{211}At compared with ^{131}I (Petrich *et al.*, 2002).

In conclusion, a definitive proof of therapeutically useful absorbed doses *in vivo* after transfer of the NIS gene is still lacking. Further studies are needed to examine pharmacological modulation of iodide efflux or the use of the hNIS gene as an *in vivo* reporter gene (Haberkorn *et al.*, 2002, 2003).

Another approach of a genetically modified isotope treatment is the transfer of the norepinephrine gene. ^{131}I–meta-odobenzylguanidine (MIBG), a metabolically stable false analog of norepinephrine, has been widely used for imaging and targeted radiotherapy in patients with neural crest–derived tumors such as neuroblastoma or pheochromocytoma. In the adrenal medulla and in pheochromocytoma MIBG is stored in the chromaffin neurosecretory granules (Smets *et al.*, 1989). The transport of MIBG by the human norepinephrine transporter (hNET) seems to be the critical step in the treatment of MIBG-concentrating tumors. The mechanism of MIBG uptake, which is qualitatively similar to that of norepinephrine, has been studied in a variety of cellular systems and two different uptake systems have been postulated. While most tissues accumulate MIBG by a nonspecific, nonsaturable diffusion process, cells of the neuroadrenergic tissues and malignant tumors derived thereof exhibit an active uptake of the tracer that is mediated by the noradrenaline (norepinephrine) transporter (Glowniak *et al.*, 1993;

Lode *et al.*, 1995; Mairs *et al.*, 1994; Wafelman *et al.*, 1994). The clinical use of MIBG radiotherapy is so far restricted to neural crest-derived malignancies and due to insufficient ^{131}I-MIBG uptake therapy in these tumor patients is not curative.

The effect of hNET gene transfection was investigated in a variety of cells, including COS-1 cells, HeLa cells, glioblastoma cells, or rat hepatoma cells, and a 3-fold to 36-fold increase of ^{131}I-MIBG or noradrenaline accumulation was achieved (Altmann *et al.*, 2003; Boyd *et al.*, 1999; Glowniak *et al.*, 1993; Pacholczyk *et al.*, 1991). *In vivo* experiments performed with nude mice bearing both the hNET-expressing and the wild-type tumor showed a 10-fold higher accumulation of ^{131}I-MIBG in the transfected tumors with respect to the wild-type tumors. Furthermore, in rat hepatoma cells, when compared with previous studies concerning the efflux of ^{131}I from hNIS-expressing cells (Haberkorn *et al.*, 2001b), a longer retention of MIBG in the hNET-transfected cells was observed (Altmann *et al.*, 2003). Nevertheless, 4 hours after incubation with MIBG an efflux of 43% of the radioactivity was determined for the recombinant cells, whereas wild-type cells had lost 95% of the radioactivity. In view of a MIBG radiotherapy in nonneuroectodermal tumors, an intracellular trapping of the tracer is required to achieve therapeutically sufficient doses of radioactivity in the genetically modified tumor cells. In that respect, a positive correlation has been observed between the content of chromaffin neurosecretory granules and the uptake of radiolabeled MIBG (Bomanji *et al.*, 1987).

Human glioblastoma cells transfected with the bovine NET gene were killed by doses of 0.5 to 1 MBq/ml ^{131}I-MIBG in monolayer cell culture as well as in spheroids (Boyd *et al.*, 1999). Accordingly, the authors expected the intratumoral activity in a 70-kg patient to be 0.021%. This corresponds to the range of MIBG uptake usually achieved in neuroblastoma. However, data obtained from *in vitro* experiments cannot be applied to the *in vivo* situation. In contrast to stable *in vitro* conditions, the radioactive dose delivered to the tumor *in vivo* differs due to decreasing radioactivity in the serum and heterogeneity within the tumor tissue. An *in vivo* dosimetry is superior in more precisely calculating the radiation dose in a particular tumor. Using 14.8 MBq ^{131}I-MIBG for the application in tumor-bearing mice corresponding to 2200 MBq/m^2 in humans, a radiation dose of 605 mGy in the hNET-expressing and of 75 mGy in the wild-type tumor was calculated (Altmann *et al.*, 2003). With regard to the treatment of patients with a non-neuroectodermal tumor transfected by the hNET gene, this absorbed dose is too low to evoke any tumor response. In addition, as with most gene transfer studies the *in vivo* experiments were performed with animals that had been transplanted with 100% stable hNET-expressing cells. Therefore, due to the low *in vivo* infection efficiency of virus particles, infection of tumor cells *in vivo* will result in even lower radiation doses.

Future development should comprise pharmacologic modulation of MIBG retention or interaction with competing catecholamines. Use of the recombinant hNET gene product as an *in vivo* reporter is not promising because the images showed high background and relatively faint appearance of the genetically modified tumor (Altmann *et al.*, 2003). Finally, it has been questioned whether the transfer of the NET gene into pheochromocytoma or neuroblastoma cells may enhance the efficiency of MIBG therapy (Boyd *et al.*, 1999).

IV. Design of New Biomolecules for Radioisotope-Based Diagnosis and Therapy

The principle of phage display is the display of the peptide libraries fused with the carboxy-terminal domain of the minor coat protein (gene III protein fragment) on the surface of a filamentous phage. The relevant molecule is then directly detected and screened using the target molecules and amplificated after infection of *E. coli*. This allows a rapid selection (within weeks) of particular clones from large pools and determination of the amino acid sequence of a peptide displayed on a phage by sequencing the relevant section of the phage genome. This technique has been use for searching antibodies, receptors for new drug discovery and cancer therapy, either as an antagonist or an agonist of a natural ligand-receptor interaction, (Smithgall, 1995; Zwick *et al.*, 1998), and custom-made enzymes for gene therapy.

Compared with antibodies, peptides show advantages for radionuclide imaging that are related mainly to their small size; the localization of the peptides is not limited by diffusion, and a fast clearance from the circulation occurs, resulting in low background activity. Furthermore, peptides have similar binding affinities as those observed with antibodies. High-affinity receptors for peptides, such as somatostatin, gastrin-releasing peptide, and vasoactive intestinal peptide, have been observed in a variety of tumors and can be used as molecular targets for isotope-based diagnosis and treatment of tumors (Maecke *et al.*, 2005).

In phage display, large numbers of different peptides (up to 10^{10}) are expressed on the surface of bacteriophages. In M13 phages the peptides can be expressed as fusion proteins with the coat proteins p3 or p8 leading to the presentation of 5 to more than 100 copies of the peptide. This coupling of genetic and proteomic information allows binding of the phages to target structures via the expressed peptide and enrichment of the peptide by means of phage amplification during several selection rounds. This was done in human prostate carcinoma cells, resulting in the identification of a new lead structure (DUP-1) with specific binding to prostate carcinoma cell lines *in vitro* and selective accumulation in prostate carcinomas *in vivo*

(Zitzmann *et al.*, 2005a). *In vitro*, a rapid internalization of fluorescein isothiocyanate (FITC)-labeled DUP-1 was observed. This internalization is useful for both imaging and potential therapeutic applications of DUP-1 derivatives. Therefore, this 12 amino acid lead structure can be the platform for the development of an improved delivery for radionuclides or pharmaceutical drugs to prostate tumors. The peptide DUP-1 contains a motif that facilitates binding to different prostate carcinoma cell lines but not to a benign prostate cell line or to human umbilical vein endothelial cells (HUVEC). This specificity for prostate carcinoma cells was also shown in nude mice bearing PC-3 or DU-145 tumors. DUP-1 showed enhanced uptake even in undifferentiated rat prostate adenocarcinomas (Dunning R3327 subline AT1) versus normal prostate tissue. The rat model showed a comparable tumor/muscle ratio of 2.54 and 3.02 at 15 minutes after injection, for AT1 tumors in rats and DU-145 tumors in mice, respectively. With a tumor/prostate ratio of about 3 at 15 minutes, DUP-1 is a promising molecule for the diagnosis of suspected prostate carcinoma. However, data in humans are needed to assess its potential for the differential diagnosis between tumor and benign hyperplasia. The target structure for DUP-1 will be investigated in further experiments using display cloning procedures. Since analyses of serum stability of DUP-1 *in vitro* with high-pressure liquid chromatography proved degradation of DUP-1 within 10 minutes, the lead sequence DUP-1 will be used to derivatize and optimize the *in vivo* stability by simultaneously maintaining the binding characteristics. This is expected to result in better target/nontarget ratios. Among the structure evolution steps sequence fragmentation, cyclization, D-amino acid substitution, and N- or C-terminal end modifications are considered. These modifications should result in enhanced stability as well as reduced binding to plasma proteins.

Another display system is the bacterial FliTrx system. Here a phagemid vector (pFliTrx) allows the display of peptides directly on the surface of *E. coli* by using two proteins: the major bacterial flagellar protein (FliC) and thioredoxin (TrxA). Peptide libraries are cloned in frame within the active site loop of TrxA that is inserted into the dispensable region of the *FliC* gene. The resulting fusion protein assembles into the flagella on the bacterial cell surface protruding from the cells. The dodecamer peptide library is constrained by a disulfide bridge, which should result in a stable conformation of the synthetic peptide as well as resistance to degradation. Another advantage of the localization of the peptide in a loop is its presence within a protein, while in phage display systems the peptide is predominantly attached to a terminus, which might lead to problems using the peptide within a protein structure. The TrxA loop peptide is easily integrated (i.e., in a viral capsid or other surface structures). This system was used to identify a peptide binding to the human prostate carcinoma cells PC-3 (Zitzmann *et al.*, 2005b). Several peptides were identified showing a potential consensus motif. One of them (MM-2) was used for further analysis. The peptide showed less affinity to primary endothelial cells (HUVEC) and benign prostate cells (PNT-2). Its binding could be competed by the unlabeled peptide. The synthetic peptide was stable in serum for more then 24 hours and showed accumulation into PC-3 tumors in a nude mouse model. MM-2 has no sequence similarity to bombesin, LH-RH, prostate-specific antigen, or to any other peptide or protein sequence available as confirmed by a search in different protein databases. In biodistribution studies the peptide showed decreasing accumulation in the tumor as well as in other organs with time, although the decrease in the tumor is slower than in the other organs. This led to an increase of the tumor/organ ratios for all organs. The decrease in tracer accumulation could indicate degradation of the peptide *in vivo* despite the observed serum stability. Alternatively, deiodination may occur while the peptide is still intact. Since other peptides identified by this system share some, but not all, of the amino acids with MM2 the peptide may be improved by modification of the sequence.

V. Conclusion

Many new molecular structures have been cloned and will be available as potential novel diagnostic or drug discovery targets. The target selection and validation will become the most critical component in this process. This evaluation of new therapeutic principles or new biomolecules will require informations about physiology, biochemistry, and pharmacology. The experimental approaches will apply many technologies including *in vivo* imaging with single photon emission computed tomography and PET. Nuclear medicine procedures can be applied for the determination of gene expression, gene function, and gene regulation using established or new tracers to study effects in knockout mice or in transgenic animals. The measurement of gene regulation may also be performed using *in vivo* reporter genes, such as enzymes, receptors, antigens, or transporters. Drug distribution studies for new biomolecules are needed to accelerate drug approval in preclinical stages of drug development. Finally, bioengineering will lead to the design of new biomolecules by methods such as phage display, which may be used for new approaches in isotope-based diagnosis and treatment of disease.

References

Alauddin, M. M., Shahinian, A., Kundu, R. K., Gordon, E. M., and Conti, P. S. (1999). Evaluation of 9-[(3-^{18}F-fluoro-1-hydroxy-2-propoxy)methyl]guanine (^{18}F]-FHPG) *in vitro* and *in vivo* as a probe for PET imaging of gene incorporation and expression in tumors. *Nucl. Med. Biol.* **26,** 371–376.

Altmann, A., and Haberkorn, U. (2003). Assessment of gene transfer using imaging methodology. *Current Genomics* **4**, 167–184.

Altmann, A., Kissel, M., Zitzmann, S., Kubler, W., mahmut, M., Peschke, P., and Haberkorn, U. (2003). Increased MIBG uptake after transfer of the human norepinephrine transporter gene in rat hepatoma. *J. Nucl. Med.* **44**, 973–980.

Altmann, A., Eisenhut, M., Mier, W., Karlsson, A., and Haberkorn, U. (2004). The *Drosophila melanogaster* deoxyribonucleoside kinase gene induces enhanced gemcitabine sensitivity in rat hepatoma cells. *Eur. J. Nucl. Med.* **31**, S357.

Anderson, L., and Seilhamer, J. (1977). A comparison of selected mRNA and protein abundances in human liver. *Electrophoresis* **18**, 533–537.

Berman, M., Hoff, E., and Barandes, M. (1968). Iodine kinetics in man: a model. *J. Clin. Endocrinol. Metab.* **28**, 1–14.

Boland, A., Magnon, C., Filetti, S., Bidart, J. M., Schlumberger, M., Yeh, P., and Perricaudet, M. (2002). Transposition of the thyroid iodide uptake and organification system in nonthyroid tumor cells by adenoviral vector-mediated gene transfers. *Thyroid* **12**, 19–26.

Bomanji, J., Levison, D. A., Flatman, W. D., Home, T., Bouloux, P. M., Ross, G., Britton, K. E., and Besser, G. M. (1987). Uptake of iodine-123 MIBG by pheochromocytomas, paragangliomas, and neuroblastomas: A histopathological comparison. *J. Nucl. Med.* **28**, 973–978.

Boyd, M., Cunningham, S. H., Brown, M. M., Mairs, R. J., and Wheldon, T. E. (1999). Noradrenaline transporter gene transfer for radiation cell kill by ^{131}I meta-iodobenzylguanidine. *Gene Ther.* **6**, 1147–1152.

Cammilleri, S., Sangrajrang, S., Perdereau, B., Brixy, F., Calvo, F., Bazin, H., and Magdelenat, H. (1996). Biodistribution of iodine-125 tyramine transforming growth factor alpha antisense oligonucleotide in athymic mice with a human mammary tumor xenograft following intratumoral injection. *Eur. J. Nucl. Med.* **23**, 448–452.

Carlin, S., Cunnungham, S. H., Boyd, M., McCluskey, A. G., and Mairs, R. J. (2000). Experimental targeted radioiodide therapy following transfection of the sodium iodide symporter gene: Effect on clonogenicity in both two- and three-dimensional models. *Cancer Gene Ther.* **7**, 1529–1536.

Cho, J. Y., Xing, S., Liu, X., Buckwalter, T. L., Hwa, L., Sferra, T. J., Chiu, I. M., and Jhiang, S. M. (2000). Expression and activity of human Na^+/I^- symporter in human glioma cells by adenovirus-mediated gene delivery. *Gene Ther.* **7**, 740–749.

Claverie, J. M. (2001). What if there are only 30,000 human genes? *Science* **291**, 1255–1257.

Dadachova, E., Bouzahza, B., Zuckier, L. S., and Pestell, R. G. (2002). Rhenium-188 as an alternative to Iodine-131 for treatment of breast tumors expressing the sodium/iodide symporter (NIS). *Nucl. Med. Biol.* **29**, 13–18.

Dai, G., Levy, O., and Carrasco, N. (1996). Cloning and characterization of the thyroid iodide transporter. *Nature* **379**, 458–460.

De Vries, E. F., van Waarde, A., Harmsen, M. C., Mulder, N. H., Vaalburg, W., and Hospers, G. A. (2000). [^{11}C]FMAU and [^{18}F]FHPG as PET tracers for herpes simplex virus thymidine kinase enzyme activity and human cytomegalovirus infections. *Nucl. Med. Biol.* **27**, 113–119.

Dewanjee, M. K., Ghafouripour, A. K., Kapadvanjwala, M., Dewanjee, S., Serafini, A. N., Lopez, D. M., and Sfakianakis, G. N. (1994). Noninvasive imaging of c-*myc* oncogene messenger RNA with indium-111-antisense probes in a mammary tumor-bearing mouse model. *J. Nucl. Med.* **35**, 1054–1063.

Fred, S., Wouters, P., Verveer, J., and Phillipe, I. H. (2001). Bastiaens: Imaging biochemistry inside cells. *Trends Cell. Biol.* **11**, 203–211.

Futcher, B., Latter, G. I., Monardo, P., McLaughlin, C. S., and Garrels, J. I. (1999). A sampling of the yeast proteome. *Mol. Cell. Biol.* **19**, 7357–7368.

Gambhir, S. S., Barrio, J. R., Phelps, M. E., Iyer, M., Namavari, M., and herschmann, H. R. (1999). Imaging adenoviral-directed reporter gene expression in living animals with positron emission tomography. *Proc. Natl. Acad. Sci. USA* **96,** 2333–2338.

Gambhir, S. S., Bauer, E., Black, M. E., Liang, Q., Kokoris, M. S., Barrio, J. R., Iyer, M., Namavari, M., Phelps, M., and Herschmann, H. R. (2000). A mutant herpes simplex virus type 1 thymidine kinase reporter gene shows improved sensitivity for imaging reporter gene expression with positron emission tomography. *Proc. Natl. Acad. Sci. USA* **97,** 2785–2790.

Gati, W. P., Misra, H. K., Knaus, E. E., and Wiebe, L. E. (1984). Structural modifications at the 2′ and 3′ positions of some pyrimidine nucleosides as determinants of their interaction with the mouse erythrocyte nucleoside transporter. *Biochem. Pharmacol.* **33,** 3325–3331.

Germann, C., Shields, A. F., Grierson, J. R., Morr, I., and Haberkorn, U. (1998). 5-Fluoro-1-(2′-deoxy-2′-fluoro-β-D-ribofuranosyl)uracil trapping in Morris hepatoma cells expressing the herpes simplex virus thymidine kinase gene. *J. Nucl. Med.* **39,** 1418–1423.

Gershengorn, M. C., Izumi, M., and Robbins, J. (1976). Use of lithium as an adjunct to radioiodine therapy of thyroid carcinoma. *J. Clin. Endocrinol. Metab.* **42,** 105–111.

Glowniak, J. V., Kilty, J. E., Amara, S. G., Hoffman, B. J., and Turner, F. E. (1993). Evaluation of metaiodobenzylguanidine uptake by the norepinephrine, dopamine and serotonin transporters. *J. Nucl. Med.* **34,** 1140–1146.

Gygi, S. P., Rist, B., Gerber, S. A., Turecek, F., Gelb, M. H., and Aebersold, R. (1999). Quantitative analysis of complex protein mixtures using isotope-coded affinity tags. *Nat. Biotechnol.* **17,** 994–999.

Haberkorn, U. (1999). Monitoring of gene transfer for cancer therapy with radioactive isotopes. *Ann. Nucl. Med.* **13,** 369–377.

Haberkorn, U., and Altmann, A. (2001). Imaging methods in gene therapy of cancer. *Current Gene Ther.* **1,** 163–182.

Haberkorn, U., Altmann, A., and Eisenhut, M. (2002). Functional genomics and proteomics-the role of nuclear medicine. *Eur. J. Nuc. Med.* **29,** 115–132.

Haberkorn, U., Altmann, A., Jiang, S., Morr, I., Mahmut, M., Peschke, P., Kübler, W., Debus, J., and Eisenhut, M. (2001a). Iodide uptake in human anaplastic thyroid carcinoma cells after transfer of the human thyroid peroxidase gene. *Eur. J. Nucl. Med.* **28,** 633–638.

Haberkorn, U., Altmann, A., Morr, I., Knopf, K. W., Germann, C., Haeckel, R., Oberdorfer, F., and van Kaick, G. (1997). Gene therapy with herpes simplex virus thymidine kinase in hepatoma cells: Uptake of specific substrates. *J. Nucl. Med.* **38,** 287–294.

Haberkorn, U., Beuter, P., Kübler, W., Eskerski, H., Eisenhut, M., Kinscherf, R., Zitzmann, S., Strauss, L. G., Dimitrakopoulou-Strauss, A., and Altmann, A. (2004). Iodide kinetics and dosimetry *in vivo* after transfer of the human sodium iodide symporter gene in rat thyroid carcinoma cells. *J. Nucl. Med.* **45,** 827–833.

Haberkorn, U., Henze, M., Altmann, A., Jiang, S., Morr, I., Mahmut, M., and Eisenhut, M. (2001b). Transfer of the human sodium iodide symporter gene enhances iodide uptake in hepatoma cells. *J. Nucl. Med.* **42,** 317–325.

Haberkorn, U., Khazaie, K., Morr, I., Altmann, A., Müller, M., and van Kaick, G. (1998). Ganciclovir uptake in human mammary carcinoma cells expressing Herpes Simplex Virus thymidine kinase. *Nucl. Med. Biol.* **25,** 367–373.

Haberkorn, U., Kinscherf, R., Kissel, M., Kübler, W., Mahmut, M., Stephanie Sieger, Eisenhut, M., Peschke, P., and Altmann, A. (2003). Enhanced iodide transport after transfer of the human sodium iodide symporter gene is associated with lack of retention and low absorbed dose. *Gene Ther.* **10,** 774–780.

Haberkorn, U., Oberdorfer, F., Gebert, J., Morr, I., Haack, K., Weber, K., Lindauer, M., van Kaick, G., and Schackert, H. K. (1996). Monitoring of gene therapy with cytosine deaminase: *In vitro* studies using ^{3}H-5-fluorocytosine. *J. Nucl. Med.* **37,** 87–94.

Hamstra, D. A., Rice, D. J., Pu, A., Oyedijo, D., Ross, B. D., and Rehemtulla, A. (1999). Enzyme/prodrug therapy for head and neck cancer using a catalytically superior cytosine deaminase. *Hum. Gene. Ther.* **10,** 1993–2003.

Hannon, G. J. (2002). RNA interference. *Nature* **418,** 244–251.

Haubner, R., Avril, N., Hanzopoulos, P. A., Gansbacher, B., and Schwaiger, M. (2000). *In vivo* imaging of herpes simplex virus type 1 thymidine kinase gene expression: Early kinetics of radiolabelled FIAU. *Eur. J. Nuc. Med.* **27,** 283–291.

Hospers, G. A., Calogero, A., van Waarde, A., Doze, P., Vaalburg, W., Mulder, N. H., and de Vries, E. F. (2000). Monitoring of herpes simplex virus thymidine kinase enzyme activity using positron emission tomography. *Cancer Res.* **60,** 1488–1491.

Huang, M., Batra, R. K., Kogai, T., Lin, Y. Q., Herschman, J. M., Lichtenstein, A., Sharma, S., and Dubinett, S. M. (2001). Ectopic expression of the thyroperoxidase gene augments radioiodide uptake and retention mediated by the sodium iodide symporter in non-small cell lung cancer. *Cancer Gene Ther.* **8,** 612–618.

Hustinx, R., Shiue, C. Y., Alavi, A., McDonald, D., Shiue, G. G., Zhuang, H., Lanuti, M., Lambright, E., Karp, J. S., and Eck, S. (2001). Imaging *in vivo* herpes simplex virus thymidine kinase gene transfer to tumour-bearing rodents using positron emission tomography and (^{18}F)FHPG. *Eur. J. Nucl. Med.* **28,** 5–12.

Iversen, P. L., Zhu, S., Meyer, A., and Zon, G. (1992). Cellular uptake and subcellular distribution of phosphorothioate oligonucleotides into cultured cells. *Antisense Res. Dev.* **2,** 211–222.

Jacobs, A., Voges, J., Reszka, R., Lercher, M., Gossmann, A., Kracht, L., Kaestle, C., Wagner, R., Wienhardt, K., and Heiss, W. D. (2001). Positron-emission tomography of vector-mediated gene expression in gene therapy for gliomas. *Lancet* **358,** 727–729.

Kobori, N., Imahori, Y., Mineura, K., Ueda, S., and Fujii, R. (1999). Visualization of mRNA expression in CNS using ^{11}C-labeled phosphorothioate oligodeoxynucleotide. *Neuroreport* **10,** 2971–2974.

Koong, S., Reynolds, J. C., Movius, E. G., Keenan, A. M., Ain, K. B., and Robbins, J. (1999). Lithium as a potential adjuvant to ^{131}I therapy of metastatic, well differentiated thyroid carcinoma. *J. Clin. Endocrinol. Metab.* **84,** 912–916.

La Perle, K. M. D., Shen, D., Buckwalter, T. L., Williams, B., Hayman, A., Hinkle, G., Pozderac, R., Capen, C. C., and Jhiang, S. M. (2002). *In vivo* expression and function of the sodium iodide symporter following gene transfer in the MATLyLu rat model of metastatic prostate cancer. *Prostate* **50,** 170–117.

Lazarus, J. H. (1998). The effects of lithium therapy on thyroid and thyrotropin-releasing hormone. *Thyroid* **8,** 909–913.

Liang, Q., Satyamurthy, N., Barrio, J. R., Toyokuni, T., Phelps, M., Gambhir, S. S., and Herschmann, H. R. (2001). Noninvasive and quantitative imaging, in living animals, of a mutant dopamine D2 receptor reporter gene in which ligand binding is uncoupled from signal transduction. *Gene Ther.* **19,** 1490–1498.

Lode, H. N., Bruchelt, G., Seitz, G., Gebhard, S., Gekeler, V., Niethammer, D., and Beck, J. (1995). Reverse transcriptase-polymerase chain reaction (RT-PCR) analysis of monoamine transporters in neuroblastoma cell lines: Correlations to meta-iodobenzylguanidine (MIBG) uptake and tyrosine hydroxylase gene expression. *Eur. J. Cancer* **31A,** 586–590.

Loke, S. L., Stein, C. A., Zhang, X. H., Mori, K., Nakanishi, M., Subasinghe, C., Cohen, J. S., and Neckers, L. M. (1989). Characterization of oligonucleotide transport into living cells. *Proc. Natl. Acad. Sci. USA* **86,** 3474–3478.

Mac Laren, D. C., Gambhir, S. S., Satyamurthy, N., Barrio, J. R., Sharfstein, S., Toyokuni, T., Wu, L., Green, L. A., Bauer, E., and Herschmann, H. R. (1999). Repetitive non-invasive imaging of the dopamine D2 receptor as a reporter gene in living animals. *Gene Ther.* **6,** 785–791.

Maecke, H. R., Hofmann, M., and Haberkorn, U. (2005). Gallium-68 labeled peptides in tumor Imaging. *J. Nucl. Med.* **46,** S172–S178.

Mahony, W. B., Domin, P. A., and Zimmermann, T. P. (1988). Acyclovir transport into human erythrocytes. *J. Biol. Chem.* **263,** 9285–9291.

Mairs, R. J., Livingstone, A., Gaze, M. N., Wheldon, T. E., and Barrett, A. (1994). A prediction of accumulation of ^{131}I-labelled meta-iodobenzylguanidine in neuroblastoma cell lines by means of reverse transcription and polymerase chain reaction. *Br. J. Cancer.* **70,** 97–101.

Mandell, R. B., Mandell, L. Z., and Link, C. J. (1999). Radioisotope concentrator gene therapy using the sodium/iodide symporter gene. *Cancer Res.* **59,** 661–668.

Marcocci, C., Cohen, J. L., and Grollman, E. F. (1984). Effect of actinomycin D on iodide transport in FRTL-5 thyroid cells. *Endocrinology* **115,** 2123–2132.

Maxon, H. R., Thomas, S. R., Hertzberg, V. S., Kereiakes, J. G., Chen, I. W., Sperling, M. I., and Saenger, E. L. (1983). Relation between effective radiation dose and outcome of radioiodine therapy for thyroid cancer. *N. Engl. J. Med.* **309,** 937–941.

Monclus, M., *et al.* (1995). Development of PET radiopharmaceuticals for gene therapy: Synthesis of 9-((1-(^{18}F)fluoro-3-hydroxy-2-propoxy)methyl)guanine. *J. Label. Comp. Radiopharm.* **37,** 193–195.

Moss, E. G. (2003). Silencing unhealthy alleles naturally. *Trends Biotechnol.* **21,** 185–187.

Mukhopadhyay, T., Tainsky, M., Cavender, A. C., and Roth, J. A. (1991). Specific inhibition of K-*ras* expression and tumorigenicity of lung cancer cells by antisense RNA. *Cancer Res.* **51,** 1744–1748.

Nakamoto, Y., Luxen, A., and Goldman, S. (2000). Establishment and characterization of a breast cancer cell line expressing Na+/I– symporters for radioiodide concentrator gene therapy. *J. Nucl. Med.* **41,** 1898–1904.

Nakamura, Y., Ohtaki, S., and Yamazaki, I. (1988). Molecular mechanism of iodide transport by thyroid plasmalemmal vesicles: Cooperative sodium activation and asymmetrical affinities for the ions on the outside and inside of the vesicles. *J. Biochem.* **104,** 544–549.

Nakamura, Y., Kotani, T., and Ohtaki, S. (1990). Transcellular ioide transport and iodination on the apical plasma membrane by monolayer porcine thyroid cells cultured on collagen-coated fibers. *J. Endocrinol.* **126,** 275–281.

Oliver, F. J., Collins, M. K. L., and Lopez-Rivas, A. (1997). Overexpression of a heterologous thymidine kinase delays apoptosis induced by factor deprivation and inhibitors of deoxynucleotide metabolism. *J. Biol. Chem.* **272,** 10624–10630.

Pacholczyk, T., Blakely, R. D., and Amara, S. G. (1991). Expression cloning of a cocaine- and antidepressant-sensitive human noradrenaline transporter. *Nature* **350,** 350–354.

Paire, A., Bernier-Valentin, F., Selmi-Ruby, S., and Rousset, B. (1997). Characterization of the rat thyroid iodide transporter using anti-peptide antibodies. *J. Biol. Chem.* **272,** 18245–18249.

Petrich, T., Helmeke, H. J., Meyer, G., Knapp, W. H., and Potter, E. (2002). Establishment of radioactive astatine and iodine uptake in cancer cell lines expressing the human sodium iodide symporter. *Eur. J. Nucl. Med.* **29,** 842–854.

Raben, D., Buchsbaum, D. J., Khazaeli, M. B., Rosenfeld, M. E., Gillespie, G. Y., Grizzle, W. E., Liu, T., and Curiel, D. T. (1996). Enhancement of radiolabeled antibody binding and tumor localization through adenoviral transduction of the human carcinoembryonic antigen gene. *Gene Ther.* **3,** 567–580.

Saito, Y., Price, R. W., Rottenberg, D. A., Fox, J. J., Su, T. L., Watanabe, K. A., and Philipps, F. S. (1982). Quantitative autoradiographic mapping of herpes simplex virus encephalitis with radiolabeled antiviral drug. *Science* **217,** 1151–1153.

Sedvall, G., Jonsson, B., Petterson, U., and Levin, K. (1968). Effects of lithium salts on plasma protein bound iodine and uptake of ^{131}I in thyroid gland of man and rat. *Life Sci.* **7,** 1257–1264.

Shi, N., Boada, R. J., and Pardridge, W. M. (2000). Antisense imaging of gene expression in the brain *in vivo*. *Proc. Natl. Acad. Sci. USA* **97**, 14709–14714.

Shimura, H., Haraguchi, K., Miyazaki, A., Endo, T., and Onaya, T. (1997). Iodide uptake and experimental [131]J therapy in transplanted undifferentiated thyroid cancer cells expressing the Na+/I-symporter gene. *Endocrinology* **138**, 4493–4496.

Sieger, S., Jiang, S., Schönsiegel, F., Eskerski, H., Kübler, W., Altmann, A., and Haberkorn, U. (2003). Tumour specific activation of the sodium/iodide symporter gene under control of the glucose transporter gene 1 promoter (GTI-1.3). *Eur. J. Nucl. Med.* **30**, 748–756.

Smanik, P. A., Liu, Q., Forminger, T. L., Ryu, K., Xing, S., Mazzaferi, L., and Jhinag, S. M. (1996). Cloning of the human sodium iodide symporter. *Biochem. Biophys. Res. Commun.* **226**, 339–345.

Smets, L. A., Loesberg, C., Janssen, M., Metwally, E. A., and Huiscamp, R. (1989). Active uptake and extravesicular storage of m-iodobenzyl guanidine in human neuroblastoma. *Cancer Res.* **49**, 2941–2944.

Smit, J. W., Schroeder van der Elst, J. P., Karperien, M., Que, I., van der Pluim, G., Goslings, B., Romijn, J. A., and van der Heide, D. (2000). Reestablishment of *in vitro* and *in vivo* iodide uptake by transfection of the human sodium iodide symporter (hNIS) in a hNIS defective human thyroid carcinoma cell line. *Thyroid* **10**, 939–943.

Smit, J. W., Schroder-van der Elst, J. P., Karperien, M., Que, I., Stokkel, M., van der Heide, D., and Romijn, J. A. (2002). Iodide kinetics and experimental (131)I therapy in a xenotransplanted human sodium-iodide symporter-transfected human follicular thyroid carcinoma cell line. *J. Clin. Endocrinol. Metab.* **87**, 1247–1253.

Smithgall, T. E. (1995). SH2 and SH3 domains: Potential targets for anti-cancer drug design. *J. Pharmacol. Toxicol. Methods* **34**, 125–132.

Spitzweg, C., O'Connor, M. K., Bergert, E. R., Tindall, D. J., Young, C. Y., and Morris, J. C. (2000). Treatment of prostate cancer by radioiodine therapy after tissue-specific expression of the sodium iodide symporter. *Cancer Res.* **60**, 6526–6530.

Spitzweg, C., Dietz, A. B., O'Connor, M. K., Bergert, E. R., Tindall, D. J., Young, C. Y., and Morris, J. C. (2001). *In vivo* sodium iodide symporter gene therapy of prostate cancer. *Gene Ther.* **8**, 1524–1531.

Stegman, L. D., Rehemtulla, A., Beattie, B., Kievitt, E., Lawrence, T. S., Blasberg, R., Tjuvajev, J. G., and Ross, B. D. (1999). Noninvasive quantitation of cytosine deaminase transgene expression in human tumor xenografts with *in vivo* magnetic resonance spectroscopy. *Proc. Natl. Acad. Sci. USA* **96**, 9821–9826.

Sui, G., Soohoo, C., Affar el, B., Gay, F., Shi, Y., Forrester, W. C., and Shi, Y. (2002). A DNA vector-based RNAi technology to suppress gene expression in mammalian cells. *Proc. Natl. Acad. Sci. USA* **99**, 5515–5520.

Tavitian, B., Terrazzino, S., Kühnast, B., Marzabal, S., Stettler, O., Dolle, F., Deverre, J. R., Jobert, A., Hinnen, F., Bendriem, B., Crouzel, C., and Di Giamberardino, L. (1998). *In vivo* imaging of oligonucleotides with positron emission tomography. *Nature Med.* **4**, 467–471.

Temple, R., Berman, M., Robbins, J., and Wolff, J. (1972). The use of lithium in the treatment of thyrotoxicosis. *J. Clin. Invest.* **51**, 2746–2756.

Tjuvajev, J. G., Finn, R., Watanabe, K., Joshi, R., Oku, T., Kennedy, J., Beattie, B., Larson, S. M., and Blasberg, R. (1995). Imaging the expression of transfected genes *in vivo*. *Cancer Res.* **55**, 6126–6132.

Tjuvajev, J. G., Avril, N., Oku, T., Sasayama, T., Miyagawa, T., Joshi, R., Safer, M., Beattie, B., DiResta, G., Daghighian, F., Augensen, F., Koutcher, J., Zweit, J., Humm, J., Larson, S. M., Finn, R., and Blasberg, R. (1998). Imaging herpes virus thymidine kinase gene transfer and expression by positron emission tomography. *Cancer Res.* **58**, 4333–4341.

Urabe, M., Hershmann, J. M., Pang, X. P., Murakami, S., and Sugawara, M. (1991). Effect of lithium on function and growth of thyroid cells *in vitro*. *Endocrinology* **129**, 807–814.

Urbain, J. L., Shore, S. K., Vekemans, M. C., Cosenza, S. C., De Riel, K., Patel, G. V., Charkes, N. D., Talmud, L. S., and Reddy, E. P. (1995). Scintigraphic imaging of oncogenes with antisense probes: Does it make sense? *Eur. J. Nucl. Med.* **22,** 499–504.

Wafelman, A. R., Hoefnagel, C. A., Maes, R. A. A., and Beijnen, J. H. (1994). Radioiodinated metaiodo-benzylguanidine: A review of its distribution and pharmacokinetics, drug interactions, cytotoxicity and dosimetry. *Eur. J. Nucl. Med.* **21,** 545–559.

Watanabe, N., Sawai, H., Endo, K., Shinozuka, K., Ozaki, H., Tanada, S., Murata, H., and Sasaki, Y. (1999). Labeling of phosphorothioate antisense oligonucleotides with yttrium-90. *Nucl. Med. Biol.* **26,** 239–243.

Weiss, S. J., Philp, N. J., and Grollman, E. F. (1984). Iodide transport in a continuous line of cultured cells from rat thyroid. *Endocrinology* **114,** 1090–1098.

Wiebe, L. I., Knaus, E. E., and Morin, K. W. (1999). Radiolabelled pyrimidine nucleosides to monitor the expression of HSV-1 thymidine kinase in gene therapy. *Nucleosides Nucleotides* **18,** 1065–1066.

Woolf, T. M., Melton, D. A., and Jennings, C. G. B. (1992). Specificity of antisense oligonucleotides *in vivo. Proc. Natl. Acad. Sci. USA* **89,** 7305–7309.

Zamecnik, P. C., and Stephenson, M. L. (1978). Inhibition of Rous sarcoma virus replication and cell transformation by a specific oligodeoxynucleotide. *Proc. Natl. Acad. Sci. USA* **75,** 280–285.

Zeng, Y., Wagner, E. J., and Cullen, B. R. (2002). Both natural and designed micro RNAs can inhibit the expression of cognate mRNAs when expressed in human cells. *Mol. Cell.* **9,** 1327–1333.

Zinn, K. R., Buchsbaum, D. J., Chaudhuri, T. R., Mountz, J. M., Grizzle, W. E., and Rogers, B. E. (2000). Noninvasive monitoring of gene transfer using a reporter receptor imaged with a high-affinity peptide radiolabeled with ^{99m}Tc or ^{188}Re. *J. Nucl. Med.* **41,** 887–895.

Zitzmann, S., Krämer, S., Mier, W., Mahmut, M., Fleig, J., Altmann, A., Eisenhut, M., and Haberkorn, U. (2005b). Identification of a new prostate specific cyclic peptide with the bacterial FLITRX system. *J. Nucl. Med.* **46,** 782–785.

Zitzmann, S., Mier, W., Schad, A., Kinscherf, R., Askoxylakis, V., Krämer, S., Altmann, A., Eisenhut, M., and Haberkorn, U. (2005a). A new prostate carcinoma binding peptide (DUP-1) for tumor imaging and therapy. *Clin. Cancer. Research* **11,** 139–146.

Zwick, M. B., Shen, J., and Scott, J. K. (1998). Phage-displayed peptide libraries. *Curr. Opin. Biotechnol.* **9,** 427–436.

Further Reading

Pradet-Balade, B., Boulme, F., Beug, H., Müllner, E. W., and Garcia-Sanz, J. A. (2001). Translation control: Bridging the gap between genomics and proteomics? *Trends Biochem. Sci.* **26,** 225–229.

8

Amyloid Imaging: From Benchtop to Bedside

*Chunying Wu,**,† *Victor W. Pike,*‡ *and Yanming Wang**
*Department of Medicinal Chemistry, College of Pharmacy
University of Illinois at Chicago, Chicago, Illinois 60612
†National Laboratory of Nuclear Medicine, Jiangsu Institute of Nuclear Medicine
Jiangsu 214063, China
‡Molecular Imaging Branch, National Institute of Mental Health
National Institutes of Health, Bethesda, Maryland 20892

Tremendous efforts have been made in the search for a cure or effective treatment of Alzheimer's disease (AD) to develop therapies aimed at halting or reversing amyloid plaque deposition in the brain. This necessitates *in vivo* detection and quantification of amyloid plaques in the brain for efficacy evaluation of anti-amyloid therapies. For this purpose, a wide array of amyloid-imaging probes has been developed, mainly for *in vivo* studies based on positron emission tomography and single photon emission computed tomography. This review provides a full account of the development of amyloid-imaging agents. The *in vitro* binding properties and *in vivo* pharmacokinetic profiles of all amyloid-imaging agents so far reported are comprehensively and uniquely surveyed. Emphasis is placed on the development of small-molecule probes based on amyloid dyes, such as Congo red and thioflavin T. Compared to large biomolecules, these small-molecule probes have been systematically investigated through extensive structure activity relationship studies. Many of the probes show favorable properties for *in vivo* studies. As a result, three lead compounds, termed PIB (Pittsburgh-Compound B, [^{11}C]6-OH-BTA-1), FDDNP (2-(1-[6-[(2-[^{18}F]fluoroethyl)(methyl)amino]-2-naphthyl]ethylidene) malononitrile), and SB-13 (4-*N*-methylamino-4′-hydroxystilbene), have been

Current Topics in Developmental Biology, Vol. 70
Copyright 2005, Elsevier Inc. All rights reserved.

171

0070-2153/05 \$35.00
DOI: 10.1016/S0070-2153(05)70008-9

identified and evaluated in human subjects. Preliminary studies have indicated that these lead compounds exhibit a characteristic retention in AD subjects that is consistent with the AD pathology, thus proving the concept that amyloid deposits in the brain can be readily detected and quantified *in vivo*. The progress to date paves the way for further investigation in various aspects of AD research. Once developed, these amyloid-imaging agents could be used as biomarkers to aid in early and definitive diagnosis of AD, facilitate drug discovery and development, and allow pathophysiological studies of the disease mechanism. Furthermore, the success in the development of amyloid-imaging agents helps with the development of imaging agents for *in vivo* studies of other AD pathologies in particular and of neurodegenerative disorders in general. © 2005, Elsevier Inc.

I. Introduction

Alzheimer's disease (AD), which was first described in 1907 by Alois Alzheimer (Alzheimer, 1907), is a chronic, neurodegenerative disorder that currently affects more than 4 million Americans and costs the nation approximately $100 billion annually. With the aging of the population, it is projected that 14 million people will be affected by the middle of the twenty-first century unless a cure or prevention is found (Olshansky *et al.*, 1993; Price and Sisodia, 1998). AD is characterized by a common set of clinical and pathological features. The clinical impairments include cognitive dysfunction and behavioral abnormality with increasing mortality. The pathological features characteristic of AD are the presence of neuron and synapse loss, extracellular neuritic plaques (NPs), and intracellular neurofibrillary tangles (NFTs) (Trojanowski *et al.*, 1997). Epidemiologically, AD can be familial and sporadic. The primary causes of familial AD are mutations in the gene for presenilin-1 on chromosome 14, presenilin-2 on chromosome 1, or the amyloid precursor protein (APP) on chromosome 21 (Clark and Goate, 1997). The risk of sporadic AD is significantly increased by the presence of the $\varepsilon 4$ allele of apolipoprotein E, an effect that is related to gene dosage (Ishii *et al.*, 1997). In most cases, aging is the most significant risk factor. The prevalence rates double every 5 years among the population of individuals 60 years or older and reach nearly 50% after 80 years of age (Katzman, 1993).

Currently there are no definitive treatments to cure AD. Much recent interest has focused on the development of anti-amyloid therapies aimed at halting and reversing amyloid formation and deposition. Therapies currently under development include inhibition of β- and γ-secretases and anti-amyloid immunotherapy approaches (Schenk *et al.*, 1999, 2000, 2001). Although the active immunization (vaccine) approach has led to significant side effects such as meningoencephalitis (Nordberg, 2003; Robinson *et al.*,

2003), many other immunotherapeutic approaches remain, including passive immunization with antibodies to β-amyloid ($A\beta$). For efficacy evaluation, all these anti-$A\beta$ therapies require a noninvasive method that can quantitate effects on $A\beta$ deposition in the brain.

II. Aβ Deposits as a Biological Marker for Alzheimer's Disease

A biological marker of AD would help physicians and researchers to study the disease quantitatively and monitor its progression closely. Considering that postmortem brain tissue staining is the only method to provide a definite diagnosis of AD (Delacourte, 1998), the call for such a biological marker is great, particularly for definite diagnosis of AD at an early stage and for further studies on the causative mechanism of AD. Therefore, a biomarker should reflect an early and specific event in AD pathology that progresses through at least the early course of the clinical disease. It should also be subject to quantitative analysis.

One pathologic hallmark characteristic of AD is the presence of extracellular NPs. The high-density, nondiffusible plaque substance is frequently found in the hippocampus and associated cortex (where neurons are damaged and lost). The major component of amyloid plaque has been isolated and characterized as a group of peptides called amyloid-β ($A\beta$) (Hilbich *et al.*, 1991). These $A\beta$ peptides, which contain 40–42 amino acid residues, are metabolic products of APP arising from cleavage by β- and γ-secretases (De Strooper and Konig, 1999). Under physiological conditions, $A\beta$ peptides adopt a β-pleated sheet structure and aggregate into fibrils *in vitro*—a process thought to be related to deposition in the brain. Genetic studies show that all mutations that cause AD are closely related to APP metabolism into $A\beta$. This strongly suggests that $A\beta$ deposition is an early and specific event in the pathogenesis of AD (Selkoe, 2000).

Despite the fact that fibrillar NPs (FNPs) are common in AD, the identity of the neurotoxic species associated with $A\beta$ has not yet been clearly revealed. This is due to the fact that $A\beta$ monomers, dimers, and oligomers coexist with amyloid plaques, including diffuse plaques and fibrillar plaques in the brain. Which specific form(s) of $A\beta$ are responsible for the neuronal damage is still under debate. Previous studies leading to the amyloid cascade hypothesis mainly pointed to NPs as the mediator of neurotoxicity (Hardy, 1992), but recent studies focus on soluble, protofibrillar $A\beta$ oligomers as true pathogenic species (Selkoe, 2002; Walsh *et al.*, 2002). The structural basis of the neurotoxicity of $A\beta$ oligomers has also been investigated (Lashuel *et al.*, 2002). Furthermore, levels of insoluble $A\beta$ correlate with degree of cognitive impairment in AD (Naslund *et al.*, 2000).

Regardless of the origin of neurotoxicity, different $A\beta$ species likely exist in equilibrium, and $A\beta$ plaques could serve as a reservoir of these $A\beta$ oligomers. Mounting evidence has also shown that the level of FNPs is a strong indication of the process underlying AD. Unlike diffuse plaques that are found in low amounts in the brains of all aged human subjects and many mammalian species, FNPs are found predominantly in AD (Joachim *et al.*, 1989). While the total number of amyloid plaques is relatively stable over the clinical course of AD, the percentage of NPs increases with the progression of dementia (Morris *et al.*, 1996). Human studies have also indicated that amyloid deposition could begin years before the onset of AD symptoms (Price, 1997).

The concentration of $A\beta$ in AD brain has been reported to be more than 2000 pmol/g wet wt ($\sim$2 μM) (Naslund *et al.*, 1994). This concentration is more than 10-fold higher than that found in age-matched control brains and about 100-fold higher than the concentrations of many neuroreceptors (e.g., the dopamine D_2 receptor) currently imaged with positron emission tomography (PET) or single photon emission computed tomography (SPECT). With an average diameter of 100 μm and a density of 20–30 plaques/mm^2 in brain regions with high densities of amyloid deposition (Price, 1997), NPs can occupy more than 25% of the cortical area in AD (Mochizuki *et al.*, 1996). Together, these findings make amyloid deposits a good candidate for a biological marker of AD.

III. *In Vivo* Imaging Tool for Drug Screening

Traditionally, drug discovery and development have been largely based on *in vitro* and *ex vivo* screening techniques to select promising lead candidates for potential human studies. These *in vitro* and *ex vivo* techniques allow accurate determination of pharmacokinetic and pharmacodynamic properties of drug candidates. Such properties are measured by invasive biological experiments, which require a large quantity of animals. These biological studies help determine drug action at certain time points but do not continuously monitor the drug action over time in a complex living environment. However, it is highly desirable to evaluate biological properties of a drug continuously in a living organism following *in vitro* or *ex vivo* studies, since the following questions might additionally be answered:

1. How much drug can be delivered to the target tissue relative to normal tissue?
2. How strongly does the drug bind to the target and with what potency?
3. How effective is the drug (Gupta *et al.*, 2002)?

Ideally, these questions need to be addressed quantitatively in individual subjects so that their answers can be an invaluable addition to the process of screening drug candidates. Noninvasive molecular imaging techniques have now become a powerful means to address and answer such questions.

Over the past decade, molecular imaging techniques have been widely used to study disease pathology and drug actions both in the clinical and preclinical stages. These studies not only facilitate the screening of novel drug candidates at early stages of drug development (Gee, 2003), but they also reduce the substantial development costs (Gupta *et al.*, 2002). Among these imaging techniques, PET and SPECT play a unique and important role due to the fact that they are functional imaging techniques. Furthermore, they are used in combination with only a trace mass dose of radiotracer that only rarely exerts any pharmacological or toxicological effect (Gee, 2003). The use of PET and SPECT depends on the development of radiotracers with high sensitivity and specificity for certain molecular targets associated with disease in both animal models and human subjects. As external imaging methodology PET and SPECT can be used in all phases of drug discovery and development for *in vivo* studies of the pharmacodynamics, pharmacokinetics, and mechanisms of drug actions. To date, PET and SPECT have been widely used for drug screening in animal models (Eckelman, 2003). They have also become an indispensable tool in the drug discovery process to bridge the gap between laboratory research and clinical applications. Recent examples include evaluation of tumor metabolism and proliferation (Brock *et al.*, 2000; Shields *et al.*, 1998a), drug receptor interaction (Anderson *et al.*, 2001; Hume *et al.*, 1997; Kapur *et al.*, 1997a,b,c), modulation of multidrug resistance (Chen *et al.*, 1997; Hendrikse *et al.*, 1999a,b; Levchenko *et al.*, 2000; Vecchio *et al.*, 1997), quantitating angiogenensis and antivascular activity (Vavere and Lewis, 2003; Weber *et al.*, 2001), detection of apoptosis (Blankenberg and Strauss, 2001; Blankenberg *et al.*, 2001; Narula *et al.*, 2001), tumor hypoxia (Casciari *et al.*, 1995; Foo *et al.*, 2004; Lewis *et al.*, 2002), therapy response, and gene expression (Blasberg, 2002; Blasberg and Gelovani-Tjuvajev, 2002; Dehdashti *et al.*, 1999; Gambhir *et al.*, 1999; Herschman, 2004; Shields *et al.*, 1998b; Walker and Zigler, 2003; Walker *et al.*, 2004).

In AD, anti-amyloid therapies are currently under development to halt or reverse the progressive accumulation of Aβ deposit in AD brain. To date, brain amyloid deposition can be detected and quantitated only at the time of autopsy. Lack of *in vivo* imaging techniques hampers the efficacy evaluation of novel anti-amyloid therapies. So far, applications of PET and SPECT for amyloid imaging have been limited by the lack of suitable radiotracers that can readily enter the brain and selectively bind to amyloid deposits. Over the past decade, tremendous effects have been made to develop amyloid probes suitable for PET and SPECT studies in human subjects. This review comprehensively surveys and discusses the structures of these amyloid probes and their *in vitro* binding properties and *in vivo* pharmacokinetic profiles. Once developed, these amyloid probes can be applied as a powerful imaging tool to many aspects of AD research, ranging from pathophysiological studies to efficacy evaluation of therapeutic treatment.

IV. Introduction to SPECT and PET

PET and SPECT are two molecular imaging techniques developed to use trace amounts of radiotracers to noninvasively image and quantify cellular and tissue changes in living human subjects. Originating in the mid-1970s, PET and SPECT are imaging techniques that provide functional information about living subjects (Nutt, 2002). PET imaging is based on the coincident detection of a pair of γ-rays (photons) as a result of annihilation between an emitted positron and a nearby electron. Upon annihilation, the mass of each particle is completely transformed into energy, resulting in the emission of a pair of photons. The two photons are emitted from the annihilation site at almost exactly $180°$ apart and each has an energy of 0.511 MeV. The coincident detection of the pair of photons by opposed crystal detectors of the scanner approximately locates the parent positron emitter to the axis between the two detecting crystals. Using an image reconstruction algorithm, sequential three-dimensional images of the specific distribution of radiotracers within the body are produced (Cherry, 2001; Walker *et al.*, 2004). Since positron-emitting tracers may be taken up at different rates by healthy tissues relative to tissues undergoing a disease process, abnormal activity may be detected in terms of the accumulation of radiotracers in a specific region.

Frequently used positron emitters are the short-lived isotopes of elements found in organic compounds, namely ^{11}C, ^{13}N, and ^{15}O, with half lives ($t_{1/2}$) of 20.4, 9.97, and 2.03 minutes, respectively. In addition, ^{18}F, with a $t_{1/2}$ of 109.7 minutes, is also commonly used. ^{18}F-labeled tracers are sufficiently long-lived to allow their distribution to sites remote from cyclotron production of the radionuclide.

SPECT complements PET in terms of costs and performance. Different isotopes, such as ^{99m}Tc, ^{123}I, and ^{111}In, are used. SPECT imaging is based on external detection of a single photon emitted from the radionuclide during decay. Hence, special collimators are needed to acquire the image data from multiple views around the body. The use of collimators results in an enormous decrease in detection efficiency. In contrast, PET imaging requires no collimators for localization of radioactivity. On the one hand, PET generally has higher resolution, higher sensitivity, and better quantitative capability than SPECT and is therefore well suited to the development and validation of new radiotracers. On the other hand, SPECT is more practical, economical, and available, which makes it more appealing for routine clinical studies.

V. Development of Amyloid Probes Based on Biomolecules

In the search for amyloid-imaging agents, the use of antibodies against $A\beta$ amyloid was first explored with the expectation that antibodies might be radiolabeled and then function as imaging agents with high sensitivity and specificity (Ikeda *et al.*, 1987). Thus Pardridge *et al.* developed a cationized monoclonal antibody (IgG) against $A\beta$ proteins (Bickel *et al.*, 1994). Because the antibody itself does not cross the blood–brain barrier (BBB) because of its high molecular weight, they modified it by cationization. It has been shown that cationization of the antibody (IgG) enables BBB penetration following peripheral administration (Friedland *et al.*, 1994). They first explored the potential of using murine monoclonal antibodies (mAbs) against $A\beta$ as *in vivo* probes of amyloid deposition in the brain. After labeling mAbs with ^{99m}Tc ($t_{1/2} = 6.02$ hours) (Table I, entry 6), they screened them for their *in vitro* binding properties and for visualizing amyloid deposits and NPs in postmortem AD brain (Bickel *et al.*, 1994; Majocha *et al.*, 1992). However, this sort of antibody did not readily cross the BBB. Even if the antibodies entered the brain, the free antibodies tended to be retained in the absence of amyloid deposits. The slow clearance of the free antibodies in the brain tends to reduce the signal to noise ratios. This effect may be countered by using a radioisotope with long half-life $t_{1/2}$ over long image acquisition times. In a further study, they cationized a specific antibody (AMY33, Table I, entry 3) at selected site(s) and radiolabeled with ^{111}In ($t_{1/2} = 2.8$ days) for potential SPECT imaging. ^{111}In was chosen because peripheral radioactive metabolites are low in blood and not readily available to the brain (Friedland *et al.*, 2000). *In vitro* binding assays showed that the modified antibody displayed a high binding affinity. Further *in vivo* studies, however, did not lead to promising results (Walker *et al.*, 1994).

Meanwhile, Walker *et al.* (1994) evaluated a murine mAb (10D5) for *in vivo* binding properties to $A\beta$ deposits in aged nonhuman primates. To circumvent the BBB, they injected unlabeled antibody into the cerebrospinal fluid of the cisterna magna. *Ex vivo* studies showed a fraction of amyloid deposits in cerebral cortex. However, no *in vivo* experiments have been performed with any radiolabeled analogs of this antibody.

Due to the difficulty of brain uptake associated with antibodies, investigators turned to $A\beta$ peptides and explored their potential as amyloid-imaging agents. This approach is based on the observation that $A\beta$ peptides have a strong tendency to aggregate into amyloid deposits. Thus, Maggio *et al.* (1992) studied the *in vitro* binding properties of a radioiodinated $A\beta$ 1–40 (Table I, entry 7). They found that radioiodinated human $A\beta$ 1–40 is rapidly deposited *in vitro* onto neuritic and diffused plaques and cerebrovascular amyloid in AD brain tissues. This deposition did not occur in the

Table I Biomolecular Amyloid Imaging Agents

No.	Abbreviation	Tg Mice	Autoradiography Immunostaining	Baboon Study	Human Study	Reference
1	I-SAP	ND	ND	ND	Peripherally	Hawkins *et al.*, 1988a; Lovat *et al.*, 1998b
2	I-Aβ 1–40-8D3(Mab)	PS1/APP	+	ND	ND	Lee *et al.*, 2002
3	^{111}In-AMY33	ND	+	ND	ND	Bickel *et al.*, 1994
4	I-bFGF	Aβ PP	+	ND	ND	Friedland *et al.*, 2000; Shi *et al.*, 2002
5	^{111}In-DTPA-Aβ 3–40	ND	+	ND	ND	Kurihara and Pardridge, 2000; Marshall *et al.*, 2002
6	^{99m}Tc-10H3(Mab)	ND	+	ND	Peripherally	Friedland *et al.*, 1994
7	I-Aβ 1–40	PS1/APP	+	+	ND	Ghilardi *et al.*, 1996; Majocha *et al.*, 1992
8	I-Aβ 1–40/SA-OX26	ND	+	ND	ND	Saito *et al.*, 1995
9	I-Aβ 1–40/SA-8314	ND	+	+	ND	Wu *et al.*, 1997
10	I-PUT-Aβ 1–40	PS1/APP	+	ND	ND	Wengenack *et al.*, 2000

ND: Not determined.

absence of amyloid plaques. The process is also reversible. Encouraged by this finding, Ghilardi *et al.* (1996) evaluated the binding properties of [^{125}I] Aβ 1–40 *ex vivo* in the aged primate's brain through intra-arterial infusion. They demonstrated that brain amyloid deposits could be detected sufficiently by [^{125}I]Aβ 1–40 at 2 hours after infusion. Autoradiography of the anterior frontal and temporal cortices showed [^{125}I]Aβ 1–40 could selectively label Aβ deposits in a pattern consistent with that of thioflavin S (ThS) or anti-Aβ antibodies.

To test the binding properties of radiolabeled Aβ peptide *in vivo*, Marshall *et al.* (2002) further developed an ^{111}In-labeled Aβ 1–40 derivative with exquisite specificity for both naturally occurring and synthetic Aβ amyloid. That agent, termed [^{111}In]DTPA-Aβ 3–40 (Table I, entry 5), was administered intravenously in a rat model in which synthetic Aβ amyloid is preimplanted in muscle tissue. The labeling of the synthetic Aβ amyloid was imaged with a planar γ-camera using high-energy parallel and pinhole collimators. The preimplanted synthetic Aβ amyloid could be detected as early as 5 minutes after radioligand injection and the signal lasted up to 7 days with the optimal signal to noise ratio attained at about 48 hours after injection.

Despite the promising *in vitro* binding properties of radiolabeled Aβ peptides as amyloid-imaging agents, *in vivo* application of these peptides is hampered by negligible ability to cross the BBB. In addition, many of these peptides could undergo rapid metabolism or degradation in plasma. It has been shown that the radioactivity in the brain was largely due to radiolabeled metabolites rather than the parent radiolabeled peptides (Lee and Pardridge, 2001; Lee *et al.*, 2002). In fact, studies in a transgenic mouse model indicated that radiolabeled Aβ peptides alone were difficult to localize in the area of amyloid plaques *in vivo* due to the lack of a mechanism for transport across the BBB (Wengenack *et al.*, 2000). Therefore, efforts have been made to conjugate the radiolabeled peptides with BBB-associated vectors (Lee *et al.*, 2002; Saito *et al.*, 1995; Wu and Pardridge, 1998; Wu *et al.*, 1997) or transporters (Wengenack *et al.*, 2000). Thus, conjugation of radioiodinated Aβ 1–40 with a vector-mediated drug delivery system (SA-OX26) (Table I, entry 8) led to increased brain uptake and concomitant decrease of peripheral metabolism (Saito *et al.*, 1995). A similar study in nonhuman primates using a vector-conjugated [^{125}I]Aβ 1–40 also showed a marked increase of brain uptake and accelerated clearance of nonspecifically bound or free peptides (Table I, entry 9) (Wu *et al.*, 1997). For the same purpose Wengenack *et al.* (2000) conjugated the [^{125}I]Aβ 1–40 with a naturally occurring polyamine, putrescine (Table I, entry 10), a growth factor associated with cell division. Modification of [^{125}I]Aβ 1–40 with putrescine increased the BBB permeability. The conjugated peptides could label amyloid deposits *in vivo* in a transgenic mouse model as well as *in vitro* in postmortem AD brain tissue sections.

Another problem associated with Aβ peptides in amyloid imaging is slow clearance of nonspecific binding. Study *in vivo* in rhesus monkey had displayed a $t_{1/2}$ for clearance of brain radioactivity of 16 hours (Kurihara and Pardridge, 2000; Lee *et al.*, 2002). Thus, extended imaging acquisition time may be acquired, particularly for potential human studies. For this reason, Kurihara and Pardridge (2000) developed a biotin-conjugated Aβ 1–40 chelated with longer-lived ^{111}In for potential SPECT studies.

Hawkins *et al.* (1988a,b) explored ^{123}I-labeled human serum amyloid P (SAP) component (Table I, entry 1), a plasma protein that exists in peripheral and cerebral amyloid deposits. Since then, efforts have been made to use the purified SAP as a specific probe for amyloid imaging. Previous studies were largely concentrated on imaging peripheral amyloid deposits (Hirschfield and Hawkins, 2003). Scintigraphic studies with [^{123}I]SAP in AD revealed no detectable accumulation of the radiotracer within the brain (Lovat *et al.*, 1998a,b). The lack of brain permeability hampers its application for brain amyloid imaging in AD. To circumvent this difficulty, Shi *et al.* (2002) investigated a new route of ligand delivery through intranasal injection (Table I, entry 4). With this noninvasive method SAP entered mouse brain and selectively detected cerebral Aβ deposits *in vivo* by staining neurons around the rim of Aβ deposits. This was in sharp contrast to the observation of no significant staining of neurons without intranasal injection of the SAP.

In summary, radiolabeled Aβ peptides and SAP have been studied extensively as amyloid-imaging agents. Efforts have been made to modify their structures and develop different drug delivery methods suitable for brain studies *in vivo*. However, significant problems still hindered the potential application in human subjects. To expedite the progress in this direction, researchers turned to the development of small molecular probes as amyloid-imaging agents.

VI. Development of Amyloid Probes Based on Histological Stains

Although the use of biomolecules proved the concept that amyloid deposits can be imaged *in vivo*, its clinical application seems to be hampered by the poor brain permeability of the imaging agents. To circumvent this difficulty, investigators have explored the use of small molecular agents for *in vivo* amyloid imaging. This is due to the fact that small molecules are capable of passive penetration across the BBB in the absence of ligand transporters or vectors (Dishino *et al.*, 1983; Levin, 1980).

Small molecules must meet the following criteria to penetrate the BBB:

1. Molecular weight less than 700 (Levin, 1980)
2. Lipophilicity in terms of octanol–water partition coefficient ($LogP_{oct}$) between 1.0 and 3.5
3. Resistance to metabolism in plasma

These criteria for brain entry are only primitive requirements for *in vivo* amyloid imaging. The key challenge is to identify pharmacophores that can selectively bind to amyloid deposits with high affinity.

One strategy to develop amyloid-binding agents has been to use amyloid dyes as prototypical compounds for the development of *in vivo* imaging agents. Such amyloid dyes include Congo Red (CR, Table II entry 62), thioflavin T (ThT, Table II, entry 1) and ThS (Table II, entry 2). These histological dyes are either positively or negatively charged and are thus incapable of crossing the BBB. Uncharged and neutral analogs of these amyloid dyes were needed to pursue structure–activity relationship (SAR) studies for *in vitro* and *in vivo* properties.

A. Neutral and Lipophilic Congo Red Derivatives

CR was first used about 100 years ago to stain amyloid plaques in postmortem AD brain section (Puchtler and Sweat, 1962). Proteins or protein aggregates that bind to CR are often called "Congophilic." CR has been widely used in AD pathologic studies. A spectrophotometric assay also has been developed to allow quantification of $A\beta$ peptide aggregation *in vitro* based on CR binding (Klunk *et al.*, 1999).

A binding model at the molecular level has been proposed by Klunk *et al.* (1989). According to this model, CR binds to the amyloid aggregates through electrostatic interaction. The amyloid properties exist in a β-sheet conformation, aggregated into a fibrillar structure. The two negatively charged sulfonate groups of CR are 19 Å apart. This distance matches the spacing of every fifth peptide strand. A binding model was thus proposed in which the negatively charged sulfonate groups interact with the equally spaced, positively charged amino acid residues on the amyloid peptide aggregates. This model explains why CR binds only to aggregated β-sheets and not to peptide monomers.

Based on this model, the structure of CR was systematically modified (Mathis *et al.*, 2004). Its immediate analog, chrysamine G (CG, Table II, entry 63) was studied first (Klunk *et al.*, 1995). It was found that CG bound to $A\beta$ aggregates with high affinity but weakly stained the postmortem AD brain tissue sections in a pattern consistent with AD pathology. Further modification of CG led to the development of diazo-free analogs termed X-34 (Table III, entry 33), where the N=N bonds were replaced with C=C

Table II Amyloid Imaging Agents for SPECT Study

| | | | In Vitro Studies | | | | In Vivo Studies | | | | |
| | | | K_i (nM) | LogP (Oct, C18) | Postmortem or Tissue Staining | Tg Mouse Studies | Brain Uptake (in Mice) | | Baboon Studies | Human Studies | Ref. |
No.	Structures	Abbreviations					2 min	30 min			
1	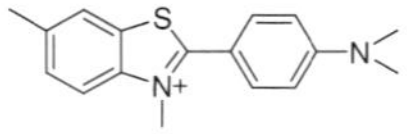	ThT	890	0.57	+	PS1/APP	ND	ND	ND	ND	1
2	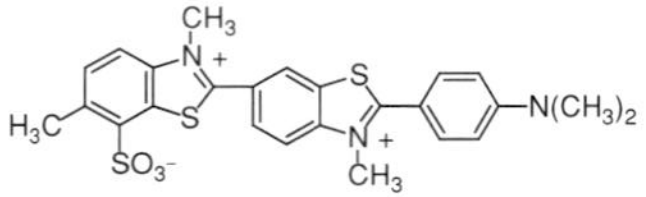	ThS (major component)	ND	ND	+	Tg2576	ND	ND	ND	ND	2
3	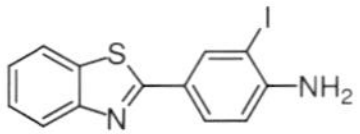	BTA-0-3'-I	8.32	3.17	ND	ND	9.08[a]	3.4[a]	ND	ND	3
4	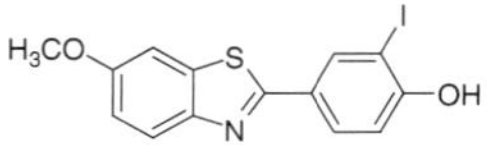	6-MeO-BTP-0-3'-I	15.8	2.31	ND	ND	ND	ND	ND	ND	3
5	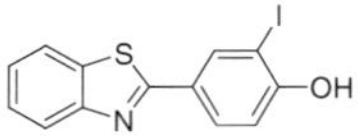	BTP-0-3'-I	19.1	2.22	ND	ND	ND	ND	ND	ND	3
6	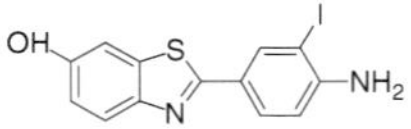	6-OH-BTA-0-3'-I	11.1	1.65	+	ND	6.43[a]	0.40[a]	ND	ND	3, 4
7	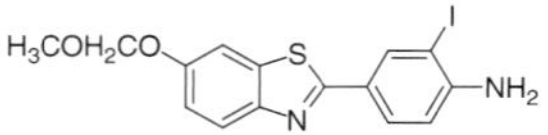	6-MOMO-BTA-0-3'-I	15.1	3.03	ND	ND	ND	ND	ND	ND	3

No.	Structure										
8	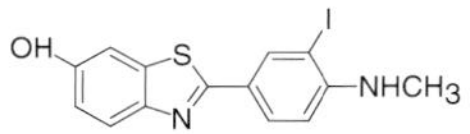 6-OH-BTA-1-3′-I	7.1*	2.35	+	PS1/APP	7.76[a]	2.66[a]	ND	ND	3, 5
9	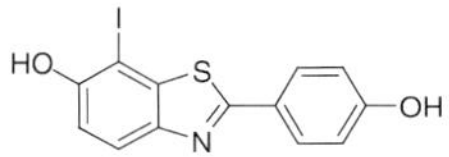 6-OH-7-I-BTP-0	34.5	1.01	ND	ND	ND	ND	ND	ND	3
10	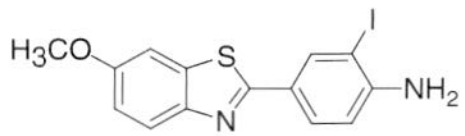 6-MeO-BTA-0-3′-I	4.4	3.08	ND	ND	ND	ND	ND	ND	3
11	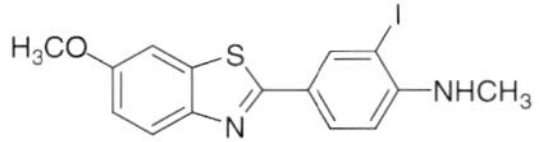 6-MeO-BTA-1-3′-I	1.93	3.80	ND	ND	ND	ND	ND	ND	3
12	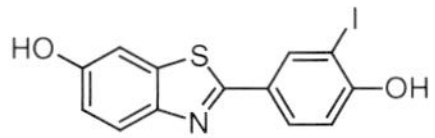 6-OH-BTP-0-3′-I	71.2	1.09	ND	ND	ND	ND	ND	ND	3
13	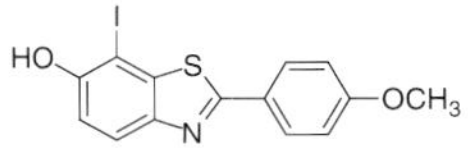 6-OH-7-I-BTP-1	7.1	2.49	ND	ND	ND	ND	ND	ND	3
14	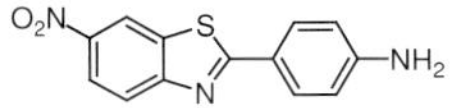 6-NO$_2$-BTA-0	17.4	2.21	ND	ND	ND	ND	ND	ND	3
15	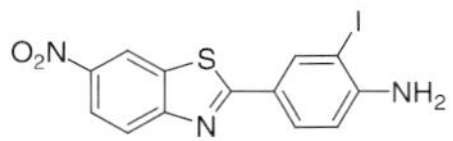 6-NO$_2$-BTA-0-3′-I	4.6	3.33	ND	ND	ND	ND	ND	ND	3
16	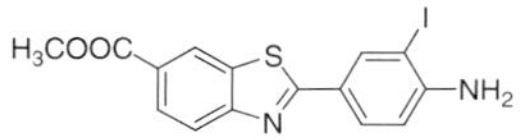 6-COOH-BTA-0-3′-I	3.34	3.29	ND	ND	ND	ND	ND	ND	3
17	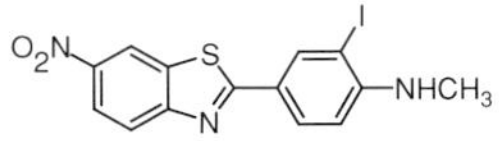 6-NO$_2$-BTA-1-3′-I	1	4.08	ND	ND	ND	ND	ND	ND	3
18	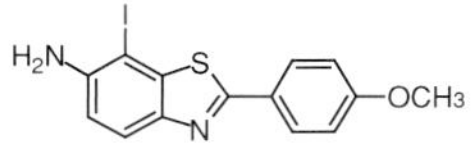 6-NH$_2$-7-I-BTP-1	3.6	2.95	ND	ND	ND	ND	ND	ND	3

(Continued)

Table II *Continued*

No.	Structures	Abbreviations	*In Vitro* Studies			Tg Mouse Studies	*In Vivo* Studies				
			K_i (nM)	LogP (Oct, C18)	Postmortem or Tissue Staining		Brain Uptake (in Mice)		Baboon Studies	Human Studies	Ref.
							2 min	30 min			
19	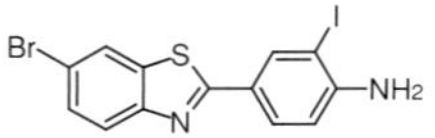	6-Br-BTA-0-3′-I	0.67	4.11	ND	ND	ND	ND	ND	ND	3
20	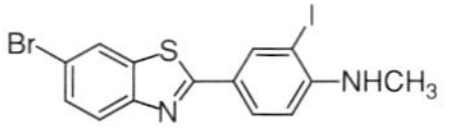	6-Br-BTA-1-3′-I	1.6	4.86	ND	ND	ND	ND	ND	ND	3
21	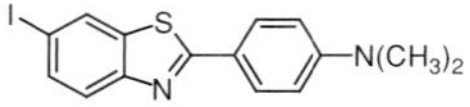	TZDM	1.9	1.85	+	Tg2576	0.6^b	0.9^b	ND	ND	6, 7
22	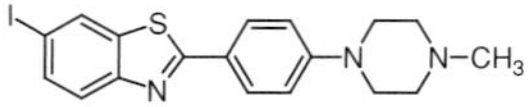	TZPI	0.13	2.49	+	ND	1.50^b	1.59^b	ND	ND	6
23	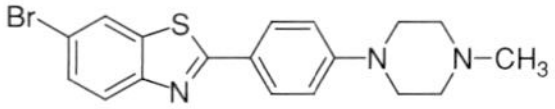	—	1.6	ND	ND	ND	ND	ND	ND	ND	6
24	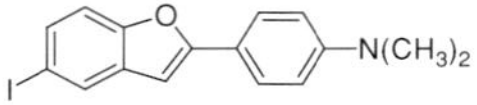	—	7.7	2.35	ND	ND	0.51^b	$0.90b^a$	ND	ND	8
25	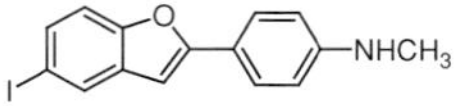	—	1.1	2.04	ND	ND	0.78^b	1.19^b	ND	ND	8
26	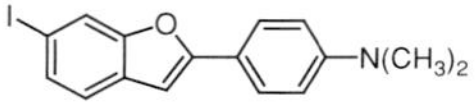	—	0.4	2.12	ND	ND	0.48^b	0.80^b	ND	ND	8
27	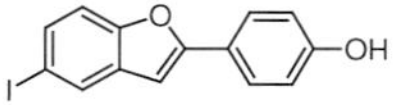	—	6.5	2.97	ND	ND	1.40^b	1.83^b	ND	ND	8

No.	Structure	Name									Ref.
28	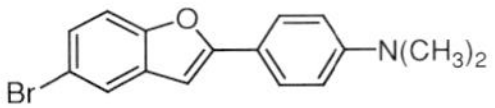 5-Br-benzofuran–C₆H₄–N(CH₃)₂	—	1.6	ND	ND	ND	ND	ND	ND	ND	8
29	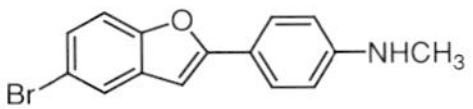 5-Br-benzofuran–C₆H₄–NHCH₃	—	2.7	ND	ND	ND	ND	ND	ND	ND	8
30	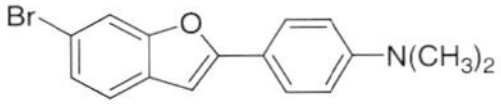 6-Br-benzofuran–C₆H₄–N(CH₃)₂	—	0.6	ND	ND	ND	ND	ND	ND	ND	8
31	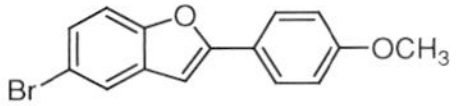 5-Br-benzofuran–C₆H₄–OCH₃	—	1.3	ND	ND	ND	ND	ND	ND	ND	8
32	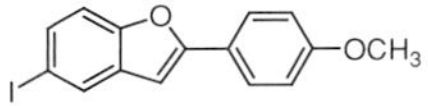 5-I-benzofuran–C₆H₄–OCH₃	—	4.2	ND	ND	ND	ND	ND	ND	ND	8
33	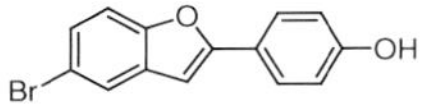 5-Br-benzofuran–C₆H₄–OH	—	9.0	ND	ND	ND	ND	ND	ND	ND	8
34	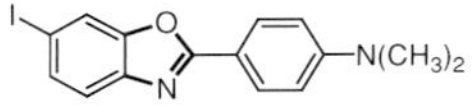 6-I-benzoxazole–C₆H₄–N(CH₃)₂	IBOX	0.8	2.09	+	Tg2576	1.43^b	2.08^b	ND	ND	8, 9
35	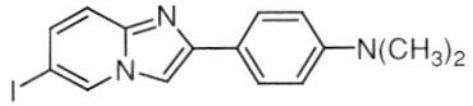 6-I-imidazopyridine–C₆H₄–N(CH₃)₂	IMPY	15.0	2.19	+	Tg2576	2.88^b	0.26^b	ND	ND	10, 11
36	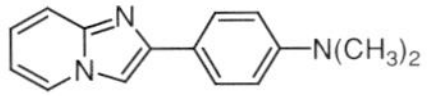 imidazopyridine–C₆H₄–N(CH₃)₂	—	>2000 1242	ND	ND	ND	ND	ND	ND	ND	10, 11
37	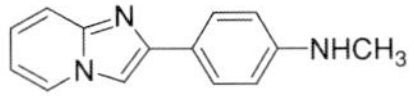 imidazopyridine–C₆H₄–NHCH₃	—	>1000	ND	ND	ND	ND	ND	ND	ND	10
38	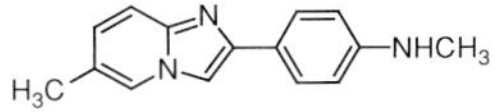 6-CH₃-imidazopyridine–C₆H₄–NHCH₃	—	>2000	ND	ND	ND	ND	ND	ND	ND	10
39	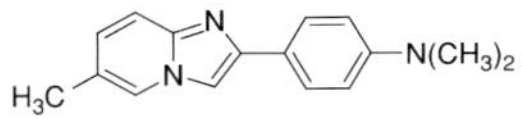 6-CH₃-imidazopyridine–C₆H₄–N(CH₃)₂	—	242	ND	ND	ND	ND	ND	ND	ND	10
40	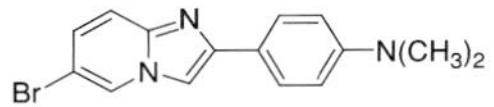 6-Br-imidazopyridine–C₆H₄–N(CH₃)₂	—	10.3	ND	ND	ND	ND	ND	ND	ND	10
41	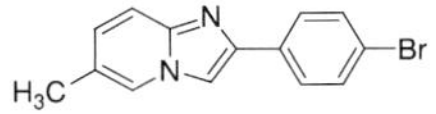 6-CH₃-imidazopyridine–C₆H₄–Br	—	638	ND	ND	ND	ND	ND	ND	ND	10

(Continued)

Table II *Continued*

| No. | Structures | Abbreviations | *In Vitro* Studies | | | *In Vivo* Studies | | | | | |
| | | | K_i (nM) | LogP (Oct, C18) | Postmortem or Tissue Staining | Tg Mouse Studies | Brain Uptake (in Mice) | | Baboon Studies | Human Studies | Ref. |
							2 min	30 min			
42		—	339	ND	ND	ND	ND	ND	ND	ND	10
43		—	>2000	ND	ND	ND	ND	ND	ND	ND	10
44		—	>2000	ND	ND	ND	ND	ND	ND	ND	10
45		BF-180	6.8	ND	ND	ND	ND	ND	ND	ND	12
46		BF-208	>5000	ND	ND	ND	ND	ND	ND	ND	12
47		BF-191	>5000	ND	ND	ND	ND	ND	ND	ND	12
48		BF-164	0.38	ND	ND	ND	ND	ND	ND	ND	12
49		BF-169	7.1	ND	ND	ND	ND	ND	ND	ND	12
50		BF-165	1.8	ND	ND	ND	ND	ND	ND	ND	12

No.	Structure	Name									Ref.
51	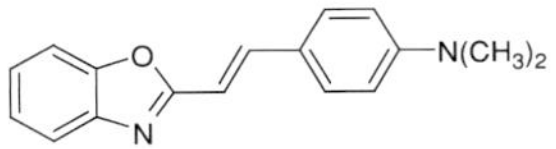 benzoxazole–CH=CH–C₆H₄–N(CH₃)₂	N-282	4.3	ND	ND	ND	ND	ND	ND	ND	12
52	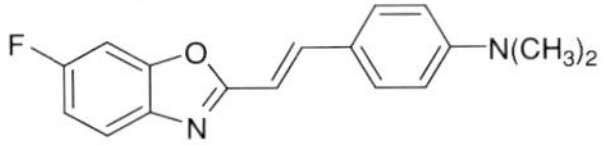 6-F-benzoxazole–CH=CH–C₆H₄–N(CH₃)₂	BF-148	4.2	ND	ND	ND	ND	ND	ND	ND	12
53	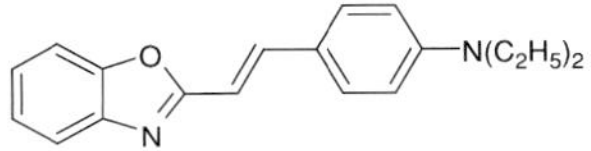 benzoxazole–CH=CH–C₆H₄–N(C₂H₅)₂	BF-125	4.9	ND	ND	ND	ND	ND	ND	ND	12
54	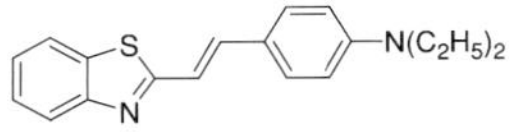 benzothiazole–CH=CH–C₆H₄–N(C₂H₅)₂	BF-124	10.9	ND	ND	ND	ND	ND	ND	ND	12
55	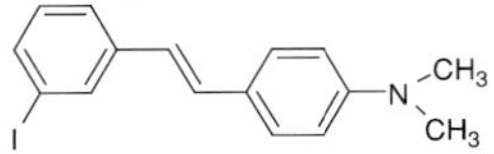 3-I-C₆H₄–CH=CH–C₆H₄–N(CH₃)CH₃	m-I-stilbene	0.19*	2.62	+	ND	0.72^b	1.12^b	ND	ND	10, 13
56	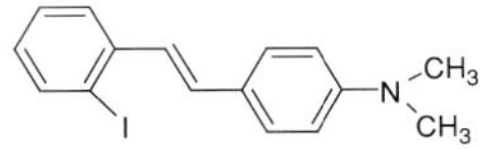 2-I-C₆H₄–CH=CH–C₆H₄–N(CH₃)CH₃	o-I-stilbene	7.7	ND	ND	ND	ND	ND	ND	ND	13
57	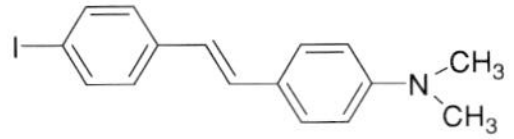 4-I-C₆H₄–CH=CH–C₆H₄–N(CH₃)CH₃	p-I-stilbene	2.0	ND	ND	ND	ND	ND	ND	ND	13
58	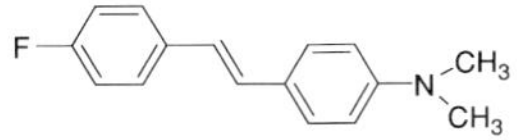 4-F-C₆H₄–CH=CH–C₆H₄–N(CH₃)CH₃	p-F-stilbene	22	ND	ND	ND	ND	ND	ND	ND	13
59	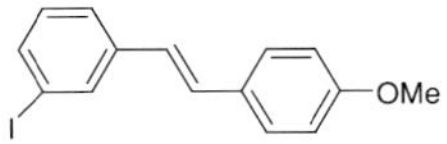 3-I-C₆H₄–CH=CH–C₆H₄–OMe	—	22	ND	ND	ND	ND	ND	ND	ND	13
60	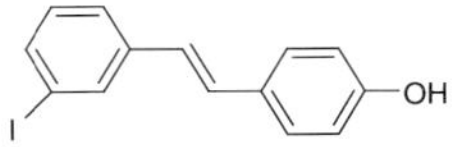 3-I-C₆H₄–CH=CH–C₆H₄–OH	—	32	ND	ND	ND	ND	ND	ND	ND	13
61	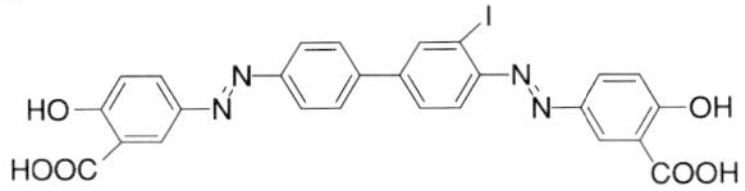 HO–(HOOC)C₆H₃–N=N–C₆H₄–C₆H₃(I)–N=N–C₆H₃(OH)(COOH)	3′-I-CG	ND	3.2	ND	ND	3^d	ND	ND	ND	14

(Continued)

Table II *Continued*

No.	Structures	Abbreviations	*In Vitro* Studies				*In Vivo* Studies				
			K_i (nM)	LogP (Oct, C18)	Postmortem or Tissue Staining	Tg Mouse Studies	Brain Uptake (in Mice) 2 min	30 min	Baboon Studies	Human Studies	Ref.
62		CR	ND	−0.18	+	ND	ND	ND	ND	ND	15
63		CG	2.7	1.8	+	ND	ND	ND	ND	ND	16
64		ISB	0.08	35	ND	ND	0.27^b (5 min)	0.06^b	ND	ND	6
65		BSB	400	ND	+	Tg2576	ND	ND	ND	ND	17, 18
66		IMSB	0.13	1.1	ND	Tg2576	0.14^b (5 min)	0.03^b	ND	ND	6, 19
67		I-Fluorene	0.92	2.47	+	Tg2576	1.13^b	1.26^b	ND	ND	7, 20

68	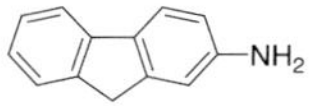	—	>1000	ND	ND	ND	ND	ND	ND	ND	*20*
69	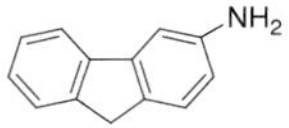	—	>1000	ND	ND	ND	ND	ND	ND	ND	*20*
70	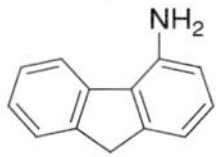	—	>1000	ND	ND	ND	ND	ND	ND	ND	*20*
71	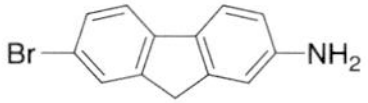	—	56	ND	ND	ND	ND	ND	ND	ND	*20*
72	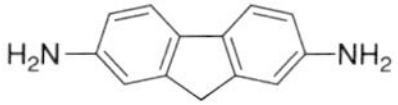	—	>1000	ND	ND	ND	ND	ND	ND	ND	*20*
73	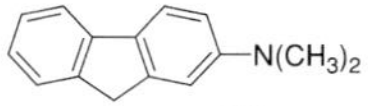	—	>1000	ND	ND	ND	ND	ND	ND	ND	*20*
74	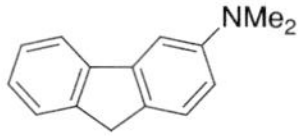	—	23.5	ND	ND	ND	ND	ND	ND	ND	*20*
75	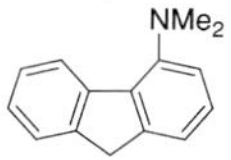	—	>1000	ND	ND	ND	ND	ND	ND	ND	*20*
76	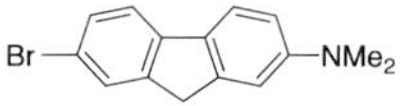	—	0.85	ND	ND	ND	ND	ND	ND	ND	*20*
77	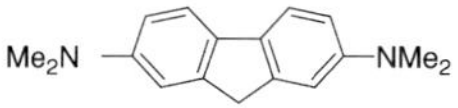	—	15.4	ND	ND	ND	ND	ND	ND	ND	*20*
78	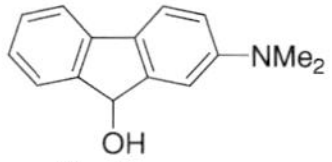	—	>1000	ND	ND	ND	ND	ND	ND	ND	*20*
79	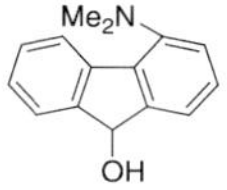	—	>1000	ND	ND	ND	ND	ND	ND	ND	*20*

(*Continued*)

Table II *Continued*

| No. | Structures | Abbreviations | *In Vitro* Studies | | | Tg Mouse Studies | *In Vivo* Studies | | Baboon Studies | Human Studies | Ref. |
| | | | K_i (nM) | LogP (Oct, C18) | Postmortem or Tissue Staining | | Brain Uptake (in Mice) | | | | |
							2 min	30 min			
80		—	88	ND	ND	ND	ND	ND	ND	ND	20
81		—	>1000	ND	ND	ND	ND	ND	ND	ND	20
82		—	>1000	ND	ND	ND	ND	ND	ND	ND	20
83		—	>1000	ND	ND	ND	ND	ND	ND	ND	20
84		—	>1000	ND	ND	ND	ND	ND	ND	ND	20
85		—	16.5	ND	ND	ND	ND	ND	ND	ND	20

No.	Structure	Name									Ref.
86	(Tc isocyanide complex; But, tBu, N≡C, Tc; HO, HOOC —N=N— pyridyl ... pyridyl —N=N— OH, COOH)	—	160	ND	ND	ND	ND	ND	ND	ND	21
87	(Tc isocyanide complex; But, tBu, N≡C, Tc; NaO₃S, NH₂, naphthyl —N=N— pyridyl ... pyridyl —N=N— naphthyl, SO₃Na, H₂N)	—	630	ND	ND	ND	ND	ND	ND	ND	21, 22
88	(Tc–MAMA chelate, S, S, O, N, Tc; HO, HOOC —N=N— phenyl ... phenyl —N=N— OH, COOH; NH, O)	^{99m}Tc-MAMA-CG	ND	1.08	ND	ND	0.3^a	ND	ND	ND	23
89	(Tc–MAMA chelate, S, S, O, N, Tc; HO, HOOC —N=N— naphthyl ... naphthyl —N=N— OH, COOH; HN, O)	—	830	0.7	ND	ND	ND	ND	ND	ND	22

^{a}%ID/g; ^{b}%ID/organ; ^{c}%ID-kg/g; ^{d}%IDI; ND: not determined.

[1]Klunk et al., 2001; [2]Kung et al., 2002b; [3]Wang et al., 2003; [4]Wang et al., 2002a; [5]Wang et al., 2004; [6]Zhuang et al., 2001b; [7]Kung et al., 2003; [8]Ono et al., 2002; [9]Zhuang et al., 2001a; [10]Zhuang et al., 2003; [11]Kung et al., 2002a; [12]Okamura et al., 2004; [13]Kung et al., 2001; [14]Mathis et al., 2004; [15]Tubis et al., 1960; [16]Klunk et al., 1995; [17]Schmidt et al., 2001; [18]Ishikawa et al., 2004; [19]Lee et al., 2002; [20]Lee et al., 2003; [21]Han et al., 1996; [22]Zhen et al., 1999; [23]Dezutter et al., 1999.

*Highest affinity selected from several reported literature values.

Table III Amyloid Imaging Agents for PET Studies

No.	Structures	Abbreviations	*In Vitro* Studies				*In Vivo* Studies				
			K_i (nM)	LogP (Oct, C18)	Auto-radiography or Tissue Staining	Tg Mouse Studies	Brain Uptake (in Mice) 2 min	30 min	Baboon Studies	Human Studies	Ref.
1	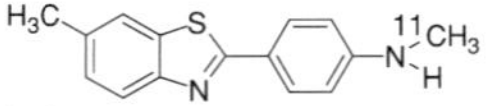	6-Me-BTA-1	10*	3.4	+	ND	7.61[a]	2.76[a]	+	ND	1, 2
2	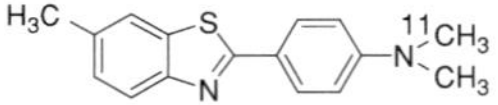	6-Me-BTA-2	64*	3.8	ND	ND	0.078[c]	0.15[c]	ND	ND	1, 3
3	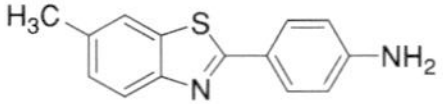	6-Me-BTA-0	9.5*	2.4	ND	ND	ND	ND	ND	ND	1, 3
4	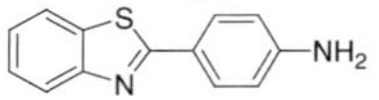	BTA-0	36.8	1.98	ND	ND	ND	ND	ND	ND	3, 4
5	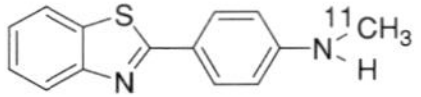	BTA-1	7.1*	2.69	+	PS1/APP	12.9[a]	1.7[a]	+	ND	4, 5
6	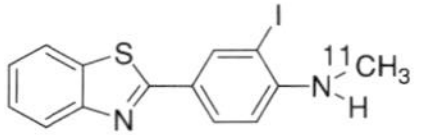	BTA-1-3'-I	4.94	3.90	ND	ND	4.40[a]	2.68[a]	ND	ND	4
7	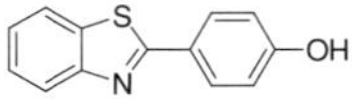	BTP-0	5.68	1.86	ND	ND	ND	ND	ND	ND	4
8	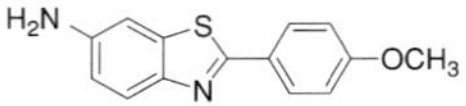	6-NH$_2$-BTP-1	6.9	1.76	ND	ND	ND	ND	ND	ND	4
9	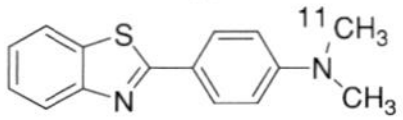	BTA-2	4.0	3.4	ND	ND	0.19[c]	0.078[c]	ND	ND	3
10	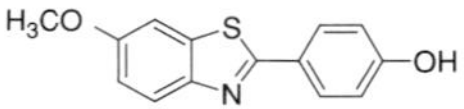	6-MeO-BTP-0	4.2	1.8	ND	ND	ND	ND	ND	ND	4
11	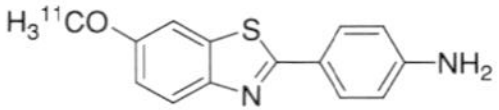	6-MeO-BTA-0	7.00	1.87	ND	ND	0.32[c]	0.084[c]	ND	ND	3, 4

No.	Structure	Compound									Ref.
12	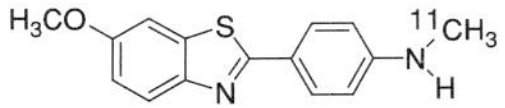 H₃CO–benzothiazole–C₆H₄–N(H)(¹¹CH₃)	6-MeO-BTA-1	4.9	2.58	ND	ND	0.33^c	0.10^c	+	ND	3, 4
13	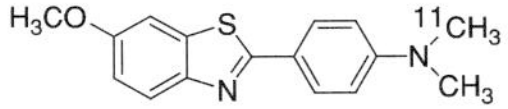 H₃CO–benzothiazole–C₆H₄–N(¹¹CH₃)(CH₃)	6-MeO-BTA-2	1.9	3.3	ND	ND	0.16^c	0.14^c	ND	ND	3
14	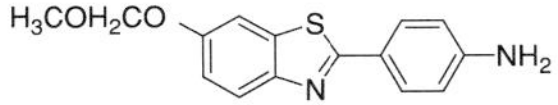 H₃COH₂CO–benzothiazole–C₆H₄–NH₂	6-MOMO-BTA-0	53.6	1.86	ND	ND	ND	ND	ND	ND	4
15	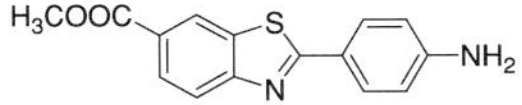 H₃COOC–benzothiazole–C₆H₄–NH₂	6-COOH-BTA-0	17.9	2.07	ND	ND	ND	ND	ND	ND	4
16	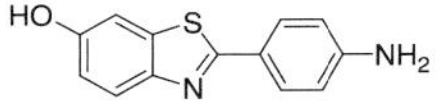 HO–benzothiazole–C₆H₄–NH₂	6-OH-BTA-0	45.6	0.66	ND	ND	ND	ND	ND	ND	3, 4
17	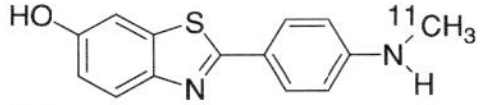 HO–benzothiazole–C₆H₄–N(H)(¹¹CH₃)	6-OH-BTA-1	4.3	1.23	+	ND	0.21^c	0.018^c	+	+	4, 6
18	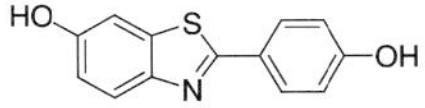 HO–benzothiazole–C₆H₄–OH	6-OH-BTP-0	16.8	0.39	ND	ND	ND	ND	ND	ND	4
19	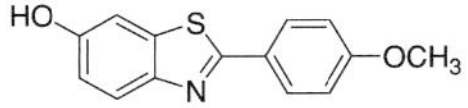 HO–benzothiazole–C₆H₄–OCH₃	6-OH-BTP-1	6.3	1.75	ND	ND	ND	ND	ND	ND	4
20	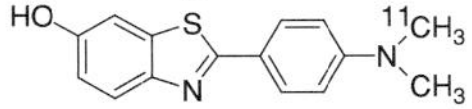 HO–benzothiazole–C₆H₄–N(¹¹CH₃)(CH₃)	6-OH-BTA-2	4.4	2.0	ND	ND	0.32^c	0.1^c	ND	ND	3
21	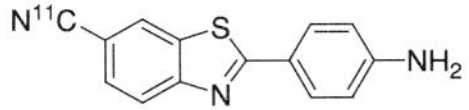 N¹¹C–benzothiazole–C₆H₄–NH₂	6-CN-BTA-0	64	1.8	ND	ND	ND	ND	ND	ND	3
22	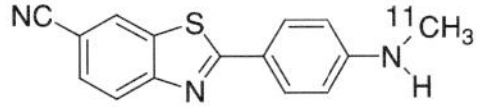 NC–benzothiazole–C₆H₄–N(H)(¹¹CH₃)	6-CN-BTA-1	8.6	2.5	ND	ND	0.32^c	0.063^c	+	ND	3
23	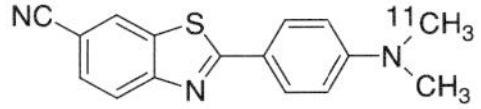 NC–benzothiazole–C₆H₄–N(¹¹CH₃)(CH₃)	6-CN-BTA-2	11	3.2	ND	ND	0.24^c	0.097^c	ND	ND	3
24	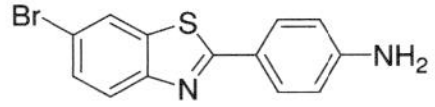 Br–benzothiazole–C₆H₄–NH₂	6-Br-BTA-0	7.22	2.87	ND	ND	ND	ND	ND	ND	3, 4
25	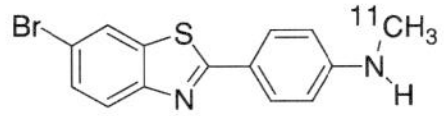 Br–benzothiazole–C₆H₄–N(H)(¹¹CH₃)	6-Br-BTA-1	1.70	3.64	ND	ND	0.12^c	0.12^c	+	ND	3, 4
26	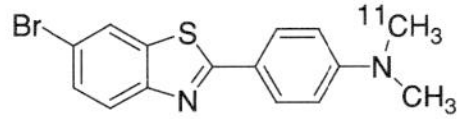 Br–benzothiazole–C₆H₄–N(¹¹CH₃)(CH₃)	6-Br-BTA-2	1.9*	4.4	ND	ND	0.054^c	0.11^c	ND	ND	3, 7

(*Continued*)

Table III *Continued*

No.	Structures	Abbreviations	K_i (nM)	LogP (Oct, C18)	Auto-radiography or Tissue Staining	Tg Mouse Studies	Brain Uptake (in Mice) 2 min	Brain Uptake (in Mice) 30 min	Baboon Studies	Human Studies	Ref.
							In Vivo Studies				
27		6-NO$_2$-BTA-1	2.75	2.96	ND	ND	ND	ND	ND	ND	*4*
28		FEM-IMPY	40	4.41	+	ND	6.4^a (1.2 m)	ND	+	ND	*8*
29		FPM-IMPY	27	4.60	ND	ND	5.7^a (0.8 m)	ND	ND	ND	*8*
30		X04-2′-MeO	26.8	2.6	+	PS1/APP NOR-beta	81^d	50^d	ND	ND	*9*
31		X04-3-OMe	38	ND	+	PS1/APP NOR-beta	7.16^a	0.71^a	ND	ND	*10*
32		X04-4-OMe	15.7	ND	ND	ND	15^d	ND	ND	ND	*11*
33		X-34	18	0.42	+	Tg2576 APP23	ND	ND	ND	ND	*12, 13*
34		—	0.81	ND	ND	ND	ND	ND	ND	ND	*11*

No.	Compound									
35	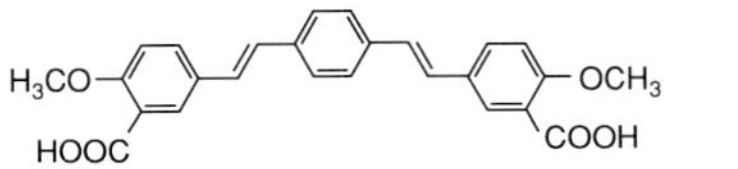 X-34-4,4-di-MeO	47	−0.95	ND	ND	ND	ND	ND	ND	11
36	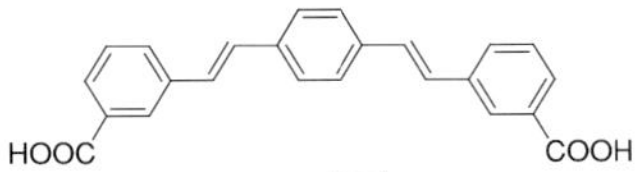 X-30	135	0.39	ND	ND	ND	ND	ND	ND	11
37	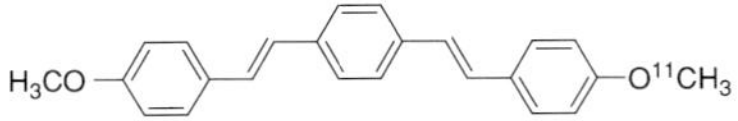 X-40-di-MeO	No inhibition	2.3	ND	ND	ND	ND	ND	ND	11
38	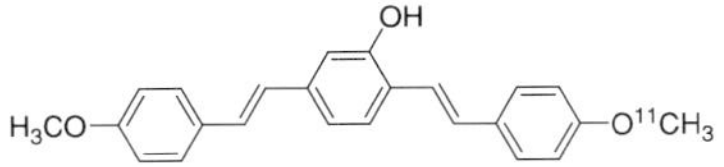 OH-X-04-di-MeO	No inhibition	ND	ND	ND	ND	ND	ND	ND	11
39	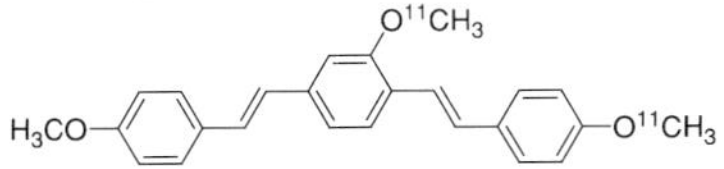 [^{11}C]MeO-X-04-di-MeO	No inhibition	ND	ND	ND	ND	ND	ND	ND	11
40	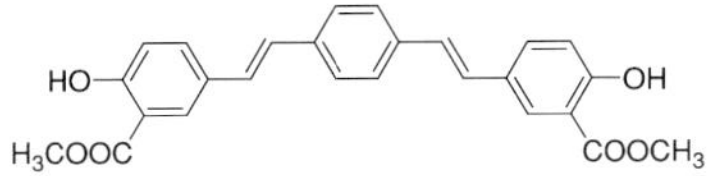 X34-diester	119	3.4	ND	ND	ND	ND	ND	ND	11
41	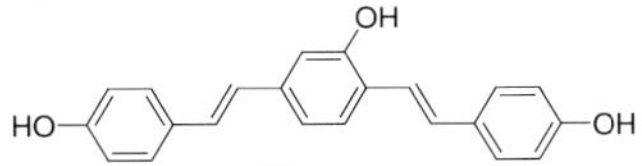 X04-2′-OH	9	ND	ND	ND	ND	ND	ND	ND	11
42	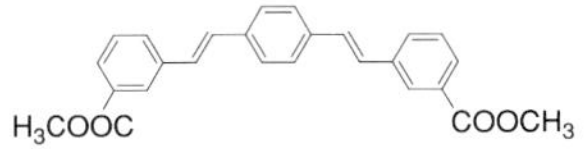 X-30-diester	No inhibition	2.5	ND	ND	ND	ND	ND	ND	11
43	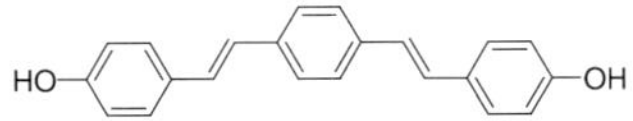 X04	3100	2.0	ND	ND	ND	ND	ND	ND	11
44	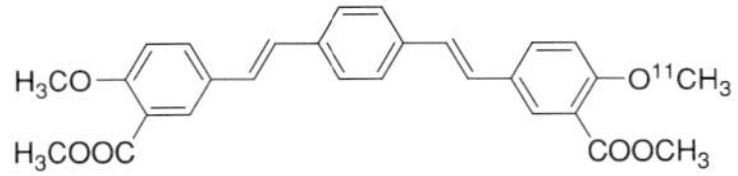 X-34-di-MeO-diester	No inhibition	1.2	ND	ND	ND	ND	ND	ND	11

(*Continued*)

Table III *Continued*

No.	Structures	Abbreviations	K_i (nM)	LogP (Oct, C18)	Auto-radiography or Tissue Staining	Tg Mouse Studies	Brain Uptake (in Mice) 2 min	Brain Uptake (in Mice) 30 min	Baboon Studies	Human Studies	Ref.
			In Vitro Studies				*In Vivo* Studies				
45	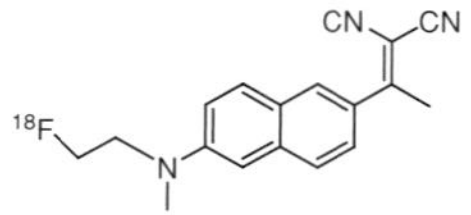	FDDNP	0.12(H) 1.86(L)	3.92	+	HuAPP	ND	ND	+	+	14
46	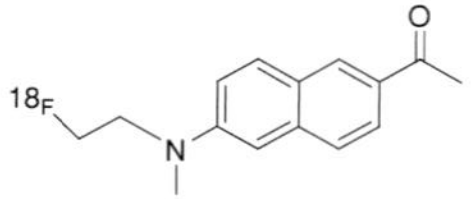	FENE	0.16(H) 71.2(L)	3.13	+	HuAPP	ND	ND	ND	ND	14, 15
47	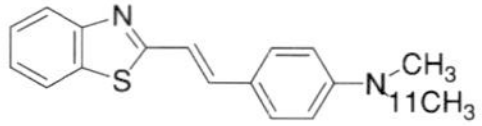	—	2.3	ND	ND	ND	ND	ND	ND	ND	16
48	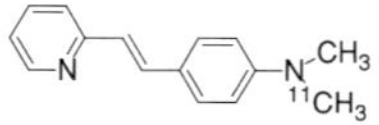	—	>3000	ND	ND	ND	ND	ND	ND	ND	16
49	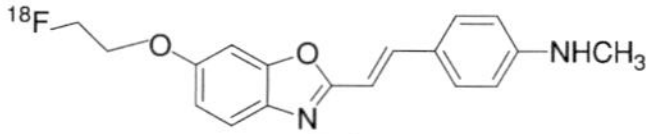	BF-168	6.4	ND	+	PS1/APP	3.9^a	1.6^a	ND	ND	17
50	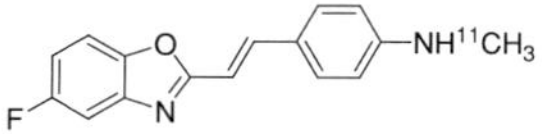	BF-145	ND	ND	+	APP23	ND	ND	ND	ND	17
51	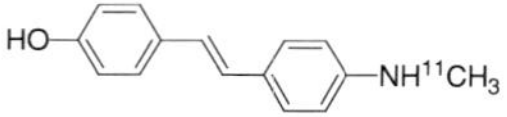	SB-13	6.0	2.36	ND	CRND8	$1.51^{e'}$	0.42^e	ND	+	18, 19
52	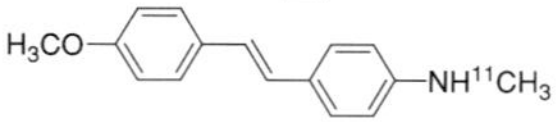	MeO-stilbene	1.2	ND	ND	ND	ND	ND	ND	ND	18

53	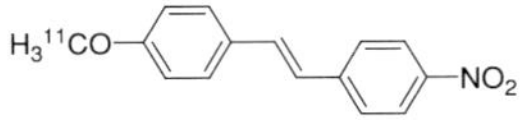 $H_3{}^{11}CO$–CH=CH–NO_2	—	151	ND	ND	ND	ND	ND	ND	ND	*18*
54	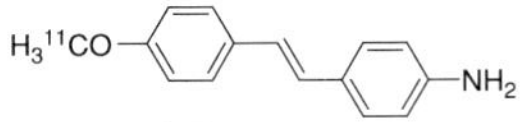 $H_3{}^{11}CO$–CH=CH–NH_2	—	36	ND	ND	ND	ND	ND	ND	ND	*18*
55	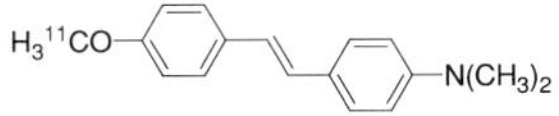 $H_3{}^{11}CO$–CH=CH–$N(CH_3)_2$	—	1.3	ND	ND	ND	ND	ND	ND	ND	*18*
56	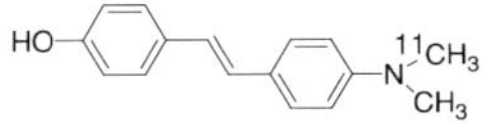 HO–CH=CH–$N({}^{11}CH_3)CH_3$	—	2.2	ND	ND	ND	ND	ND	ND	ND	*18*
57	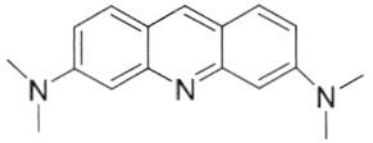 (acridine, dimethylamino)	Acridine orange	32	1.77	ND	ND	ND	ND	ND	ND	*20*
58	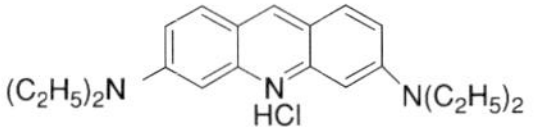 $(C_2H_5)_2N$–(acridine, HCl)–$N(C_2H_5)_2$	BF009	167	3.01	+	ND	ND	ND	ND	ND	*20*
59	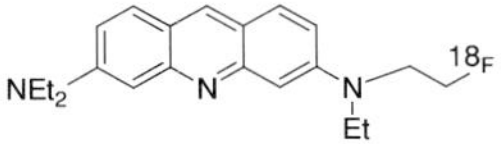 NEt_2–(acridine)–N(Et)$CH_2CH_2{}^{18}F$	BF-108	135	2.56	+	APP23	0.42[a]	1.53[a]	ND	ND	*20, 21*

[a]%ID/g; [b]%ID/organ; [c]%ID-kg/g; [d]%IDI; [e]ID%/g (cortex); ND: not determined.

[29]Klunk *et al.*, 2001; [30]Mathis *et al.*, 2002; [31]Mathis *et al.*, 2003; [32]Wang *et al.*, 2003; [33]Ishikawa *et al.*, 2004; [34]Klunk *et al.*, 2004; [35]Zhuang *et al.*, 2001b; [36]Cai *et al.*, 2004; [37]Klunk *et al.*, 2002; [38]Wang *et al.*, 2002a; [39]Mathis *et al.*, 2004; [40]Link *et al.*, 2001; [41]Styren *et al.*, 2000; [42]Agdeppa *et al.*, 2001; [43]Agdeppa *et al.*, 2003; [44]Kung *et al.*, 2001; [45]Okamura *et al.*, 2004; [46]Ono *et al.*, 2003; [47]Verhoeff *et al.*, 2004; [48]Suemoto *et al.*, 2004; [49]Shimadzu *et al.*, 2003.

*Highest affinity selected from several reported literature values.

bonds. Compared to CG, X-34 was a better histochemical stain for AD pathology (Styren *et al.*, 2000). A brominated analog of X-34, termed BSB (Table II, entry 65), was found to label a diverse array of β-pleated sheet structures in postmortem human brain in diseases (Schmidt *et al.*, 2001). Systematic injection of BSB in a transgenic mouse model indicated that BSB could stain amyloid deposits *ex vivo* (Skovronsky *et al.*, 2000). Fluorinated and ^{13}C-labeled BSB analogs have also been synthesized and evaluated for use as a histochemical stain. Such analogs have the potential for use as contrast agents for MRI (Sato *et al.*, 2004).

However, *in vivo* application of X-34 was suboptimal because its carboxylic acid groups are detrimental to brain uptake. More lipophilic analogs were needed. This led to the development of acid-free analogs by substitution of the salicylic acid groups with phenols (Klunk *et al.*, 2001) or catechols (Wang *et al.*, 2002b). Both of these analogs exhibited enhanced brain entry and selective binding to amyloid deposits with respect to future *in vivo* studies in human subjects. However, the level of brain entry of these neutral and uncharged CR derivatives was still insufficient. Further SAR studies were severely limited by the rigid scaffold of the *bis*-styrylbenzene structure.

To further increase the flexibility of structural modification, Kung *et al.* (2001) developed a series of stilbene analogs, the semi-analogs of CR derivative previously developed. These stilbene derivatives readily entered the brain and so potentially could bind to amyloid deposits. Interestingly, these stilbene analogs bound to the sites in amyloid aggregates in a different manner than did the CR analogs, as shown in competitive binding assays.

B. Neutral and Lipophilic Thioflavin T Derivatives

ThT (MW = 283) is another fluorescent dye that has been used infrequently as a histological stain for amyloid (Burns *et al.*, 1967). Levine *et al.* (1993) pioneered the use of ThT as a method of measuring amyloid aggregation. The binding mechanism of ThT to amyloid fibrils is unknown but appears to be specific and saturable (LeVine, 1999). In comparison, ThS is a fluorescent dye widely used for histologic studies of amyloid deposits but has many disadvantages for development of *in vivo* amyloid probes compared with ThT. These specific limitations are as follows:

1. ThS is not chemically defined but is a mixture of at least six components.
2. The molecular weight of ThS is about twice of that of ThT.
3. ThT is structurally simpler than CR, lending itself to more efficient chemical derivatization.

For these reasons, efforts have been focused on the development of lipophilic ThT analogs.

The positive charge of ThT can be eliminated by removal of the methyl group of the benzothiazolium nitrogen. This approach led to a generation of a series of 2-aryl–substituted benzothiazole derivatives that were neutral at physiologic pH and more lipophilic than ThT. Depending on the substituents, there are two synthetic approaches to benzothiazole derivatives. One straightforward route is based on coupling between substituted aminothiophenols and benzoic acid chlorides. If aminothiophenols are not readily available or very unstable, a more general, but lengthy approach can be taken. ^{11}C labeling can be readily achieved with [^{11}C]iodomethane. Some compounds can also be labeled with ^{18}F by nucleophilic substitution with [^{18}F]fluoride in appropriately activated precursors (Kilbourn *et al.*, 1990; Mathis *et al.*, 2002).

After their synthesis ThT derivatives have been examined for the following properties, which may be considered as criteria to be met in successful imaging agents:

1. Specificity for staining Aβ deposits in postmortem AD brain
2. Quantitative binding affinity and binding stoichiometry for synthetic Aβ fibrils
3. Reversibility of binding
4. Quantitative differentiation of binding to homogenates of AD, in control and non-AD dementia brain
5. Lack of peripheral and brain metabolism and satisfactory pharmacokinetics in animals
6. Lack of binding to other central nervous system receptor sites in wide assay screens
7. Efficacy in *ex vivo* and micro-PET studies in transgenic mice that deposit Aβ in the brain
8. Lack of toxicity determined by standard toxicological studies normally required for Food and Drug Administration approval

Neutral ThT derivatives can be obtained by removal of the positive charge from the quaternary heterocyclic nitrogen of ThT without affecting its ability to bind Aβ 1–40 fibrils and NFTs (Klunk *et al.*, 2001). These uncharged ThT derivatives exhibited up to 45-fold higher affinity than ThT (Ki = 890 nM) itself. At nanomolar concentrations, these neutral ThT analogs bind better to amyloid plaque than to NFTs as examined by tissue staining of well-confirmed AD brain section. Further studies showed that they could penetrate the BBB very well. The mouse brain uptake reached a level that is considered sufficient for *in vivo* PET imaging of amyloid deposit in human subjects. Further optimization of the benzothiazole derivatives led to the development of a series of promising candidates suitable for PET imaging.

The *in vivo* pharmacokinetic profiles of these compounds were systematically evaluated in nonhuman primates. The binding properties were also quantitatively analyzed in brain tissue from AD subjects, controls, and subjects with non-AD dementias (Klunk *et al.*, 2003; Mathis *et al.*, 2003). The first *in vivo* binding study was performed with multiphoton fluorescence microscopy in living PS1/APP transgenic mice, which demonstrated that modification of ThT dyes would lead to a valuable *in vivo* amyloid-imaging agent. Encouraged by these findings, comprehensive SAR studies were conducted (Mathis *et al.*, 2003). The structures of these ThT analogs were derived by introducing different functional groups in various positions of the 2-aryl benzothiazole system. All these compounds exhibited high affinities for Aβ 1–40 aggregates, and the binding affinities generally increased with lipophilicity (Wang *et al.*, 2003). After comparison of *in vitro* and *in vivo* properties such as lipophilicity, binding affinity and specificity, brain entry, retention, and clearance, a lead compound was identified for *in vivo* PET imaging in human subjects following further investigation of *in vivo* pharmacokinetic profiles in baboons (Mathis *et al.*, 2003). The selected PET ligand, termed [^{11}C]6-OH-BTA-1 (PIB; Table III, entry 17), exhibited a rapid clearance of nonspecific binding and displayed a time–radioactivity course very similar to those PET radioligands currently used in clinical studies. Peripheral and brain metabolism of [^{11}C]6-OH-BTA-1 in mice and baboons indicated that the radioactive metabolites of plasma were polar and unable to cross the BBB. PIB has been successfully applied to PET studies in AD subjects (Klunk *et al.*, 2004).

For SPECT studies, different radionuclides, such as ^{99m}Tc or ^{123}I, should be introduced into the candidate ligand. No ^{99m}Tc-labeled ThT derivative has been reported. Encouraged by the success in the development of PET amyloid-imaging agents, radioiodinated ThT derivatives were developed for potential SPECT imaging. SAR studies of these iodinated derivatives led to identification of two lead compounds (termed 6-OH-BTA-0-3′-I [Table II, entry 6] and 6-OH-BTA-1–3′-I [Table II, entry 8]) (Wang *et al.*, 2003). Both exhibited *in vitro* binding properties and *in vivo* pharmacokinetic profiles similar to those of PET radioligands. In particular, 6-OH-BTA-1-3′-I has the advantage that it can be labeled with either ^{11}C as a PET radioligand or ^{123}I as a SPECT radioligand. This unique structural feature allows the combination of the quantitative ability of PET with the clinical availability of SPECT. The dual agent would permit direct comparison of the clinic data from the two imaging modalities due to identical *in vivo* pharmacodynamic and pharmacokinetic properties (Wang *et al.*, 2004).

In the interim, other groups have also developed a wide array of iodinated ThT analogs for amyloid imaging represented by two neutral ThT derivatives, termed TZDM (2-[4′-(dimethylamino)phenyl]-6-iodobenzothiazole, Table II, entry 21) and TZPI (2-[4′-(4′′′-methylpiperazin-1-yl)phenyl]-6-iodobenzothiazole, Table II, entry 22) (Zhuang *et al.*, 2001a). Both TZDM and TZPI exhibited high affinity for Aβ fibrils. *Ex vivo* autoradiography demonstrated distinctive labeling of plaques. Despite the promising *in vitro* binding properties, *in vivo* application of these two compounds is hampered by slow brain clearance of the radioactivity in mice and limited brain uptake. Further studies include the replacement of a benzothiazole ring by a benzofuran ring (Ono *et al.*, 2002). The obtained compounds as represented by IBOX (2-(4'-dimethylaminophenyl)-6-iodobenzoxazole; Table II, entry 34) were found to have excellent *in vitro* binding affinity for Aβ aggregates. However, the nonspecific binding in the mouse brain was high, indicating the unsuitability of IBOX for *in vivo* plaque imaging. Further SAR studies to improve the pharmacokinetics of brain uptake led to the development of a novel ligand, termed IMPY (6-iodo-2-(4'-dimethylamino-) phenyl-imidazo[1,2]pyridine; Table II, entry 35), and its series derivatives (Zhuang *et al.*, 2003). IMPY, with an imidazol[1,2-*a*]pyridine ring and *N,N*-dimethylaminophenyl group in its structure, displayed high binding affinity for Aβ aggregates and selective amyloid plaque labeling in postmortem AD brain sections (Kung *et al.*, 2002a, 2003, 2004; Zhuang *et al.*, 2003). *In vivo* brain uptake showed that the initial uptake of [^{125}I]IMPY in normal mice was sufficiently high for potential human studies. Compared with TZDM and IBOX, IMPY exhibited a rapid brain clearance from normal mouse brain. *Ex vivo* labeling of amyloid plaques in Tg2576 transgenic mice showed selective retention of radioactivity in Tg mouse brain relative to aged-matched control litter mates. The plaques labeled by [^{125}I]IMPY were identical to those stained with ThS. These promising results suggested that IMPY might be a good candidate as a SPECT imaging agent for amyloid plaque. However, detailed quantitative validation studies and pharmacological studies in primate brain are needed to confirm the potential of IMPY. In addition, two ^{18}F-labeled IMPY analogs, termed FPM-IMPY (Table III, entry 29) and FEM-IMPY (Table III, entry 28), have also been developed for potential PET imaging (Cai *et al.*, 2004).

Most recently, novel styrylbenzoxazole derivatives for *in vivo* imaging of amyloid plaques have been reported (Okamura *et al.*, 2004). The structures of these compounds contain functional groups necessary for binding, such as benzoxazole and styryl, as well as *N*-methylaminophenyl or *N,N*-dimethyla-minophenyl. Moreover, these structures can be labeled with ^{11}C, ^{18}F, or ^{123}I. The so-designed compounds showed high affinity for Aβ aggregates. Of these compounds, [^{18}F]BF-168 (Table III, entry 49) was identified as the lead compound. *In vivo* biodistribution studies showed that [^{18}F]BF-168 displays a high initial brain uptake in normal mice at early time points. Neuropathological staining of senile plaques (SPs) and NFTs in AD brain section showed that BF-168 clearly stained both neuric and diffuse amyloid plaques. Moreover, *ex vivo* studies using PS1/APPsw and APP23 transgenic mice indicated that [^{18}F]BF-168 could visualize early amyloid deposition in the brain, which makes it another useful candidate for both PET and SPECT imaging of brain amyloid plaques.

C. Acridine Derivatives

Recently acridine orange has also been explored for the development of *in vivo* amyloid imaging. Acridine has previously been used to locate the active DNA templates. At physiological pH, acridine is neutral, but very hydrophilic, with limited brain permeability. Shimadzu *et al.* (2003) synthesized and screened neutral derivatives of acridine orange for binding to Aβ aggregates, which led to the identification of a novel uncharged compound [^{18}F]BF-108 (Suemoto *et al.*, 2004) (Table III, entry 59). *Ex vivo* studies in transgenic mice and AD brain sections showed that [^{18}F]BF-108 exhibited high affinity for both SPs and NFTs. Further studies are currently underway to optimize the *in vitro* binding properties and *in vivo* pharmacokinetic profile.

VII. Human Studies of Selected Amyloid-Imaging Agents

To date, several amyloid-imaging agents have been evaluated in human subjects for potential clinical applications. [^{18}F]FDDNP (Table III, entry 45) was the first probe that was studied for imaging amyloid plaques in living subjects (Agdeppa *et al.*, 2001; Barrio *et al.*, 1999; Shoghi-Jadid *et al.*, 2002). Its structure is based on a highly lipophilic, solvent-sensitive, and fluorescent probe, termed DDNP, which is able to permeably cross membrane barriers (Jacobson *et al.*, 1996). *Ex vivo* autoradiography of AD brain sections indicated that [^{18}F]FDDNP also labeled NFTs and prion plaques (Bresjanac *et al.*, 2003). When administrated to human subjects, [^{18}F]FDDNP displayed good brain uptake. Selective retention in the brains of AD subjects relative to controls was observed following data analyses based on relative residence time (RRT) (Shoghi-Jadid *et al.*, 2002). The RRT was higher in the hippocampus of AD subjects, which also correlated with memory performance scores. Furthermore, the accumulations of [^{18}F]FDDNP corresponded to the regions of decreased glucose metabolism and atrophy (Fig. 1).

Among the lipophilic ThT derivatives, extensive SAR studies have led to identification of a lead radioligand, PIB, for human PET studies. Preliminary results indicated that PIB entered the human brain very well. The initial distribution appeared to be proportional to blood flow, as expected (Klunk *et al.*, 2004). The PET studies were carried out in 9 control subjects and 15 AD patients. As a group, the healthy control subjects showed rapid entry and clearance of PIB in all cortical and subcortical gray matter areas, including cerebellar cortex (Fig. 2). The uptake and clearance of PIB in the

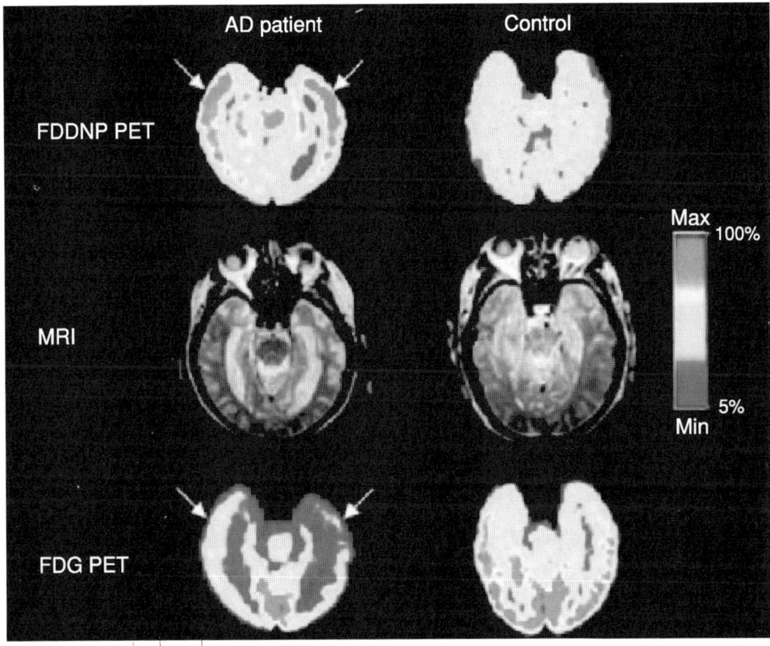

Figure 1 [^{18}F]FDDNP-PET, MRI, and ^{18}F-labeled deoxyglucose (FDG)-PET images of a patient with AD and a normal subject. The [^{18}F]FDDNP and FDG images of each stage are co-registered to their respective MR images. Areas of FDG hypometabolism are matched with the localization of neurofibrillary tangles and amyloid plaques (APs) resulting from [^{18}F]FDDNP binding (arrows). The [^{18}F]FDDNP images represent activity 25–54 minutes after ^{18}F-FDDNP administration. The FDG images represent activity 20–60 minutes after FDG injection. Reprinted with permission from the American Journal of Geriatric Psychiatry. Copyright 2002, American Psychiatric Association. (See Color Insert.)

cerebellum were nearly identical in the control and AD subjects. Relatively lower entry and slower clearance were observed in the white matter, but PIB retention was very similar in both groups. In contrast, PIB retention in AD subjects was significantly different from that in control subjects in areas of the brain known to contain large amounts of amyloid deposits in AD, such as the parietal and frontal cortices. The AD patients showed a marked retention of PIB compared with control subjects, indicating a quantifiable discrimination between patients with mild cognitive impairment and AD and control subjects. In control subjects, there was very little retention of PIB in cortical regions. In AD subjects, the absolute amount of PIB retained in the frontal cortex was more than 90% higher than that retained in the control frontal cortex or cerebellum of either controls or AD patients.

As the time–activity data would predict, the topographical pattern of PIB retention was clearly different in AD patients compared with the control

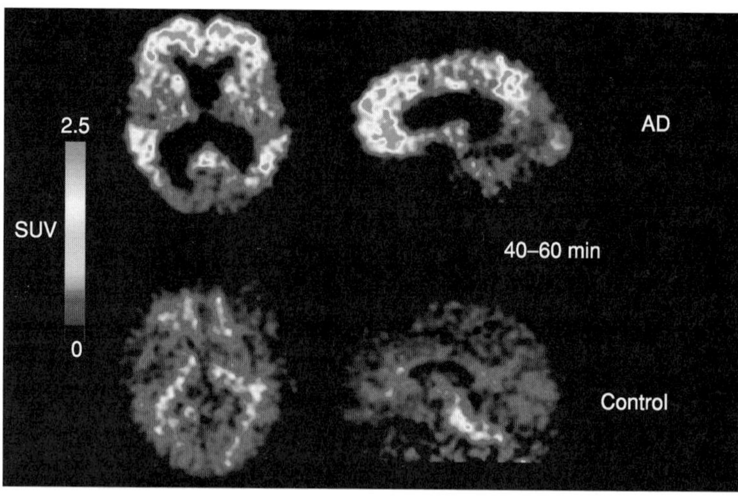

Figure 2 Transaxial (left column) and sagittal (right column) PIB-PET images expressed as standardized uptake value (SUV) in both a patient with suspected AD (top) and an age-matched control subject (bottom). In the AD image, note the relative intensity of retention in the frontal and temporoparietal cortices and the relative lack of retention in the visual cortex. Reprinted with permission. (See Color Insert.)

subjects. In AD patients as a group, PIB retention was most prominent in cortical association areas and lower in the white matter area. PIB images from control subjects showed little or no PIB retention in cortical areas, leaving the subcortical white matter regions highest in relative terms. But in absolute terms, the accumulation of PIB in white matter was essentially the same in AD and control subjects. This pattern of distribution of PIB in AD subjects is very consistent with the known topology of amyloid plaque distribution in AD brain.

Quantitative comparison of AD and control subjects showed that, in cortical areas, the mean PIB standard uptake value (SUV) of AD patients was significantly greater than the mean PIB SUV value of control subjects (Klunk *et al.*, 2004). This indicates increased retention of PIB in areas known to have extensive amyloid deposition in AD. The average PIB SUV values in the control subjects were low and similar to each other in all cortical and subcortical gray matter areas. In both control subjects and AD patients, the retention also was similar in the cerebellar gray matter, indicating the lack of PIB retention in control cortex and in the cerebellum of both AD and controls, brain areas that would not be expected to have significant amyloid deposition. In the white matter, the mean PIB SUV value of control subjects was similar to the SUV values found in AD subjects. These values were higher than those found in cortical areas of the control subjects. This suggests higher, nonspecific retention of PIB in white matter than in gray matter areas.

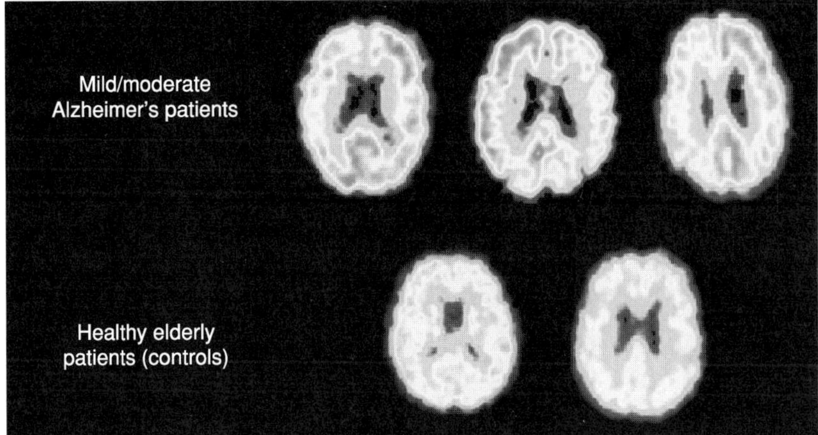

Figure 3 Parametric images of standardized uptake values obtained by normalizing tissue concentration (nCi/mL) by injected dose per body mass (nCi/g) of PET images summed over 40–120 minutes after injection of 10 mCi of [^{11}C]SB-13. Data are shown for representative Alzheimer's disease patients and comparison subjects. (Courtesy N. P. Verhoeff.) (See Color Insert.)

Most recently, a stilbene derivative, termed ^{11}C-SB-13 (Table III, entry 51) has been evaluated in human AD and control subjects compared with ^{11}C-PIB (Verhoeff *et al.*, 2004). As shown in Figure 3, ^{11}C-SB-13 and ^{11}C-PIB display a similar BBB permeability. Like ^{11}C-PIB, ^{11}C-SB-13 showed increased retention in frontal and posterior temporal–inferior parietal association cortices in AD compared with the control subjects. These studies indicated that ^{11}C-SB-13 could be used to differentiate patients with AD from healthy controls.

VIII. Conclusion

A wide array of amyloid-imaging agents has been developed, ranging from biomolecules to small-molecule compounds. Each type of compounds displayed unique *in vitro* binding properties and *in vivo* pharmacokinetic profiles. Over the past decade, significant progress has been made to meet the challenge of *in vivo* detection of amyloid deposits in the brain. To date, the concept of amyloid imaging has materialized and crosses from benchtop to the bedside of AD patients. It is hoped that this comprehensive survey of these amyloid-imaging agents may serve not only as a complete reference but rather as a gateway for future endeavors in the development of many new imaging agents for studies of neurodegerative disorders.

Acknowledgments

This work is supported in part by grants from the Institute for the Study of Aging (Y.W.) and the National Institute on Aging (grant No. AG22048, Y.W.).

References

Agdeppa, E. D., Kepe, V., Liu, J., Flores-Torres, S., Satyamurthy, N., Petric, A., Cole, G. M., Small, G. W., Huang, S. C., and Barrio, J. R. (2001). Binding characteristics of radiofluorinated 6-dialkylamino-2-naphthylethylidene derivatives as positron emission tomography imaging probes for beta-amyloid plaques in Alzheimer's disease. *J. Neurosci.* **21,** RC189.

Alzheimer, A. (1907). Ueber eine eigenartige Erkrankung der Hirnrinde. *Allgemeine Zeitschrift der Psychiatrie* **64,** 146–148.

Anderson, C. J., Dehdashti, F., Cutler, P. D., Schwarz, S. W., Laforest, R., Bass, L. A., Lewis, J. S., and McCarthy, D. W. (2001). ^{64}Cu-TETA-octreotide as a PET imaging agent for patients with neuroendocrine tumors. *J. Nucl. Med.* **42,** 213–221.

Barrio, J. R., Huang, S. C., Cole, G. M., Satyamurthy, N., Petric, A., Phelps, M. E., and Small, G. W. (1999). PET imaging of tangles and plaques in Alzheimer's disease with a highly hydrophobic probes. *J. Label. Radiopharm.* **42,** S194–S195.

Bickel, U., Lee, V. M., Trojanowski, J. Q., and Pardridge, W. M. (1994). Development and *in vitro* characterization of a cationized monoclonal antibody against beta A4 protein: A potential probe for Alzheimer's disease. *Bioconjug.Chem.* **5,** 119–125.

Blankenberg, F. G., Naumovski, L., Tait, J. F., Post, A. M., and Strauss, H. W. (2001). Imaging cyclophosphamide-induced intramedullary apoptosis in rats using ^{99}mTc-radiolabeled annexin V. *J. Nucl. Med.* **42,** 309–316.

Blankenberg, F. G., and Strauss, H. W. (2001). Noninvasive strategies to image cardiovascular apoptosis. *Cardiol. Clin.* **19,** 165–172.

Blasberg, R. (2002). PET imaging of gene expression. *Eur. J. Cancer* **38,** 2137–2146.

Blasberg, R. G., and Gelovani-Tjuvajev, J. (2002). *In vivo* molecular-genetic imaging. *J. Cell Biochem.* **39**(Suppl.), 172–183.

Bresjanac, M., Smid, L. M., Vovko, T. Z., Petric, A., Barrio, J. R., and Popovic, M. (2003). Molecular-imaging probe 2-(1-{6-[(2-fluoroethyl)(methyl) amino]-2-naphthyl}ethylidene) malononitrile labels prion plaques *in vitro*. *J. Neurosci.* **23,** 8029–8033.

Brock, C. S., Young, H., O'Reilly, S. M., Matthews, J., Osman, S., Evans, H., Newlands, E. S., and Price, P. M. (2000). Early evaluation of tumour metabolic response using [^{18}F]fluorodeoxyglucose and positron emission tomography: A pilot study following the phase II chemotherapy schedule for temozolomide in recurrent high-grade gliomas. *Br. J. Cancer* **82,** 608–615.

Burns, J., Pennock, C. A., and Stoward, P. J. (1967). The specificity of the staining of amyloid deposits with thioflavine T. *J. Pathol. Bacteriol.* **94,** 337–344.

Cai, L., Chin, F. T., Pike, V. W., Toyama, H., Liow, J. S., Zoghbi, S. S., Modell, K., Briard, E., Shetty, H. U., Sinclair, K., Donohue, S., Tipre, D., Kung, M. P., Dagostin, C., Widdowson, D. A., Green, M., Gao, W., Herman, M. M., Ichise, M., and Innis, R. B. (2004). Synthesis and evaluation of two ^{18}F-labeled 6-iodo-2-(4'-N,N-dimethylamino)phenylimidazo[1,2-*a*] pyridine derivatives as prospective radioligands for beta-amyloid in Alzheimer'′s disease. *J. Med. Chem.* **47,** 2208–2218.

Casciari, J. J., Graham, M. M., and Rasey, J. S. (1995). A modeling approach for quantifying tumor hypoxia with [F-18]fluoromisonidazole PET time-activity data. *Med. Phys.* **22,** 1127–1139.

Chen, C. C., Meadows, B., Regis, J., Kalafsky, G., Fojo, T., Carrasquillo, J. A., and Bates, S. E. (1997). Detection of *in vivo* P-glycoprotein inhibition by PSC 833 using Tc-99m sestamibi. *Clin. Cancer Res.* **3**, 545–552.

Cherry, S. R. (2001). Fundamentals of positron emission tomography and applications in preclinical drug development. *J. Clin. Pharmacol.* **41**, 482–491.

Clark, R. F., and Goate, A. M. (1997). Pharmacological treatment of Alzheimer's disease. *In* "Molecular and Neurobiological Foundations" (J. D. Brioni and M. W. Decker, Eds.), pp. 193–216. Wiley-Liss, New York.

De Strooper, B., and Konig, G. (1999). Alzheimer's disease. A firm base for drug development. *Nature* **402**, 471–472.

Dehdashti, F., Flanagan, F. L., Mortimer, J. E., Katzenellenbogen, J. A., Welch, M. J., and Siegel, B. A. (1999). Positron emission tomographic assessment of "metabolic flare" to predict response of metastatic breast cancer to antiestrogen therapy *Eur. J. Nucl. Med.* **26**, 51–56.

Delacourte, A. (1998). Diagnosis of Alzheimer's disease [in French]. *Ann. Biol. Clin. (Paris)* **56**, 133–142.

Dishino, D. D., Welch, M. J., Kilbourn, M. R., and Raichle, M. E. (1983). Relationship between lipophilicity and brain extraction of C-11-labeled radiopharmaceuticals. *J. Nucl. Med.* **24**, 1030–1038.

Eckelman, W. C. (2003). The use of PET and knockout mice in the drug discovery process. *Drug Discov. Today* **8**, 404–410.

Foo, S. S., Abbott, D. F., Lawrentschuk, N., and Scott, A. M. (2004). Functional imaging of intratumoral hypoxia. *Mol. Imaging Biol.* **6**, 291–305.

Friedland, R. P., Majocha, R. E., Reno, J. M., Lyle, L. R., and Marotta, C. A. (1994). Development of an anti-A beta monoclonal antibody for *in vivo* imaging of amyloid angiopathy in Alzheimer's disease. *Mol. Neurobiol.* **9**, 107–113.

Friedland, R. P., Shi, J., Lamanna, J. C., Smith, M. A., and Perry, G. (2000). Prospects for noninvasive imaging of brain amyloid beta in Alzheimer's disease. *Ann. NY Acad. Sci.* **903**, 123–128.

Gambhir, S. S., Barrio, J. R., Herschman, H. R., and Phelps, M. E. (1999). Assays for noninvasive imaging of reporter gene expression. *Nucl. Med. Biol.* **26**, 481–490.

Gee, A. D. (2003). Neuropharmacology and drug development. *Br. Med. Bull.* **65**, 169–177.

Ghilardi, J. R., Catton, M., Stimson, E. R., Rogers, S., Walker, L. C., Maggio, J. E., and Mantyh, P. W. (1996). Intra-arterial infusion of [^{125}I]A beta 1-40 labels amyloid deposits in the aged primate brain *in vivo*. *Neuroreport* **7**, 2607–2611.

Gupta, N., Price, P. M., and Aboagye, E. O. (2002). PET for *in vivo* pharmacokinetic and pharmacodynamic measurements. *Eur. J. Cancer* **38**, 2094–2107.

Hardy, J. (1992). An "anatomical cascade hypothesis" for Alzheimer's disease. *Trends Neurosci.* **15**, 200–201.

Hawkins, P. N., Myers, M. J., Epenetos, A. A., Caspi, D., and Pepys, M. B. (1988a). Specific localization and imaging of amyloid deposits *in vivo* using ^{123}I-labeled serum amyloid P component. *J. Exp. Med.* **167**, 903–913.

Hawkins, P. N., Myers, M. J., Lavender, J. P., and Pepys, M. B. (1988b). Diagnostic radionuclide imaging of amyloid: Biological targeting by circulating human serum amyloid P component. *Lancet* **1**, 1413–1418.

Hendrikse, N. H., de Vries, E. G., Eriks-Fluks, L., van der Graaf, W. T., Hospers, G. A., Willemsen, A. T., Vaalburg, W., and Franssen, E. J. (1999). A new *in vivo* method to study P-glycoprotein transport in tumors and the blood-brain barrier. *Cancer Res.* **59**, 2411–2416.

Hendrikse, N. H., Franssen, E. J., van der Graaf, W. T., Vaalburg, W., and de Vries, E. G. (1999). Visualization of multidrug resistance *in vivo*. *Eur. J. Nucl. Med.* **26**, 283–293.

Herschman, H. R. (2004). Noninvasive imaging of reporter gene expression in living subjects. *Adv. Cancer Res.* **92,** 29–80.

Hilbich, C., Kisters-Woike, B., Reed, J., Masters, C. L., and Beyreuther, K. (1991). Aggregation and secondary structure of synthetic amyloid beta A4 peptides of Alzheimer's disease. *J. Mol. Biol.* **218,** 149–163.

Hirschfield, G. M., and Hawkins, P. N. (2003). Amyloidosis: New strategies for treatment. *Int. J. Biochem. Cell Biol.* **35,** 1608–1613.

Hume, S. P., Brown, D. J., Ashworth, S., Hirani, E., Luthra, S. K., and Lammertsma, A. A. (1997). *In vivo* saturation kinetics of two dopamine transporter probes measured using a small animal positron emission tomography scanner. *J. Neurosci. Methods* **76,** 45–51.

Ikeda, S., Wong, C. W., Allsop, D., Landon, M., Kidd, M., and Glenner, G. G. (1987). Immunogold labeling of cerebrovascular and neuritic plaque amyloid fibrils in Alzheimer's disease with an anti-beta protein monoclonal antibody. *Lab. Invest.* **57,** 446–449.

Ishii, K., Tamaoka, A., Mizusawa, H., Shoji, S., Ohtake, T., Fraser, P. E., Takahashi, H., Tsuji, S., Gearing, M., Mizutani, T., Yamada, S., Kato, M., St George-Hyslop, P. H., Mirra, S. S., and Mori, H. (1997). Abeta1-40 but not Abeta1-42 levels in cortex correlate with apolipoprotein E epsilon4 allele dosage in sporadic Alzheimer's disease. *Brain Res.* **748,** 250–252.

Jacobson, A., Petric, A., Hogenkamp, D., Sinur, A., and Barrio, J. R. (1996). 1,1-Dicyano-2-[6-(dimethylamino)naphthalen-2-yl]propene (DDNP): A solvent polarity and viscosity sensitive fluorophore for fluorescence microscopy. *J. Am. Chem. Soc.* **118,** 5572–5579.

Joachim, C. L., Morris, J. H., and Selkoe, D. J. (1989). Diffuse senile plaques occur commonly in the cerebellum in Alzheimer's disease. *Am. J. Pathol.* **135,** 309–319.

Kapur, S., Jones, C., Da Silva, J., Wilson, A., and Houle, S. (1997a). Reliability of a simple non-invasive method for the evaluation of 5-HT$_2$ receptors using [^{18}F]-setoperone PET imaging. *Nucl. Med. Commun.* **18,** 395–399.

Kapur, S., Zipursky, R., Remington, G., Jones, C., McKay, G., and Houle, S. (1997b). PET evidence that loxapine is an equipotent blocker of 5-HT$_2$ and D$_2$ receptors: Implications for the therapeutics of schizophrenia. *Am. J. Psychiatry* **154,** 1525–1529.

Kapur, S., Zipursky, R., Roy, P., Jones, C., Remington, G., Reed, K., and Houle, S. (1997c). The relationship between D$_2$ receptor occupancy and plasma levels on low dose oral haloperidol: A PET study. *Psychopharmacology (Berl)* **131,** 148–152.

Katzman, R. (1993). Education and the prevalence of dementia and Alzheimer's disease. *Neurology* **43,** 13–20.

Kilbourn, M. R., Pavia, M. R., and Gregor, V. E. (1990). Synthesis of fluorine-18 labeled GABA uptake inhibitors. *Int. J. Rad. Appl. Instrum. [A]* **41,** 823–828.

Klunk, W. E., Debnath, M. L., and Pettegrew, J. W. (1995). Chrysamine-G binding to Alzheimer and control brain: Autopsy study of a new amyloid probe. *Neurobiol. Aging* **16,** 541–548.

Klunk, W. E., Engler, H., Nordberg, A., Wang, Y., Blomqvist, G., Holt, D. P., Bergstrom, M., Savitcheva, I., Huang, G. F., Estrada, S., Ausen, B., Debnath, M. L., Barletta, J., Price, J. C., Sandell, J., Lopresti, B. J., Wall, A., Koivisto, P., Antoni, G., Mathis, C. A., and Langstrom, B. (2004). Imaging brain amyloid in Alzheimer's disease with Pittsburgh Compound-B. *Ann. Neurol.* **55,** 306–319.

Klunk, W. E., Jacob, R. F., and Mason, R. P. (1999). Quantifying amyloid beta-peptide (Abeta) aggregation using the Congo red-Abeta (CR-abeta) spectrophotometric assay. *Anal. Biochem.* **266,** 66–76.

Klunk, W. E., Pettegrew, J. W., and Abraham, D. J. (1989). Quantitative evaluation of congo red binding to amyloid-like proteins with a beta-pleated sheet conformation. *J. Histochem. Cytochem.* **37,** 1273–1281.

Klunk, W. E., Wang, Y., Huang, G. F., Debnath, M. L., Holt, D. P., and Mathis, C. A. (2001). Uncharged thioflavin-T derivatives bind to amyloid-beta protein with high affinity and readily enter the brain. *Life Sci.* **69**, 1471–1484.

Klunk, W. E., Wang, Y., Huang, G. F., Debnath, M. L., Holt, D. P., Shao, L., Hamilton, R. L., Ikonomovic, M. D., De Kosky, S. T., and Mathis, C. A. (2003). The binding of 2-(4'-methylaminophenyl)benzothiazole to postmortem brain homogenates is dominated by the amyloid component. *J. Neurosci.* **23**, 2086–2092.

Kung, H. F., Kung, M. P., Zhuang, Z. P., Hou, C., Lee, C. W., Plossl, K., Zhuang, B., Skovronsky, D. M., Lee, V. M., and Trojanowski, J. Q. (2003). Iodinated tracers for imaging amyloid plaques in the brain. *Mol. Imaging Biol.* **5**, 418–426.

Kung, H. F., Lee, C. W., Zhuang, Z. P., Kung, M. P., Hou, C., and Plossl, K. (2001). Novel stilbenes as probes for amyloid plaques. *J. Am. Chem. Soc.* **123**, 12740–12741.

Kung, M. P., Hou, C., Zhuang, Z. P., Skovronsky, D., and Kung, H. F. (2004). Binding of two potential imaging agents targeting amyloid plaques in postmortem brain tissues of patients with Alzheimer's disease. *Brain Res.* **1025**, 98–105.

Kung, M. P., Hou, C., Zhuang, Z. P., Skovronsky, D. M., Zhang, B., Gur, T. L., Trojanowski, J. Q., Lee, V. M., and Kung, H. F. (2002a). Radioiodinated styrylbenzene derivatives as potential SPECT imaging agents for amyloid plaque detection in Alzheimer's disease. *J. Mol. Neurosci.* **19**, 7–10.

Kurihara, A., and Pardridge, W. M. (2000). Abeta(1-40) peptide radiopharmaceuticals for brain amyloid imaging: [111]In chelation, conjugation to poly(ethylene glycol)-biotin linkers, and autoradiography with Alzheimer's disease brain sections. *Bioconjug. Chem.* **11**, 380–386.

Lashuel, H. A., Hartley, D. M., Balakhaneh, D., Aggarwal, A., Teichberg, S., and Callaway, D. J. (2002). New class of inhibitors of amyloid-beta fibril formation. Implications for the mechanism of pathogenesis in Alzheimer's disease. *J. Biol. Chem.* **277**, 42881–42890.

Lee, H. J., and Pardridge, W. M. (2001). Pharmacokinetics and delivery of tat and tat-protein conjugates to tissues *in vivo*. *Bioconjug. Chem.* **12**, 995–999.

Lee, H. J., Zhang, Y., Zhu, C., Duff, K., and Pardridge, W. M. (2002). Imaging brain amyloid of Alzheimer disease *in vivo* in transgenic mice with an Abeta peptide radiopharmaceutical. *J. Cereb. Blood Flow Metab.* **22**, 223–231.

Levchenko, A., Mehta, B. M., Lee, J. B., Humm, J. L., Augensen, F., Squire, O., Kothari, P. J., Finn, R. D., Leonard, E. F., and Larson, S. M. (2000). Evaluation of ^{11}C-colchicine for PET imaging of multiple drug resistance. *J. Nucl. Med.* **41**, 493–501.

Levin, V. A. (1980). Relationship of octanol/water partition coefficient and molecular weight to rat brain capillary permeability. *J. Med. Chem.* **23**, 682–684.

Levine, L. E., Reiss, G., and Smith, D. A. (1993). *In situ* scanning-tunneling-microscopy studies of early-stage electromigration in Ag. *Physical Review. B. Condensed Matter.* **48**, 858–863.

Lewis, J. S., Herrero, P., Sharp, T. L., Engelbach, J. A., Fujibayashi, Y., Laforest, R., Kovacs, A., Gropler, R. J., and Welch, M. J. (2002). Delineation of hypoxia in canine myocardium using PET and copper(II)-diacetyl-bis(N(4)-methylthiosemicarbazone). *J. Nucl. Med.* **43**, 1557–1569.

Lovat, L. B., O'Brien, A. A., Armstrong, S. F., Madhoo, S., Bulpitt, C. J., Rossor, M. N., Pepys, M. B., and Hawkins, P. N. (1998a). Scintigraphy with ^{123}I-serum amyloid P component in Alzheimer disease. *Alzheimer Dis. Assoc. Disord.* **12**, 208–210.

Lovat, L. B., Persey, M. R., Madhoo, S., Pepys, M. B., and Hawkins, P. N. (1998b). The liver in systemic amyloidosis: Insights from ^{123}I serum amyloid P component scintigraphy in 484 patients. *Gut* **42**, 727–734.

Maggio, J. E., Stimson, E. R., Ghilardi, J. R., Allen, C. J., Dahl, C. E., Whitcomb, D. C., Vigna, S. R., Vinters, H. V., Labenski, M. E., and Mantyh, P. W. (1992). Reversible *in vitro* growth of Alzheimer disease beta-amyloid plaques by deposition of labeled amyloid peptide. *Proc. Natl. Acad. Sci. USA* **89**, 5462–5466.

Majocha, R. E., Reno, J. M., Friedland, R. P., Van Haight, C., Lyle, L. R., and Marotta, C. A. (1992). Development of a monoclonal antibody specific for beta/A4 amyloid in Alzheimer's disease brain for application to *in vivo* imaging of amyloid angiopathy. *J. Nucl. Med.* **33**, 2184–2189.

Marshall, J. R., Stimson, E. R., Ghilardi, J. R., Vinters, H. V., Mantyh, P. W., and Maggio, J. E. (2002). Noninvasive imaging of peripherally injected Alzheimer's disease type synthetic A beta amyloid *in vivo*. *Bioconjug. Chem.* **13**, 276–284.

Mathis, C. A., Bacskai, B. J., Kajdasz, S. T., McLellan, M. E., Frosch, M. P., Hyman, B. T., Holt, D. P., Wang, Y., Huang, G. F., Debnath, M. L., and Klunk, W. E. (2002). A lipophilic thioflavin-T derivative for positron emission tomography (PET) imaging of amyloid in brain. *Bioorg Med. Chem. Lett.* **12**, 295–298.

Mathis, C. A., Wang, Y., Holt, D. P., Huang, G. F., Debnath, M. L., and Klunk, W. E. (2003). Synthesis and evaluation of ^{11}C-labeled 6-substituted 2-arylbenzothiazoles as amyloid imaging agents. *J. Med. Chem.* **46**, 2740–2754.

Mathis, C. A., Wang, Y., and Klunk, W. E. (2004). Imaging beta-amyloid plaques and neurofibrillary tangles in the aging human brain. *Curr. Pharm. Des.* **10**, 1469–1492.

Mochizuki, A., Peterson, J. W., Mufson, E. J., and Trapp, B. D. (1996). Amyloid load and neural elements in Alzheimer's disease and nondemented individuals with high amyloid plaque density. *Exp. Neurol.* **142**, 89–102.

Morris, J. C., Storandt, M., McKeel, D. W., Jr., Rubin, E. H., Price, J. L., Grant, E. A., and Berg, L. (1996). Cerebral amyloid deposition and diffuse plaques in "normal" aging: Evidence for presymptomatic and very mild Alzheimer's disease. *Neurology* **46**, 707–719.

Narula, J., Acio, E. R., Narula, N., Samuels, L. E., Fyfe, B., Wood, D., Fitzpatrick, J. M., Raghunath, P. N., Tomaszewski, J. E., Kelly, C., Steinmetz, N., Green, A., Tait, J. F., Leppo, J., Blankenberg, F. G., Jain, D., and Strauss, H. W. (2001). Annexin-V imaging for noninvasive detection of cardiac allograft rejection. *Nat. Med.* **7**, 1347–1352.

Naslund, J., Haroutunian, V., Mohs, R., Davis, K. L., Davies, P., Greengard, P., and Buxbaum, J. D. (2000). Correlation between elevated levels of amyloid beta-peptide in the brain and cognitive decline. *JAMA* **283**, 1571–1577.

Naslund, J., Schierhorn, A., Hellman, U., Lannfelt, L., Roses, A. D., Tjernberg, L. O., Silberring, J., Gandy, S. E., Winblad, B., Greengard, P., Nordstedt, C., and Terenius, L. (1994). Relative abundance of Alzheimer A beta amyloid peptide variants in Alzheimer disease and normal aging. *Proc. Natl. Acad. Sci. USA* **91**, 8378–8382.

Nordberg, A. (2003). Toward an early diagnosis and treatment of Alzheimer's disease. *Int. Psychogeriatr.* **15**, 223–237.

Nutt, R. (2002). For: Is LSO the future of PET? *Eur. J. Nucl. Med. Mol. Imaging* **29**, 1523–1525.

Okamura, N., Suemoto, T., Shimadzu, H., Suzuki, M., Shiomitsu, T., Akatsu, H., Yamamoto, T., Staufenbiel, M., Yanai, K., Arai, H., Sasaki, H., Kudo, Y., and Sawada, T. (2004). Styrylbenzoxazole derivatives for *in vivo* imaging of amyloid plaques in the brain. *J. Neurosci.* **24**, 2535–2541.

Olshansky, S. J., Carnes, B. A., and Cassel, C. K. (1993). The aging of the human species. *Sci. Am.* **268**, 46–52.

Ono, M., Kung, M. P., Hou, C., and Kung, H. F. (2002). Benzofuran derivatives as Abeta-aggregate-specific imaging agents for Alzheimer's disease. *Nucl. Med. Biol.* **29**, 633–642.

Price, D. L., and Sisodia, S. S. (1998). Mutant genes in familial Alzheimer's disease and transgenic models. *Annu. Rev. Neurosci.* **21**, 479–505.

Price, J. L. (1997). Diagnostic criteria for Alzheimer's disease. *Neurobiol. Aging* **18**, S67–S70.

Puchtler, H., and Sweat, F. (1962). Amidoblack as a stain for hemoglobin. *Arch. Pathol.* **73**, 245–249.

Robinson, S. R., Bishop, G. M., and Munch, G. (2003). Alzheimer vaccine: Amyloid-beta on trial. *Bioessays* **25**, 283–288.

Saito, Y., Buciak, J., Yang, J., and Pardridge, W. M. (1995). Vector-mediated delivery of ^{125}I-labeled beta-amyloid peptide A beta 1-40 through the blood-brain barrier and binding to Alzheimer disease amyloid of the A beta 1-40/vector complex. *Proc. Natl. Acad. Sci. USA* **92**, 10227–10231.

Sato, K., Higuchi, M., Iwata, N., Saido, T. C., and Sasamoto, K. (2004). Fluoro-substituted and ^{13}C-labeled styrylbenzene derivatives for detecting brain amyloid plaques. *Eur. J. Med. Chem.* **39**, 573–578.

Schenk, D., Barbour, R., Dunn, W., Gordon, G., Grajeda, H., Guido, T., Hu, K., Huang, J., Johnson-Wood, K., Khan, K., Kholodenko, D., Lee, M., Liao, Z., Lieberburg, I., Motter, R., Mutter, L., Soriano, F., Shopp, G., Vasquez, N., Vandevert, C., Walker, S., Wogulis, M., Yednock, T., Games, D., and Seubert, P. (1999). Immunization with amyloid-beta attenuates Alzheimer-disease-like pathology in the pharmacodynamicAPP mouse. *Nature* **400**, 173–177.

Schenk, D., Games, D., and Seubert, P. (2001). Potential treatment opportunities for Alzheimer's disease through inhibition of secretases and Abeta immunization. *J. Mol. Neurosci.* **17**, 259–267.

Schenk, D. B., Seubert, P., Lieberburg, I., and Wallace, J. (2000). beta-Peptide immunization: A possible new treatment for Alzheimer disease. *Arch. Neurol.* **57**, 934–936.

Schmidt, M. L., Schuck, T., Sheridan, S., Kung, M. P., Kung, H., Zhuang, Z. P., Bergeron, C., Lamarche, J. S., Skovronsky, D., Giasson, B. I., Lee, V. M., and Trojanowski, J. Q. (2001). The fluorescent Congo red derivative, (*trans, trans*)-1-bromo-2,5-bis-(3-hydroxycarbonyl-4-hydroxy)styrylbenzene (BSB), labels diverse beta-pleated sheet structures in postmortem human neurodegenerative disease brains. *Am J. Pathol.* **159**, 937–943.

Selkoe, D. J. (2000). Toward a comprehensive theory for Alzheimer's disease. Hypothesis: Alzheimer's disease is caused by the cerebral accumulation and cytotoxicity of amyloid beta-protein. *Ann. NY Acad. Sci.* **924**, 17–25.

Selkoe, D. J. (2002). Alzheimer's disease is a synaptic failure. *Science* **298**, 789–791.

Shi, J., Perry, G., Berridge, M. S., Aliev, G., Siedlak, S. L., Smith, M. A., La Manna, J. C., and Friedland, R. P. (2002). Labeling of cerebral amyloid beta deposits *in vivo* using intranasal basic fibroblast growth factor and serum amyloid P component in mice. *J. Nucl. Med.* **43**, 1044–1051.

Shields, A. F., Grierson, J. R., Dohmen, B. M., Machulla, H. J., Stayanoff, J. C., Lawhorn-Crews, J. M., Obradovich, J. E., Muzik, O., and Mangner, T. J. (1998a). Imaging proliferation *in vivo* with [F-18]FLT and positron emission tomography. *Nat. Med.* **4**, 1334–1336.

Shields, A. F., Mankoff, D. A., Link, J. M., Graham, M. M., Eary, J. F., Kozawa, S. M., Zheng, M., Lewellen, B., Lewellen, T. K., Grierson, J. R., and Krohn, K. A. (1998b). Carbon-11-thymidine and FDG to measure therapy response. *J. Nucl. Med.* **39**, 1757–1762.

Shimadzu, H., Suemoto, T., Suzuki, M., Shiomitsu, T., Okamura, N., Kudo, Y., and Sawada, T. (2003). A novel probe for imaging amyloid-β: Synthesis of F-18 labelled BF-108, an acridine orange analog. *J. Labelled Comp. Radiopharm.* **46**, 765–772.

Shoghi-Jadid, K., Small, G. W., Agdeppa, E. D., Kepe, V., Ercoli, L. M., Siddarth, P., Read, S., Satyamurthy, N., Petric, A., Huang, S. C., and Barrio, J. R. (2002). Localization of neurofibrillary tangles and beta-amyloid plaques in the brains of living patients with Alzheimer disease. *Am. J. Geriatr. Psychiatry* **10**, 24–35.

Skovronsky, D. M., Zhang, B., Kung, M. P., Kung, H. F., Trojanowski, J. Q., and Lee, V. M. (2000). *In vivo* detection of amyloid plaques in a mouse model of Alzheimer's disease. *Proc. Natl. Acad. Sci. USA* **97**, 7609–7614.

Styren, S. D., Hamilton, R. L., Styren, G. C., and Klunk, W. E. (2000). X-34, a fluorescent derivative of Congo red: A novel histochemical stain for Alzheimer's disease pathology. *J. Histochem. Cytochem.* **48**, 1223–1232.

Suemoto, T., Okamura, N., Shiomitsu, T., Suzuki, M., Shimadzu, H., Akatsu, H., Yamamoto, T., Kudo, Y., and Sawada, T. (2004). *In vivo* labeling of amyloid with BF-108. *Neurosci. Res.* **48,** 65–74.

Trojanowski, J. Q., Clark, C. M., Schmidt, M. L., Arnold, S. E., and Lee, V. M. (1997). Strategies for improving the postmortem neuropathological diagnosis of Alzheimer's disease. *Neurobiol. Aging* **18,** S75–S79.

Vavere, A. L., and Lewis, J. S. (2003). Imaging the effects of anti-angiogenic treatments. *Q. J. Nucl. Med.* **47,** 163–170.

Vecchio, S. D., Ciarmiello, A., Potena, M. I., Carriero, M. V., Mainolfi, C., Botti, G., Thomas, R., Cerra, M., D' Aiuto, G., Tsuruo, T., and Salvatore, M. (1997). *In vivo* detection of multidrug-resistant (MDR1) phenotype by technetium-99m sestamibi scan in untreated breast cancer patients. *Eur. J. Nucl. Med.* **24,** 150–159.

Verhoeff, N. P., Wilson, A. A., Takeshita, S., Trop, L., Hussey, D., Singh, K., Kung, H. F., Kung, M. P., and Houle, S. (2004). *In-vivo* imaging of Alzheimer disease beta-amyloid with [^{11}C]SB-13 PET. *Am. J. Geriatr. Psychiatry* **12,** 584–595.

Walker, L. C., Price, D. L., Voytko, M. L., and Schenk, D. B. (1994). Labeling of cerebral amyloid *in vivo* with a monoclonal antibody. *J. Neuropathol. Exp. Neurol.* **53,** 377–383.

Walker, R. C., Purnell, G. L., Jones-Jackson, L. B., Thomas, K. L., Brito, J. A., and Ferris, E. J. (2004). Introduction to PET imaging with emphasis on biomedical research. *Neurotoxicology* **25,** 533–542.

Walker, R. C., and Zigler, S. S. (2003). PET practice in nuclear pharmacy. *J. Am. Pharm. Assoc. (Wash)* **43,** S42–S43.

Walsh, D. M., Klyubin, I., Fadeeva, J. V., Rowan, M. J., and Selkoe, D. J. (2002). Amyloid-beta oligomers: Their production, toxicity and therapeutic inhibition. *Biochem. Soc. Trans.* **30,** 552–557.

Wang, Y., Klunk, W. E., Debnath, M. L., Huang, G. F., Holt, D. P., Shao, L., and Mathis, C. A. (2004). Development of a PET/SPECT agent for amyloid imaging in Alzheimer's disease. *J. Mol. Neurosci.* **24,** 55–62.

Wang, Y., Mathis, C. A., Huang, G. F., Debnath, M. L., Holt, D. P., Shao, L., and Klunk, W. E. (2003). Effects of lipophilicity on the affinity and nonspecific binding of iodinated benzothiazole derivatives. *J. Mol. Neurosci.* **20,** 255–260.

Wang, Y., Mathis, C. A., Huang, G. F., Holt, D. P., Debnath, M. L., and Klunk, W. E. (2002b). Synthesis and ^{11}C-labelling of (*E,E*)-1-(3′,4′-dihydroxystyryl)-4-(3′-methoxy-4′-hydroxystyryl) benzene for PET imaging of amyloid deposits. *J. Labelled Comp. Radiopharm.* **45,** 647–664.

Weber, W. A., Haubner, R., Vabuliene, E., Kuhnast, B., Wester, H. J., and Schwaiger, M. (2001). Tumor angiogenesis targeting using imaging agents. *Q. J. Nucl. Med.* **45,** 179–182.

Wengenack, T. M., Curran, G. L., and Poduslo, J. F. (2000). Targeting Alzheimer amyloid plaques *in vivo*. *Nat. Biotechnol.* **18,** 868–872.

Wu, D., and Pardridge, W. M. (1998). Pharmacokinetics and blood-brain barrier transport of an anti-transferrin receptor monoclonal antibody (OX26) in rats after chronic treatment with the antibody. *Drug Metab. Dispos.* **26,** 937–939.

Wu, D., Yang, J., and Pardridge, W. M. (1997). Drug targeting of a peptide radiopharmaceutical through the primate blood-brain barrier *in vivo* with a monoclonal antibody to the human insulin receptor. *J. Clin. Invest.* **100,** 1804–1812.

Zhuang, Z. P., Kung, M. P., Hou, C., Plossl, K., Skovronsky, D., Gur, T. L., Trojanowski, J. Q., Lee, V. M., and Kung, H. F. (2001a). IBOX(2-(4′-dimethylaminophenyl)-6-iodobenzoxazole): A ligand for imaging amyloid plaques in the brain. *Nucl. Med. Biol.* **28,** 887–894.

Zhuang, Z. P., Kung, M. P., Wilson, A., Lee, C. W., Plossl, K., Hou, C., Holtzman, D. M., and Kung, H. F. (2003). Structure-activity relationship of imidazo[1,2-*a*]pyridines as ligands for detecting beta-amyloid plaques in the brain. *J. Med. Chem.* **46,** 237–243.

Further Reading

Agdeppa, E. D., Kepe, V., Petri, A., Satyamurthy, N., Liu, J., Huang, S. C., Small, G. W., Cole, G. M., and Barrio, J. R. (2003). *In vitro* detection of (*S*)-naproxen and ibuprofen binding to plaques in the Alzheimer's brain using the positron emission tomography molecular imaging probe 2-(1-[6-[(2-[(18)F]fluoroethyl)(methyl)amino]-2-naphthyl]ethylidene)malono nitrile. *Neuroscience* **117,** 723–730.

Dezutter, N. A., Dom, R. J., de Groot, T. J., Bormans, G. M., and Verbruggen, A. M. (1999). [99]mTc-MAMA-chrysamine G, a probe for beta-amyloid protein of Alzheimer's disease. *Eur. J. Nucl. Med.* **26,** 1392–1399.

Han, H., Cho, C. G., and Lansbury, P. T., Jr. (1996). Technetium complexes for the quantitation of brain amyloid. *J. Am. Chem. Soc.* **118,** 4506–4507.

Ishikawa, K., Doh-ura, K., Kudo, Y., Nishida, N., Murakami-Kubo, I., Ando, Y., Sawada, T., and Iwaki, T. (2004). Amyloid imaging probes are useful for detection of prion plaques and treatment of transmissible spongiform encephalopathies. *J. Gen. Virol.* **85,** 1785–1790.

Klunk, W. E., Bacskai, B. J., Mathis, C. A., Kajdasz, S. T., McLellan, M. E., Frosch, M. P., Debnath, M. L., Holt, D. P., Wang, Y., and Hyman, B. T. (2002). Imaging Abeta plaques in living transgenic mice with multiphoton microscopy and methoxy-X04, a systemically administered Congo red derivative. *J. Neuropathol. Exp. Neurol.* **61,** 797–805.

Kung, M. P., Hou, C., Zhuang, Z. P., Zhang, B., Skovronsky, D., Trojanowski, J. Q., Lee, V. M., and Kung, H. F. (2002b). IMPY: An improved thioflavin-T derivative for *in vivo* labeling of beta-amyloid plaques. *Brain Res.* **956,** 202–210.

Lee, C. W., Kung, M. P., Hou, C., and Kung, H. F. (2003). Dimethylamino-fluorenes: Ligands for detecting beta-amyloid plaques in the brain. *Nucl. Med. Biol.* **30,** 573–580.

Le Vine, H., 3rd. (1999). Quantification of beta-sheet amyloid fibril structures with thioflavin T. *Methods Enzymol.* **309,** 274–284.

Link, C. D., Johnson, C. J., Fonte, V., Paupard, M., Hall, D. H., Styren, S., Mathis, C. A., and Klunk, W. E. (2001). Visualization of fibrillar amyloid deposits in living, transgenic *Caenorhabditis elegans* animals using the sensitive amyloid dye, X-34. *Neurobiol. Aging* **22,** 217–226.

Ono, M., Wilson, A., Nobrega, J., Westaway, D., Verhoeff, P., Zhuang, Z. P., Kung, M. P., and Kung, H. F. (2003). ^{11}C-labeled stilbene derivatives as Abeta-aggregate-specific PET imaging agents for Alzheimer's disease. *Nucl. Med. Biol.* **30,** 565–571.

Tubis, M., Blahd, W. H., and Nordyke, R. A. (1960). The preparation and use of radioiodinated Congo red in detecting amyloidosis. *J. Am. Pharm. Assoc.* **49,** 422–425.

Wang, Y., Klunk, W. E., Huang, G. F., Debnath, M. L., Holt, D. P., and Mathis, C. A. (2002a). Synthesis and evaluation of 2-(3′-iodo-4′-aminophenyl)-6-hydroxybenzothiazole for *in vivo* quantitation of amyloid deposits in Alzheimer's disease. *J. Mol. Neurosci.* **19,** 11–16.

Zhen, W., Han, H., Anguiano, M., Lemere, C. A., Cho, C. G., and Lansbury, P. T., Jr. (1999). Synthesis and amyloid binding properties of rhenium complexes: Preliminary progress toward a reagent for SPECT imaging of Alzheimer's disease brain. *J. Med. Chem.* **42,** 2805–2815.

Zhuang, Z. P., Kung, M. P., Hou, C., Skovronsky, D. M., Gur, T. L., Plossl, K., Trojanowski, J. Q., Lee, V. M., and Kung, H. F. (2001b). Radioiodinated styrylbenzenes and thioflavins as probes for amyloid aggregates. *J. Med. Chem.* **44,** 1905–1914.

9

In Vivo Imaging of Autoimmune Disease in Model Systems

Eric T. Ahrens and Penelope A. Morel†*
*Department of Biological Sciences and Pittsburgh NMR Center for Biomedical Research, Carnegie Mellon University, Pittsburgh, Pennsylvania 15213
†Department of Immunology, University of Pittsburgh School of Medicine Pittsburgh, Pennsylvania 15261

Autoimmune diseases are characterized by infiltration of the target tissue with specific immune cells that ultimately leads to the destruction of normal tissue and the associated disease. There is a need for imaging tools that allow the monitoring of ongoing inflammatory disease as well as the response to therapy. We discuss new magnetic resonance imaging–based technologies that have been used to monitor inflammation and disease progression in animal models of type 1 diabetes, multiple sclerosis, and rheumatoid arthritis. Therapeutic strategies for these diseases include the transfer of immune cells, such as dendritic cells, with the aim of preventing or halting the disease course. We discuss several new MRI labeling techniques developed to allow tracking of immune cells *in vivo*. These include direct *ex vivo* labeling techniques as well as the genetic modification of cells to allow them to produce their own contrast agents. This is an area of intense recent research and can be expanded to other conditions such as cancer. © 2005, Elsevier Inc.

Current Topics in Developmental Biology, Vol. 70
Copyright 2005, Elsevier Inc. All rights reserved.

0070-2153/05 $35.00
DOI: 10.1016/S0070-2153(05)70009-0

I. Introduction

Autoimmune disease occurs when a sustained immune response is mounted against the body's own tissues. One of the hallmarks of autoimmune disease is an inappropriate trafficking of immune cells into tissues that do not come under routine surveillance from these cells in the healthy state. The mechanisms underlying the initiation of autoimmune disease are largely unknown but are believed to have genetic and environmental components. Understanding the trafficking patterns of immune cells in early and late phases of autoimmune disease is paramount in our attempts to understand the pathogenesis of autoimmunity and in designing immunotherapeutic interventions. Animal models of autoimmune disease, particularly mouse models, play a key role in elucidating aspects of these diseases. The ability to noninvasively image the trafficking of phenotypically defined populations of immune cells without killing the animal would be tremendously beneficial to these studies.

The phenotype of an immune cell is defined by the pattern and level of expression of a lexicon of cell surface molecules (i.e., CD antigens). These molecules are commonly assayed *in vitro* using sensitive techniques such as fluorescence-activated cell sorting (FACS) or immunohistochemistry. Determining which surface molecules are present, and at what level under various conditions in diseased tissues, is only one piece of the equation. A more difficult question to answer, but one that is at least as important, is "what biological role do these cell surface markers perform *in vivo*?" Vital imaging of immune cell trafficking patterns can play a key role in answering these sorts of questions. Histology can provide only a snapshot view of a dynamic process such as cell migration. A large number of snapshots would be required, followed by a methodical statistic analysis, to elucidate the migration patterns. This can be extremely time consuming and often does not reveal the true range of individual variability among subjects. A longitudinal view of individual subjects is superior in many ways. Fewer subjects are needed, which saves time and money, and kinetic and topographic information about migration patterns are read out in real-time. This in turn can reveal individual variability and biological complexity in ways that may be missed by static snapshots.

In addition to basic science, a key long-term application of immune cell imaging is monitoring the trafficking of cellular therapeutics *in vivo*. Several immunotherapeutic cell types, such as dendritic cells (DCs), T cells, and natural killer (NK) cells, are currently being studied for therapeutic use. These cells can originate from the patients themselves, from other individuals, or from immortalized cell lines. Labeling cells for imaging can be an additional cell treatment before their implantation into the patient. Visualizing therapeutic cells noninvasively can be difficult, and any approach that can speed the testing of these treatments will be extremely useful and welcomed. In the future, *in vivo* cellular imaging strategies will be closely aligned with a therapeutic agent; they will be used to help calibrate dosage and delivery efficacy.

It is important that any cell labeling scheme used for imaging does not significantly alter the immunological properties of the cells or cause significant cytotoxicity or changes in function, which may confound the interpretation of cell migration data *in vivo*. Furthermore, many immuno-therapeutic strategies rely on specific immune cell subsets (e.g., mature versus immature or different splenic subsets). Thus it is essential that the labeling process does not alter the cell's phenotype, as this may reduce the therapeutic efficacy.

This chapter describes emerging methods and applications of noninvasive imaging as applied to models of autoimmune disease. We emphasize cellular–molecular imaging approaches using magnetic resonance imaging (MRI). However, in certain instances other complementary techniques are discussed with the goal of stimulating more research in MRI in these areas. Mouse models of autoimmune disease serve as a convenient framework for our discussion. Section II of this chapter surveys prototypical murine models that are widely used in autoimmunity studies and shows how MRI and other imaging modalities have been used in these systems. The models include the non-obese diabetic (NOD) mouse; experimental allergic encephalomyeli-tis (EAE), which is a multiple sclerosis (MS) model; and rheumatoid arthritis (RA) models. After providing a brief description of the model biology, we review key imaging studies using cellular–molecular MRI in the above autoimmune systems. Section III describes several recent advances in cell labeling technologies developed in our laboratory. The first technology uses receptor-mediated endocytosis (RME) to deliver a high concentration of superparamagnetic iron oxide (SPIO) agent to immune cells with minimal effect on the cellular phenotype and function (Ahrens *et al.*, 2003). Next, we describe a promising new approach that uses genetically encoded transgenes to instruct the cell to produce its own intracellular MRI contrast agent (Genove *et al.*, 2005). This new class of agents relies on the expression of iron-binding metalloproteins that impart exogenous contrast to targeted cells. This approach may be useful for highly specific long-term immune cell labeling. Alternatively, this approach could be used for monitoring therapeutic gene delivery to tissues in autoimmune disease models.

II. Imaging Studies in Model Systems

Much of our current understanding of self-recognition in autoimmune disease is dervied from studies in mouse models. Three prototypical mouse models have been widely studied, including NOD, EAE, and RA mice. Numerous *in vivo* imaging studies use these animal models, which serve as valuable platforms for the development of therapies since aspects of their phenotypes mimic the pathogenesis of the corresponding human diseases.

All of these models share a common feature—a destructive proinflammatory response against specific tissues accompanied by an anomalous influx of cellular infiltrates. Noninvasive imaging can be used to visualize these cellular and molecular events. Conventional (^{1}H) anatomical MRI can be highly effective in visualizing nonspecific inflammation with intrinsic contrast mechanisms. As noninvasive imaging methods evolve, so will the ability to read out specific immunobiological information directly from *in vivo* images. To realize these goals, exogenous agents and labeling methods must be used that tag specific cell populations, selectively target specific proteins, or are responsive to key biomolecules, such as nucleic acids. The following brief overview summarizes key studies utilizing MRI and other complementary imaging modalities to investigate immunobiologic and therapeutic aspects of the NOD, EAE, and RA models *in vivo*.

A. Type 1 Diabetes and the Non-Obese Diabetic Mouse

Type 1 diabetes is an autoimmune disease characterized by the destruction of the insulin-producing β cells of the islets of Langerhans (Castano and Eisenbarth, 1990). Prior to the development of diabetes the islets become heavily infiltrated with lymphoid cells, including CD4$^+$, CD8$^+$ T cells, DCs, and monocytes (Jansen *et al.*, 1994; Miyazaki *et al.*, 1985). By the time diabetes appears, more than 90% of the islets have been destroyed by these infiltrates. Therapeutic interventions to prevent diabetes have been aimed at the period of insulitis during which most of the destruction takes place. The NOD mouse is a good model of human type 1 diabetes because it shares many of the genetic and immunological features of the human disease (Leiter *et al.*, 1987). NOD mice spontaneously develop diabetes (females > males) between 15 and 20 weeks of age (Leiter *et al.*, 1987). Genetic analysis of diabetes susceptibility in the NOD mouse has revealed that a minimum of 15 genes are implicated, and one of these is mapped to the major histocompatibility complex (MHC) (Todd and Wicker, 2001). Many of the other genes implicated in its pathogenesis have important functions in the immune system, including genes important in T-cell differentiation and function (Todd and Wicker, 2001).

As early as 4 weeks of age T cells infiltrate the islets of Langerhans and begin to destroy insulin-producing β cells. The disease is mediated by T lymphocytes since the disease can be prevented by treatment with anti–T cell antibodies (Chatenoud *et al.*, 1994; Koike *et al.*, 1987; Shizuru *et al.*, 1988) and can be transferred with T cells from diabetic animals (Wicker *et al.*, 1986). Adoptive transfer of the disease has been shown to require both CD4[+] and CD8[+] T cells (Bendelac *et al.*, 1987; Nagata *et al.*, 1994), but several reports have demonstrated that the disease could be transferred by CD4[+] islet-specific T-cell clones (Daniel *et al.*, 1995; Haskins and McDuffie, 1990; Healey *et al.*, 1995; Zekzer *et al.*, 1998). In addition, the depletion of CD4[+] T cells in NOD mice starting at 90–110 days of age, by which time the mice had already developed insulitis, halted the progression to overt diabetes (Shizuru *et al.*, 1988). Thus, prior to the development of overt disease there is a period of immune-mediated inflammation and tissue destruction.

In early imaging studies NOD mice were injected with radiolabeled interleukin-2 (IL-2), which binds to activated T cells present at sites of inflammation (Rolandsson *et al.*, 2001), but this was not found to be useful diagnostically since IL-2 did not accumulate preferentially in the pancreas. Several recent cellular imaging studies have used MRI to analyze insulitis in the NOD model (Denis *et al.*, 2004; Moore *et al.*, 2004). Denis *et al.* (2004) used long-circulating T2 contrast agents to probe the microvascular changes accompanying inflammation. These agents were composite particles comprised of dextran-coated superparamagnetic iron oxide (SPIO) nanoparticles with an integrated fluorophore that made it possible to track their accumulation using MRI and later histologically. The nanoparticles persist in the circulation for longer than 10 hours and accumulate in areas of inflammation. Particles left the vasculature in areas of insulitis, where they were rapidly taken up by CD11b[+]/CD11c[−] macrophages in the vicinity. This technique was capable of detecting early insulitic lesions, but once the disease had progressed to more established lesions, the differences between NOD and diabetic-resistant mice were no longer apparent.

T cells also have been used as therapeutic agents in NOD mice (Salomon *et al.*, 2000); several studies have reported the use of MRI to track the distribution of T cells following intravenous administration (Moore *et al.*, 2002, 2004). These studies focused on pathogenic T cells that would be expected to infiltrate the pancreas. In one study CLIO-Tat particles (see Chapter 1) were used to label splenocytes from a diabetic NOD mouse that were then transferred to a healthy NOD-SCID (severe combined immune deficiency) mouse. The pancreas was removed and the cells could be visualized in the islets (Moore *et al.*, 2002). A more effective study (Moore *et al.*, 2004) used the fact that a CD8[+] T cell specific for an islet antigen represents the dominant population of cells found in early insulitic lesions in NOD mice (Kita *et al.*, 2003). In this study Moore *et al.* designed a label that would specifically bind to this autoreactive T cell. This consisted of CLIO nanoparticle coupled to avidin fluorescein isothiocyanate (FITC) to which peptide–MHC complexes were attached. This takes advantage of the well-known MHC–peptide tetramer technology (Kita *et al.*, 2003) to track antigen-specific T cells since these reagents will only bind to T cells expressing the appropriate T-cell receptor. This agent labeled $CD8^+$ T cells efficiently, without altering the cells' function, and following transfer into a NOD mouse, they were visualized by MRI in the pancreas (Moore *et al.*, 2004). It remains to be seen whether this agent can be used to track spontaneous insulitis in NOD mice.

Bioluminescence has been used to track the survival of islet grafts in NOD-SCID mice (Lu *et al.*, 2004). In these studies isolated islets were engineered using lentiviral or adenoviral vectors to express a bioluminescent reporter gene. The islets were transplanted into diabetic animals, and no difference was observed between transduced or nontransduced islets in the return to normoglycemia. Lentiviral vectors were superior to adenoviral vectors in terms of imaging since strong bioluminescence signals could be detected for as long as 140 days after transplant (Lu *et al.*, 2004).

Over the years our laboratory has developed cellular therapeutics that can influence the course of insulitis in NOD mice; with the help of vital imaging we are now poised to further investigate the mechanism by which these cells perform their function *in vivo*. Our focus has been on the role of DC subsets in the pathogenesis and therapy of diabetes in NOD mice (Feili-Hariri *et al.*, 1999, 2002, 2003). We have found that a single injection of bone marrow–derived DCs can protect young prediabetic NOD mice from the development of diabetes (Feili-Hariri *et al.*, 1999, 2002, 2003). The therapeutic DC populations expressed high levels of co-stimulatory molecules (CD80, CD86, and CD40) and produced low levels of IL-12p70 following CD40 ligation, whereas a nontherapeutic bone marrow–derived DC population expresses low levels of co-stimulatory molecules (Feili-Hariri and Morel, 2001). Interestingly, the therapeutic DC expresses higher levels of several chemokines and chemokine receptors, which are likely to influence the trafficking ability of these cells. When we performed fluorescent-based imaging using confocal or two-photon microscopy, we could identify DCs in the pancreas, pancreatic lymph nodes, and spleen (Feili-Hariri *et al.*, 1999, 2003). These techniques have not allowed us to quantitate differences between the two DC populations in terms of trafficking and thus we are developing MRI-based technologies to do this (see Section III). Our study of DC trafficking in NOD is just one example of the potential use of *in vivo* imaging in the diabetic model to address fundamental questions about the disease and its treatment.

B. Experimental Allergic Encephalomyelitis

An inappropriate immune response to polypeptides found in the central nervous system (CNS) can have profoundly debilitating effects, of which the human disease MS is a clear example. MS is a demyelinating autoimmune disease involving recognition of myelin proteins by T cells followed by a cascade of destructive actions by cells of the immune system (Steinman, 1996). Although MS is associated with certain genetic markers, the etiology of the disease remains a mystery (Steinman, 1996).

Much of our current understanding about MS self-recognition in autoimmune disease has been generated from studies in EAE, an animal model with many clinical and histopathological similarities to MS (Owens and Sriram, 1995; Raine, 1984; Tuohy *et al.*, 1987; Zamvil and Steinman, 1990). EAE has been studied in numerous species (e.g., mouse, rat, guinea pig, monkeys). EAE is most commonly induced by immunizing animals with myelin proteins or their disease-inducing peptides, often referred to as encephalitogenic determinants. The most common proteins used for inducing EAE include myelin proteolipid protein (PLP), myelin basic protein (MBP), and myelin oligodendrocyte glycoprotein (MOG). Clinical disease develops when primed $CD4^+$ T cells enter the CNS and recognize their cognate self-determinant presented in the context of MHC class II molecules. The resulting perivascular and parenchymal infiltrations in the CNS often lead to clinical paralysis, demyelination, and permanent disability.

In mouse models, EAE can be induced by immunization with a variety of antigens (reviewed by Anderson and Karlsson, 2004; Martin and McFarland, 1995; Martin *et al.*, 1992; Swanborg, 1995), particularly PLP and MOG. One widely studied mouse model uses a peptide of PLP to induce the disease (Tuohy *et al.*, 1989). After immunization with the immunodominant PLP 139–151 or transfer of PLP 139–151–activated $CD4^+$ T cells into naive recipients, SJL/J and SWXJ mice develop acute EAE followed by a relapsing–remitting clinical course with each relapse progressively more severe and with each remission leaving mice progressively more impaired (Tuohy *et al.*, 1989; Yu *et al.*, 1996); severe CNS demyelination is observed histologically, particularly in the spinal cord. Inoculation with MOG or immunodominant peptides can also elicit severe relapsing–remitting EAE in the several mouse strains (Amor *et al.*, 1994; Mendel *et al.*, 1995). Another method used to induce chronic demyelinating disease in the mouse is by inoculating the brain with Theiler's murine encephalomyelitis virus (TMEV) (Miller *et al.*, 1997). This picornavirus is a natural mouse pathogen that induces a chronic demyelinating disease. The clinical symptoms and neurohistopathology are similar to that of EAE. Demyelination in TMEV-transduced SJL/J mice is initiated by an inflammatory response that is mediated by virus-specific $CD4^+$ T cells (Miller *et al.*, 1997).

Although inoculation models of EAE have formed the basis of many important findings, key aspects of the disease cannot be addressed with these models; one of the criticisms is that autoimmunity must be induced by immunization, whereas MS arises spontaneously. Motivated by these concerns, transgenic mice have been generated that express high levels of rearranged T-cell receptor α and β transgenes (Goverman *et al.*, 1993). These form a T-cell receptor (TCR) that is specific for an MBP epitope (Ac1–11). This transgenic mouse spontaneously acquires EAE in certain environments without the need for inoculation (Goverman *et al.*, 1993). The spontaneous triggering of the disease makes these transgenic animals a closer model to human MS in some respects because it offers the intriguing possibility that the environmental factors (e.g., viral factors) initiating the model disease might be identified.

As with MS, MRI is a powerful technique used to locate the sites and assess the level of activity of EAE lesions. Early studies (Karlik *et al.*, 1990; Steward *et al.*, 1985) report increases in T1 and T2 in white matter regions containing EAE lesions, and these appear hypointense and hyperintense in T1- and T2-weighted images, respectively. These observations are not cell or pathology specific and are consistent with the presence of demyelination, inflammation, and edema. Figure 1b is an example of a T2-weighted image showing lesions in the EAE mouse spinal cord. The use of intravenous contrast agents (e.g., gadolinium–diethylenetriamine penta-acetic acid [Gd-DTPA]) have been effective in elucidating active regions of breakdown in the blood–brain barrier (Hawkins *et al.*, 1990, 1991; Karlik *et al.*, 1993; Morrissey *et al.*, 1996; Namer *et al.*, 1992, 1993). Diffusion-weighted images (DWIs) and apparent diffusion coefficient (ADC) maps have been used to elucidate EAE lesions. Heide *et al.* (1993) observed changes in DWIs on or before the day lesions became apparent in T2-weighted images. Verhoye *et al.* (1996) reported a significant correlation between increased ADCs and clinical score within white matter. Ahrens *et al.* (1998) first used diffusion tensor imaging to assay the pathologic state of EAE lesions in the transgenic EAE mouse model (Fig. 1c and d).

More recently MRI has been used to gain insights into fundamental cellular and biochemical mechanisms of EAE. With the help of exogenous MRI agents, cellular and molecular aspects of EAE have been investigated by several groups. Experiments in rat monitored the infiltration of inflammatory cells into the CNS following *in situ* labeling of macrophages with SPIO particles (Dousset *et al.*, 1999; Rausch *et al.*, 2003). The SPIO, initially administered intravenously, could be detected within 24 hours *in vivo* within lesions; electron microscopy analyses in lesions revealed the presence of SPIO in cells with macrophage morphology. In related experiments, monocyte infiltration into the CNS was monitored using *in vivo* [19]F MRI following intravenous inoculation of emulsion nanoparticles of perfluoro-15-crown-5-ether (Noth *et al.*, 1997) that were then taken up by macrophages. Pirko *et al.* (2003), using the TMEV mouse model, injected SPIO conjugated to monoclonal antibodies specific for the T-cell surface markers $CD4^+$ and $CD8^+$. Specific binding to these T-cell subsets *in vivo* was claimed, resulting in selective contrast enhancement in lesional regions (Pirko *et al.*, 2003). Adoptively

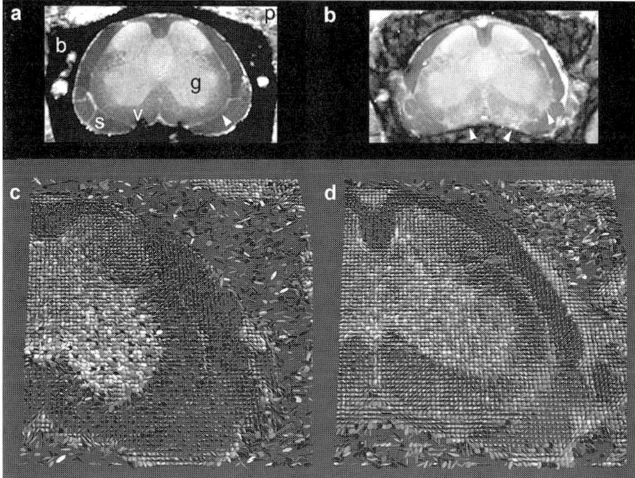

Figure 1 MRI in control and EAE mouse spinal cord using T2-weighted (a and b) and diffusion tensor imaging (DTI (c-d). Panel a, control B10.PL mice; various anatomical regions are indicated, including trabecula bone (b), ventral column white matter (v), gray matter (g), spinal cord boundary (arrow), spinal nerves (s), and phosphate-buffered saline (p) inside the capillary tube surrounding the fixed vertebral segment. In the diseased spinal cords (panel b) taken from the EAE transgenic model (stage 3), lesions are clearly identifiable within and surrounding white matter and appear as regions of hyperintensity; this is consistent with the presence of demyelination, inflammatory infiltrates, and edema. Lesions appear primarily near the ventral median fissure, ventral column, meningeal lining, and major spinal vessels. Renderings of the DTI data in the control and EAE spinal cords are shown in panels c and d, respectively. (Panels a and b are the corresponding anatomical images.) These renderings are effective in representing the relative diffusion anisotropy on a voxel-by-voxel basis. The shape of the diffusion ellipsoid represents diffusion anisotropy in three-dimensions. Its long axis is oriented in the direction of the fastest diffusion, or along the direction of the axon tract. In c and d, normal white matter and spinal nerves appear as prolate (cigar-shaped) ellipsoids, indicating highly anisotropic diffusion. These are oriented primarily along the spinal axis, as is expected from the known fiber organization. Gray matter, fluid-filled regions, and regions of white matter lesions have reduced anisotropy, and thus, the ellipsoids appear more spherical. An important observation is that regions of lesions (d) show greatly reduced anisotropy in the same regions that appear hyperintense in anatomical images (b), especially around the meningeal boundary, the ventral median fissure, and the ventral columns. This is a signature of the presence of inflammatory cells and edema (Ahrens *et al.*, 1998). Panels a and b are T2-weighted spin-echo images calculated from the two-dimensional diffusion-weighted images using to fit the DTIs with repetition time = 2000 ms, echo time = 17 ms, and 20 × 20 × 300 μm resolution. Images were acquired in fixed dorsal columns at 11.7 T (see Ahrens *et al.*, 1998, for additional details).

transferred T cells, either labeled or unlabeled with SPIO, were visualized in the EAE mouse spinal cord in studies by Anderson *et al.* (2004). These results showed that the activated T cells, which were SPIO labeled, could effectively transfer the disease to a naive recipient and could also be imaged within lesions following disease induction. During the onset of EAE, there is an upregulation of intercellular adhesion molecule-1 (ICAM-1) on the vascular endothelium. This leukocyte receptor aids in the initial adhesion and subsequent entry of immune cells into the CNS. Sipkins *et al.* (2000) constructed antibody-conjugated paramagnetic liposomes that target ICAM-1. EAE mice that were infused with this agent showed significant signal enhancement in fixed brains imaged *ex vivo* compared to controls. Overall, there is a wide range of potential MRI techniques and molecular targets that can be investigated in the EAE model.

C. Rheumatoid Arthritis

RA is a multisystem autoimmune disease characterized by destructive inflammation of the joints and circulating autoantibodies. The disease is believed to be initiated by recognition of self-antigen autoreactive $CD4^+$ T cells leading to polyclonal B-cell activation, followed by the recruitment of inflammatory mediators and cells to the joints. Recent studies have identified several genetic factors that may lead to the initiation of the autoimmune process, including the expression of particular HLA-DR alleles (Winchester *et al.*, 1992). In addition, several important inflammatory mediators, such as IL-1 and tumor necrosis factor (TNF)-α, have been implicated in the perpetuation of the inflammatory state (Firestein, 1991). A common feature of RA is the presence of autoantibodies, the best known of which is rheumatoid factor (RF). This antibody is directed towards the Fc portion of IgG and can be both of IgM and IgG isotypes (Moore and Dorner, 1993). The role of RF and IgG immune complexes in the pathogenesis of TA is not fully understood, but they are likely to interact with FcγR on the surface of neutrophils, NK cells, DCs, and monocyte/macrophages. These cells are all potent producers of inflammatory cytokines, and NK cells and DCs are also important in the regulation of the inflammatory response (Mailliard *et al.*, 2003).

Several animal models have been developed to study the pathogenesis and treatment of RA. In general, these have involved the induction of disease in susceptible mouse or rat strains through the deliberate immunization with joint constituents, such as collagen (Duris *et al.*, 1994) or the infusion of cationic immune complexes (van Lent *et al.*, 1992) or adjuvants (van Eden and Waksman, 2003). Recently, a novel spontaneous model of arthritis has been developed (Kouskoff *et al.*, 1996). In this model, TCR transgenic mice were bred onto the NOD background and the offspring was found to spontaneously develop a severe, unremitting joint disease similar to human RA. One distinguishing feature is that these mice do not develop RA, but recent studies have shown that the disease can be transferred to Rag$-$/$-$ mice with purified IgG demonstrating the arthritogenic potential of these antibodies (Korganow *et al.*, 1999). All of these models are characterized by intense inflammation in the synovial lining of the joints resulting from the accumulation of macrophages, lymphocytes, and neutrophils. Macrophages are the major source of inflammatory cytokines such as TNF-α and IL-1 that ultimately lead to the loss of cartilage and joint destruction.

One of the challenges in the treatment and management of patients with RA is the ability to assess the degree of tissue inflammation and disease severity. Traditional radiological and MRI imaging provide anatomical information concerning the degree of joint narrowing or erosions that occur late in the course of the disease. With the advent of novel immunotherapies (Moreland *et al.*, 1997) that target the inflammatory cytokines secreted by macrophages, it has become important to develop strategies to monitor inflammatory events in the joint. In the past several years new techniques for imaging affected joints have been developed. Since macrophages are important cells in the inflammatory process, several studies have described the use of MRI to label macrophage migration into the joint (Beckmann *et al.*, 2003; Dardzinski *et al.*, 2001; Lutz *et al.*, 2004). In these studies mice (Dardzinski *et al.*, 2001), rats (Beckmann *et al.*, 2003), or rabbits (Lutz *et al.*, 2004) were injected with SPIO nanoparticles following the induction of arthritis on the premise that the phagocytic macrophages would preferentially take up the particles (Dousset *et al.*, 1999). These three studies all demonstrated increased uptake of SPIO particles in arthritic joints and not in control nonarthritic joints, suggesting that this approach might be useful in the assessment of macrophage infiltration into affected tissues.

In addition to MRI, several other *in vivo* imaging techniques have been used in studies of RA. The technique of near-infrared fluorescence has been applied to the imaging of arthritic joints in recent years (Hansch *et al.*, 2004a,b; Wunder *et al.*, 2004). This technique involves the use of fluorescent probes, such as Cy5.5, that emit in the far-red region of the spectrum (~700 nm) and have less absorbance by biological tissue. One study examined infiltrations into the joint using Cy5.5-conjugated F4/80 antibody that targets macrophages (Hansch *et al.*, 2004b). This group also directly injected the Cy5.5 dye as a means to detect inflammation in the joint, arguing that increased vascular permeability in affected joints would allow the dye to enter and be taken up by resident macrophages (Hansch *et al.*, 2004a). In this study the authors observed that the Cy5.5 dye bound albumin in serum and that the presence of Cy5.5 in the affected joint was due to an increased permeability to albumin. The presence of cartilage-degrading proteases is a hallmark of the inflamed joint, and recently a probe has been developed that takes advantage of this (Wunder *et al.*, 2004). In this study a probe was created that requires proteolytic cleavage in order to release the fluorescent Cy5.5. This technique proved very effective in detecting affected joints in a mouse model of collagen-induced arthritis and was also able to detect changes in protease activity following the treatment of the mice with methotrexate (Wunder *et al.*, 2004).

MicroPET has also been used in the spontaneous murine model of arthritis (Wipke *et al.*, 2002). In this model it has been shown that disease can be transferred with an antibody specific for the autoantigen in the disease, glucose-6-phosphate isomerase (GPI). Since GPI is ubiquitously expressed, it was not clear why joints should be targeted in this autoimmune response. Using ^{64}Cu-labeled anti-GPI antibodies and microPET, Wipke *et al.* demonstrated that anti-GPI antibodies accumulate in the distal joints within 10 minutes of injection and that they persist for more than 24 hours. This distribution correlated with the pattern of arthritis seen in this model. This study changes the concept of autoimmunity by demonstrating that an autoimmune response to a antigen whose expression is not tissue specific can lead to tissue-specific autoimmune disease.

Bioluminescence has been used in the collagen-induced arthritis model to track transferred T cells that have been engineered to express potentially therapeutic genes (Tarner *et al.*, 2002). This approach takes advantage of the fact that specific T cells will traffic to sites of inflammation and can be used as delivery vehicles for therapeutic agents such as cytokines (Slavin *et al.*, 2002). This was tested in the model of collagen-induced arthritis (Tarner *et al.*, 2002). Collagen-specific T-cell hybridomas were retrovirally transduced to express IL-4 or a GFP-luciferase gene. Bioluminescence imaging of mice that had received the transduced cells demonstrated specific accumulation of these cells in the joints 3 days after infusion (Tarner *et al.*, 2002). Use of cells that expressed IL-4 also resulted in an improvement in the arthritis in the treated mice (Tarner *et al.*, 2002).

III. Recent Advances in Immune Cell Labeling with MRI

A. Receptor-Mediated Endocytosis of SPIO

SPIO agents have attracted great interest for a wide range of *in vivo* cellular MRI tracking studies. A detailed review of the SPIO labeling methods and agent compositions is described elsewhere (see Chapter 1 and (Bulte *et al.*, 2004). However, we describe one SPIO labeling method recently investigated in our laboratory using the receptor-mediated endocytosis (RME) uptake mechanism that can be applied to models of autoimmunity

(Ahrens *et al.*, 2003). This method tags phentypically defined population of immune cells in culture at high efficiency. It relies on targeting specific surface accessory molecules with SPIO agents that are conjugated to mono-clonal antibodies (mAb). After a short incubation period under physiological conditions, the mAb–SPIO complex is internalized into the cell. The cells are then transplanted into the subject.

Using the RME mechanism to label immune cells offers several key advantages. The RME approach offers a high uptake efficiency using a modest incubation time (~1–3 hours). The particle uptake efficiency is the same order of magnitude as reported for the peptide-conjugated nanoparti-cles (Josephson *et al.*, 1999) or transfection agents (Frank *et al.*, 2003; Hoehn *et al.*, 2002). Peptide- and transfection-based delivery can enter a wide range of cell types. In contrast, RME uptake is selective to a specific cell-surface phenotype, and this may be more desirable in targeting only a single, or set of phenotypes, in a mixed population of cultured cells. Importantly, RME of SPIO does not significantly alter the cell's immunological phenotype or function (Ahrens *et al.*, 2003) nor does it cause cytotoxicity. These factors can be a concern when using transfection agents to label primary immune cells. An additional advantage of using mAb–SPIO agents for immune cells is that these agents are commercially available and are widely used in magnetic cell sorting applications. Thus, a large selection of specific anti-bodies coupled to SPIO particles is readily available. Additionally, research-ers can combine positive-selection magnetic cell sorting techniques, followed by a 37 °C incubation period of the recovered cell fraction, to efficiently sort and label the cells for MRI.

Studies in our laboratory have focused on using RME to label DCs. DCs are known to be the most efficient antigen-presenting cells (APCs) and are capable of stimulating naive T cells to initiate an immune response. We are interested in visualizing the trafficking patterns of DCs in the NOD mouse, where abnormalities have been reported in the number and function of DCs (Morel and Feili-Hariri, 2001). Immature DCs actively take up soluble proteins and small particles by an active process known as macropinocyto-sis. SPIO particles are typically ~50 nm in size, and thus these would not be taken up by DCs via macropinocytosis; however, high-efficiency labeling can be achieved by using RME (Ahrens *et al.*, 2003). In our studies we targeted the CD11c surface molecule, which is expressed at high levels on DCs, using an mAb–SPIO complex. The result was a concentrated particle uptake by the DCs, as indicated by relaxation time measurements in cell pellets, with no adverse effects on immunological phenotype and function (Ahrens *et al.*, 2003). Electron microscopy was used to confirm the intracel-lular incorporation of the SPIO (Ahrens *et al.*, 2003). The labeled cells can be visualized *in vivo* by MRI for several days (Fig. 2).

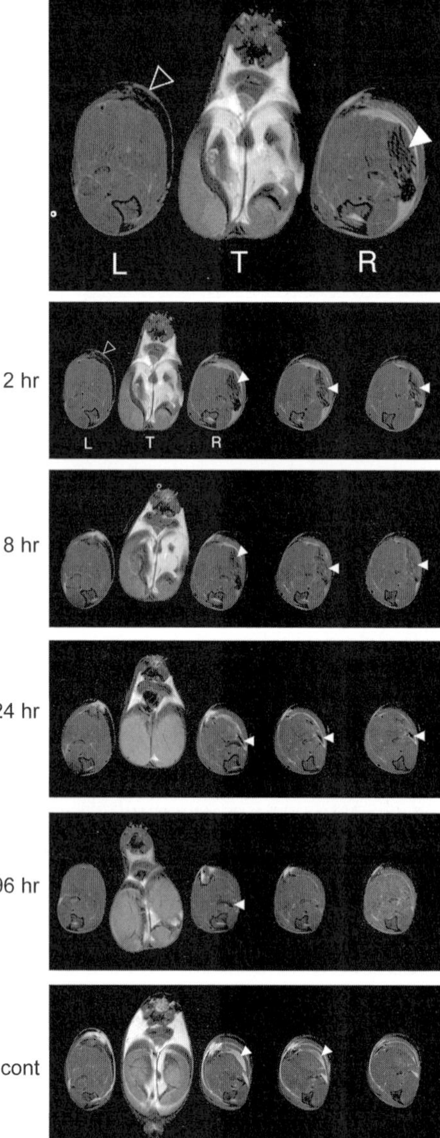

Figure 2 *In vivo* MRI data showing SPIO-labeled DCs in mouse quadriceps. DCs were labeled *ex vivo* by receptor-mediated endocytosis (RME) uptake via the CD11c surface marker and then injected intramuscularly. The top panel shows an axial slice through the left and right legs (labeled L and R, respectively) and testis (T) at 2 hours after injection. Region of hypointensity from (3×10^6) labeled cells are indicated by the solid arrow. The neighboring hyperintensity is excess phosphate-buffered saline. The left leg was not injected. This same animal was followed for 96 hours with serial MRI. The time series is shown in the lower panels. At each time point

B. Nucleic Acid–Based MRI Reporters

SPIO nanoparticles are effective in visualizing immune cell infiltration during autoimmune disease in a variety of cases. However, there are several disadvantages to using SPIO as intracellular agents:

1. Targeting the agents to specific cell populations *in situ* with high specificity can be difficult.
2. The mean intracellular agent concentration is diluted by every cell division, which tends to diminish the image contrast of the cells over time.
3. The SPIO particles can be degraded by the cell because they normally reside in low-pH vesicles (i.e. lysosomal compartments) (Okon *et al.*, 1994).
4. When SPIO-labeled cells die, the SPIO nanoparticles may be taken up by resident phagocytic cells, resulting in nonspecific macrophage labeling.

Recently intracellular labeling approaches have been devised using genetically encoded metalloproteins that avoid many of limitations imposed by exogenous metal-complexed agents such as SPIO (Cohen *et al.*, 2005; Genove *et al.*, 2005). The goal is to induce the expression of metalloproteins from the ferritin family into specific host tissues using a vector (Genove *et al.*, 2005). The ferritin protein is made superparamagnetic by sequestering endogenous iron from the organism. In this novel approach, the MRI "contrast agent" is assembled *in situ* via genetic instructions introduced by the vector (Genove *et al.*, 2005). Moreover, by combining this MRI reporter with another transgene of interest (e.g., a therapeutic gene), it is feasible to visualize transgene delivery to cells (Genove *et al.*, 2005). Conceptually, these applications are similar to those using green fluorescent protein, where fluorescent expression patterns of cell populations can be controlled by genetic means.

---

two additional contiguous slices are shown for R in proximity to the injection point. Regions of hypointensity created by labeled cells (solid arrows) are easily observed at 2, 8, and 24 hours after injection and tend to diminish over time. At later times the number of apparent labeled DCs diminished, suggesting that the cells had either migrated to a distant site or that they had died and the SPIO particles had dispersed into resident phagocytic cells. The bottom panels (cont) shows contiguous slices in a mouse injected 2 hours earlier with unlabeled DCs. The arrow indicates the hyperintensity from PBS. Due to the high magnetic field strength used in these experiments, chemical shift artifacts from fat are readily visualized; an example of this is indicated in the left leg with the open arrow (top panel). Images were acquired using an 11.7 T MRI system, a two-dimensional ferritin spin-echo sequence, repetition time = 1500 ms, echo time = 30 ms, in-plane resolution 49 μm, and 0.6-mm thick slices. For additional details see Ahrens *et al.*, 2003.

Ferritin is a natural prototype molecule for developing MRI reporters. Ferritin is ubiquitous and highly conserved throughout almost all organisms (Theil, 1987). It is primarily responsible for the storage of intracellular Fe in a nontoxic (Fe^{3+}) form (Corsi *et al.*, 1998; Epsztejn *et al.*, 1999) and for physiological homeostasis of Fe metabolism (Levi *et al.*, 1992; Santambrogio *et al.*, 1993). Ferritin has a crystalline ferrihydrite core that exhibits properties of superparamagnetism (Bulte *et al.*, 1994). Ferritin has a marked effect on solvent nuclear magnetic resonance (NMR) relaxation rates (Bulte *et al.*, 1994; Gillis and Koenig, 1987; Gottesfeld and Neeman, 1996; Vymazal *et al.*, 1996, 1998). Naturally occurring ferritin in cells is a source of intrinsic MRI contrast seen in various tissues and organs (Gossuin *et al.*, 2004; Vymazal *et al.*, 1996). Ferritin is a natural by-product of the intracellular degradation of many SPIO contrast agent compositions that are currently used for cellular imaging. Over time these agents are degraded in the low-pH environment of lysosomal compartments, and ferritin is upregulated as a consequence of this degradation (Okon *et al.*, 1994). Thus, when SPIO is used for long-term cell tracking studies *in vivo*, it is likely that at least some of the MRI contrast is produced by ferritin and not the original SPIO composition.

The advantages of using genetically encoded tags for MRI of immune cells are several-fold:

1. No exogenous metal-complexed contrast agent is required, thereby simplifying intracellular delivery because an efficient vector can be used to deliver only the transgene.
2. Because the contrast agent is genetically encoded, the cell and its progeny can be instructed to produce the agent for extended time periods; this can mitigate the effect of contrast reduction due to dilution and agent degradation.
3. With genetically encoded reporters, when the cell dies the reporter transgenes are destroyed, as are the contrasting metalloproteins via efficient proteases, thus the contrast agent will not be passed to other cells.

Recent results from our laboratory have investigated the use of human ferritin to label cells for *in vivo* MRI (Genove *et al.*, 2005). The ferritin transgenes were introduced to cells via an efficient recombinant adenovirus vector (AdV). The impact of reporter expression on the spin-spin NMR relaxation rate (1/T2) is a function of both the intracellular ferritin concentration and the amount of Fe loaded into the ferritin cores (Vymazal *et al.*, 1996). To initially detect relative Fe loading in transduced cells *in vitro*, we measured 1/T2 in pelleted A549 cells. We observed a significantly enhanced 1/T2 by approximately 2.5-fold in virus-transduced cells compared with control cells in culture. By performing T2-weighted MRI on the pellets,

we confirmed that the 1/T2 changes due to FT transgene expression correlated with significant image contrast *in vitro* (Genove *et al.*, 2005).

To evaluate the true potential of this transgene imaging approach *in vivo*,
we visualized MRI reporter expression in mouse. We injected adenovirus
containing the MRI reporter stereotactically into the striatum and then
imaged the mice ($n = 5$) at 5, 11, and 39, days after inoculation (Genove
et al., 2005). After 5 days, transduced cells displayed robust contrast in both
T2-weighted images (left arrows in Fig. 3a), and the contrast could be seen
up 39 days (Genove *et al.*, 2005). No significant contrast could be detected
on the contralateral side injected with the AdV-lacZ control vector (right
arrowsin Fig. 3a). At 5 days after transduction, we performed histologic
studies in selected brains. The X-Gal staining pattern of the AdV-lacZ
inoculation mimicked the AdV-FT–induced MRI contrast (Fig. 3b). In the
same brains used for MRI, immunohistochemistry was performed to detect
human ferritin expression in the striatum (Fig. 3c). The spatial pattern of
recombinant ferritin expression was consistent with the MRI (Fig. 3a).

Metalloprotein-based MRI reporters represent a novel approach for labeling immune and other cells types (e.g., stem cells), but more studies are
need to evaluate its true potential. This approach is appealing because it

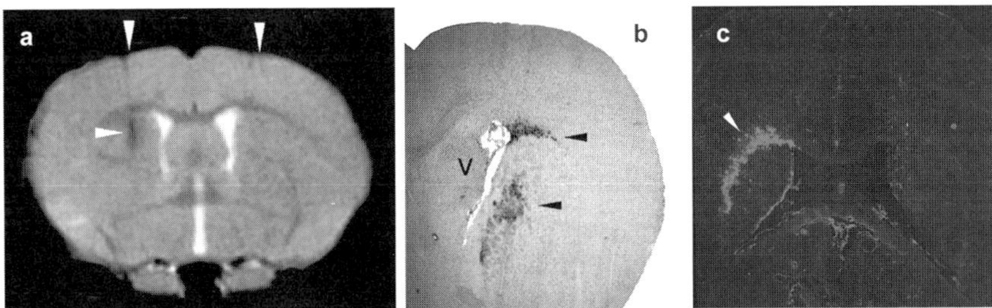

Figure 3 *In vivo* longitudinal results of MRI reporter expression in the mouse brain.
Adenovirus-containing reporter transgenes encoding ferritin subunits were injected into the
striatum. Panel a shows the T2-weighted image 5 days after injection showing the injection sites
(arrows, MRI reporters left, AdV-lacZ control right). The MRI reporter–transduced cells
appear hypointense, while the cells with LacZ show no contrast in MRI. Panel b shows the X-
Gal–stained AdV-lacZ–transduced pattern at 5 days after transduction. In b, the staining
pattern, similar to the MRI, is predominantly in white matter (top arrow) and striatum (bottom
arrow), where v denotes ventricle. Panel c shows immunohistochemistry (IHC) results of ferritin
transgene expression in mouse brain slices at 5 days after inoculation. Adenovirus containing
the MRI reporters was injected into the left striatum and appears immunopositive in IHC.
AdV-lacZ was injected in the contralateral side and shows no staining above background.
In vivo images (a) were obtained in anesthetized mice at 11.7 T. Coronal slices were acquired at
the injection site using a T2-weighted spin-echo sequence with repetition time = 1200 ms, echo
time = 35 ms, 0.75-mm slice thickness, and in-plane resolution of 98 μm. For additional details
see Genove *et al.*, 2005. (See Color Insert.)

combines powerful "off the shelf" molecular biological tools and MRI to modulate contrast in targeted cells. With these constructs long-term cell labeling is possible. Alternatively, the MRI reporters could be combined with inducible promoters to "turn on" contrast at key times, for example, after immunotherapeutic or stem cells are believed to have migrated to and reconstituted specific (e.g., diseased) tissues. Furthermore, by uniting an MRI reporter with a gene of interest, a multitude of applications exist. In the near-term these include visualizing patterns of gene expression in model systems, such as following the delivery of engineered vectors, or for use in imaging transgene expression in genetically manipulated animals.

IV. Conclusions and Future Directions

Noninvasive *in vivo* imaging techniques such as MRI, in conjunction with various emerging cell labeling techniques, will be valuable for studies of animal models of autoimmune disease and other diseases such as cancer. Autoimmune diseases are characterized by infiltration of affected tissues with immune cells. There has been rapid progress in the development of new immune cell labeling approaches that can be used to monitor disease progression in models of type 1 diabetes, MS, and RA, particularly the trafficking of specific immune cell types such as macrophages or T cells.

For cellular studies of autoimmunity, *ex vivo* labeling of immune cells can be an additional treatment to the cells prior to infusion into the animal. Labeling can achieved via direct uptake of reagents, such as by using RME of SPIO particles, or by genetic modification of cells such that they make their own contrast agents. Labeled cells are then introduced *in vivo* by a variety of routes. Delivery methods involving the direct implantation of cells into tissues or organs yield more sensitivity and may be preferable to a systemic delivery; in the later case, false positives in ^{1}H anatomical images can confound the interpretation of cell localization, particularly if the biodistribution is not know *a priori*.

The labeling schemes used in the study of autoimmunity can also be used to monitor transplanted cells used as part of a cell-based therapeutic. Cellular therapeutics are currently being applied to various diseases, including cancer, neurological, hematological, and immunological disorders. We have discussed the use of DCs as cellular therapeutics in autoimmune disease, but DCs are also being used as vaccines in cancer patients. DCs are the most efficient antigen-presenting cells and have the capacity to stimulate naive T cells to initiate an immune response. DCs pulsed with tumor antigens have been used in many clinical protocols (Figdor *et al.*, 2004). Imaging methods that will allow tracking of injected DC to the tumor sites as well as lymph nodes will be beneficial in the monitoring of the efficacy of these new therapies. In addition, newer approaches to cancer vaccines designed to target specific DC subsets *in* vivo could also benefit from cellular imaging technologies.

In conclusion, we are entering a new era where MRI will increasingly be used to monitor ongoing inflammatory processes *in vivo*. Additionally, researchers will be able to determine the trafficking patterns of different cell types currently being used, or under development, for therapies in many immune-mediated diseases.

References

Ahrens, E. T., Feili-Hariri, M., Xu, H., Genove, G., and Morel, P. A. (2003). Receptor-mediated endocytosis of iron-oxide particles provides efficient labeling of dendritic cells for *in vivo* MR imaging. *Magn. Reson. Med.* **49**, 1006–1013.

Ahrens, E. T., Laidlaw, D. H., Readhead, C., Brosnan, C. F., Fraser, S. E., and Jacobs, R. E. (1998). MR microscopy of transgenic mice that spontaneously acquire experimental allergic encephalomyelitis. *Magn. Reson. Med.* **40**, 119–132.

Amor, S., Groome, N., Linington, C., Morris, M. M., Dornmair, K., Gardinier, M. V., Matthieu, J. M., and Baker, D. (1994). Identification of epitopes of myelin oligodendrocyte glycoprotein for the induction of experimental allergic encephalomyelitis in SJL and Biozzi AB/H Mice. *J. Immunol.* **153**, 4349–4356.

Anderson, A., and Karlsson, J. (2004). Genetics of experimental autoimmune encephalomyelitis in the mouse. *Archivum Immunologiae Et Therapiae Experimentalis* **52**, 316–325.

Anderson, S. A., Shukaliak-Quandt, J., Jordan, E. K., Arbab, A. S., Martin, R., McFarland, H., and Frank, J. A. (2004). Magnetic resonance imaging of labeled T-cells in a mouse model of multiple sclerosis. *Annals of Neurology* **55**, 654–659.

Beckmann, N., Falk, R., Zurbrugg, S., Dawson, J., and Engelhardt, P. (2003). Macrophage infiltration into the rat knee detected by MRI in a model of antigen-induced arthritis. *Magn. Reson. Med.* **49**, 1047–1055.

Bendelac, A., Carnaud, C., Boitard, C., and Bach, J. F. (1987). Syngeneic transfer of autoimmune diabetes from diabetic NOD mice to healthy neonates. Requirement for both L3T4$^+$ and lyt-2$^+$ T cells. *J. Exp. Med.* **166**, 823–832.

Bulte, J. W., Douglas, T., Mann, S., Frankel, R. B., Moskowitz, B. M., Brooks, R. A., Baumgarner, C. D., Vymazal, J., Strub, M. P., and Frank, J. A. (1994). Magnetoferritin: Characterization of a novel superparamagnetic MR contrast agent. *J. Magn. Reson. Imag.* **4**, 497–505.

Bulte, J. W. M., Arbab, A. S., Douglas, T., and Frank, J. A. (2004). Preparation of magnetically labeled cells for cell tracking by magnetic resonance imaging. *Meth. Enzymol.* **386**, 275–299.

Castano, L., and Eisenbarth, G. S. (1990). Type-I diabetes: A chronic autoimmune disease of human, mouse and rat. *Annu. Rev. Immunol.* **8**, 647–679.

Chatenoud, L., Thervet, E., Primo, J., and Bach, J. F. (1994). Anti-CD3 antibody induces long term remission of overt autoimmunity in nonobese diabetic mice. *Proc. Natl. Acad. Sci. USA* **91**, 123–127.

Cohen, B., Dafni, H., Meir, G., Harmelin, A., and Neeman, M. (2005). Ferritin as an endogenous MRI reporter for noninvasive imaging of gene expression in C6 glioma tumors. *Neoplasia* **7**, 109–117.

Corsi, B., Perrone, F., Bourgeois, M., Beaumont, C., Panzeri, M. C., Cozzi, A., Sangregorio, R., Santambrogio, P., Albertini, A., Arosio, P., and Levi, S. (1998). Transient overexpression of human H- and L-ferritin chains in COS cells. *Biochem. J.* **330,** 315–320.

Daniel, D., Gill, R. G., Schloot, N., and Wegmann, D. (1995). Epitope specificity, cytokine production profile and diabetogenic activity of insulin-specific T cell clones isolated from NOD mice. *Eur. J. Immunol.* **25,** 1056–1062.

Dardzinski, B. J., Schmithorst, V. J., Holland, S. K., Boivin, G. P., Imagawa, T., Watanabe, S., Lewis, J. M., and Hirsch, R. (2001). MR imaging of murine arthritis using ultrasmall superparamagnetic iron oxide particles. *Magn. Reson. Imag.* **19,** 1209–1216.

Denis, M. C., Mahmood, U., Benoist, C., Mathis, D., and Weissleder, R. (2004). Imaging inflammation of the pancreatic islets in type 1 diabetes. *Proc. Natl. Acad. Sci. USA* **101,** 12634–12639.

Dousset, V., Delalande, C., Ballarino, L., Quesson, B., Seilhan, D., Coussemacq, M., Thiaudiere, E., Brochet, B., Canioni, P., and Caille, J. M. (1999). *In vivo* macrophage activity imaging in the central nervous system detected by magnetic resonance. *Magn. Reson. Med.* **41,** 329–333.

Duris, F. H., Fava, R. A., and Noelle, R. J. (1994). Collagen-induced arthritis as a model of rheumatoid arthritis. *Clin. Immunol. Immunopathol.* **73,** 11–18.

Epsztejn, S., Glickstein, H., Picard, V., Slotki, I. N., Breuer, W., Beaumont, C., and Cabantchik, Z. I. (1999). H-ferritin subunit overexpression in erythroid cells reduces the oxidative stress response and induces multidrug resistance properties. *Blood* **94,** 3593–3603.

Feili-Hariri, M., Dong, X., Alber, S. M., Watkins, S. C., Salter, R. D., and Morel, P. A. (1999). Immunotherapy of NOD mice with bone marrow-derived dendritic cells. *Diabetes* **48,** 2300–2308.

Feili-Hariri, M., Falkner, D. H., Gambotto, A., Papworth, G. D., Watkins, S. C., Robbins, P. D., and Morel, P. A. (2003). Dendritic cells transduced to express interleukin-4 prevent diabetes in nonobese diabetic mice with advanced insulitis. *Human Gene Therapy* **14,** 13–23.

Feili-Hariri, M., Falkner, D. H., and Morel, P. A. (2002). Regulatory Th2 response induced following adoptive transfer of dendritic cells in prediabetic NOD mice. *Eur. J. Immunol.* **32,** 2021–2030.

Feili-Hariri, M., and Morel, P. A. (2001). Phenotypic and functional characteristics of BM-derived DC from NOD and non diabetes-prone strains. *Clin. Immunol.* **98,** 133–142.

Figdor, C. G., de Vries, I. J. M., Lesterhuis, W. J., and Melief, C. J. M. (2004). Dendritic cell immunotherapy: Mapping the way. *Nat. Med.* **10,** 475–480.

Firestein, G. S. (1991). The immunopathogenesis of rheumatoid arthritis. *Curr. Opin. Rheumatol.* **3,** 398–406.

Frank, J. A., Miller, B. R., Arbab, A. S., Zywicke, H. A., Jordan, E. K., Lewis, B. K., Bryant, L. H., and Bulte, J. W. M. (2003). Clinically applicable labeling of mammalian and stem cells by combining superparamagnetic iron oxides and transfection agents. *Radiology* **228,** 480–487.

Genove, G., De Marco, U., Xu, H. Y., Goins, W. F., and Ahrens, E. T. (2005). A new transgene reporter for *in vivo* magnetic resonance imaging. *Nat. Med.* **11,** 450–454.

Gillis, P., and Koenig, S. H. (1987). Transverse relaxation of solvent protons induced by magnetized spheres – application to ferritin, erythrocytes, and magnetite. *Magn. Reson. Med.* **5,** 323–345.

Gossuin, Y., Muller, R. N., and Gillis, P. (2004). Relaxation induced by ferritin: A better understanding for an improved MRI iron quantification. *NMR Biomed.* **17,** 427–432.

Gottesfeld, Z., and Neeman, M. (1996). Ferritin effect on the transverse relaxation of water: NMR microscopy at 9.4 T. *Magn. Reson. Med.* **35,** 514–520.

Goverman, J., Woods, A., Larson, L., Weiner, L. P., Hood, L., and Zaller, D. M. (1993). Transgenic mice that express a myelin basic protein-specific T-cell receptor develop spontaneous autoimmunity. *Cell* **72,** 551–560.

Hansch, A., Frey, O., Hilger, I., Sauner, D., Haas, M., Schmidt, D., Kurrat, C., Gajda, M., Malich, A., Brauer, R., and Kaiser, W. A. (2004a). Diagnosis of arthritis using near-infrared fluorochrome Cy5.5. *Invest. Radiol.* **39,** 626–632.

Hansch, A., Frey, O., Sauner, D., Hilger, I., Haas, M., Malich, A., Brauer, R., and Kaiser, W. A. (2004b). *In vivo* imaging of experimental arthritis with near-infrared fluorescence. *Arthritis Rheum.* **50,** 961–967.

Haskins, K., and McDuffie, M. (1990). Acceleration of diabetes in young NOD mice with a CD4$^+$ islet cell-specific T cell clone. *Science* **249,** 1433–1435.

Hawkins, C. P., Mackenzie, F., Tofts, P., du Boulay, E. P., and McDonald, W. I. (1991). Patterns of blood-brain barrier breakdown in inflammatory demyelination. *Brain* **114,** 801–810.

Hawkins, C. P., Munro, P. M., Mac Kenzie, F., Kesselring, J., Tofts, P. S., du Boulay, E. P., Landon, D. N., and McDonald, W. I. (1990). Duration and selectivity of blood-brain barrier breakdown in chronic relapsing experimental allergic encephalomyelitis studied by gadolinium-DTPA and protein markers. *Brain* **113,** 365–378.

Healey, D., Ozegbe, P., Arden, S., Chandler, P., Hutton, J., and Cooke, A. (1995). *In vivo* activity and *in vitro* specificity of CD4$^+$ Th1 and Th2 cells derived from the spleens of diabetic NOD mice. *J. Clin. Invest.* **95,** 2979–2985.

Heide, A. C., Richards, T. L., Alvord, E. C., Peterson, J., and Rose, L. M. (1993). Diffusion imaging of experimental allergic encephalomyelitis. *Magn. Reson. Med.* **29,** 478–484.

Hoehn, M., Kustermann, E., Blunk, J., Wiedermann, D., Trapp, T., Wecker, S., Focking, M., Arnold, H., Hescheler, J., Fleischmann, B. K., Schwindt, W., and Buhrle, C. (2002). Monitoring of implanted stem cell migration *in vivo*: A highly resolved *in vivo* magnetic resonance imaging investigation of experimental stroke in rat. *Proc. Natl. Acad. Sci. USA* **99,** 16267–16272.

Jansen, A., Homodelarche, F., Hooijkaas, H., Leenen, P. J., Dardenne, M., and Drexhage, H. A. (1994). Immunohistochemical Characterization of Monocytes-Macrophages and Dendritic Cells Involved in the Initiation of the Insulitis and Beta-Cell Destruction in Nod Mice. *Diabetes* **43,** 667–675.

Josephson, L., Tung, C. H., Moore, A., and Weissleder, R. (1999). High-efficiency intracellular magnetic labeling with novel superparamagnetic-tat peptide conjugates. *Bioconj. Chem.* **10,** 186–191.

Karlik, S. J., Gilbert, J. J., Wong, C., Vandervoort, M. K., and Noseworthy, J. H. (1990). NMR studies in experimental allergic encephalomyelitis: Factors which contribute to T1 and T2 values. *Magn. Reson. Med.* **14,** 1–11.

Karlik, S. J., Grant, E. A., Lee, D., and Noseworthy, J. H. (1993). Gadolinium enhancement in acute and chronic-progressive experimental allergic encephalomyelitis in the guinea pig. *Magn. Reson. Med.* **30,** 326–331.

Kita, H., He, X.-S., and Gershwin, M. E. (2003). Application of tetramer technology in studies on autoimmune diseases. *Autoimmunity Reviews* **2,** 43–49.

Koike, T., Itoh, Y., Ishi, I., Tabayashi, K., Maruyama, N., Tomioka, H., and Yoshida, S. (1987). Preventive effect of monoclonal anti-L3T4 antibody on the development of diabetes in NOD mice. *Diabetes* **36,** 539–541.

Korganow, A.-S., Ji, H., Mangialaio, S., Duchatelle, V., Pelanda, R., Martin, T., Degott, C., Kikutani, H., Rajewsky, K., Pasquali, J.-L., Benoist, C., and Mathis, D. (1999). From systemic T cell self-reactivity to organ-specific autoimmune disease via immunoglobulins. *Immunity* **10,** 451–461.

Kouskoff, V., Korganow, A.-S., Duchatelle, V., Degott, C., Benoist, C., and Mathis, D. (1996). Organ-specific disease provoked by systemic autoreactivity. *Cell* **87,** 811–822.

Leiter, E. H., Prochazka, M., and Coleman, D. L. (1987). The nonobese diabetic (NOD) mouse. *Am. J. Pathol.* **28,** 380–383.

Levi, S., Yewdall, S. J., Harrison, P. M., Santambrogio, P., Cozzi, A., Rovida, E., Albertini, A., and Arosio, P. (1992). Evidence that H-chains and L-chains have cooperative roles in the iron-uptake mechanism of human ferritin. *Biochem. J.* **288,** 591–596.

Lu, Y., Dang, H., Middleton, B., Zhang, Z., Washburn, L., Campbell-Thompson, M., Atkinson, M. A., Gambhir, S. S., Tian, J., and Kaufman, D. L. (2004). Bioluminescent monitoring of islet graft survival after transplantation. *Mol. Ther.* **9,** 428–435.

Lutz, A. M., Seemayer, C., Corot, C., Gay, R. E., Goepfert, K., Michel, B. A., Marincek, B., Gay, S., and Weishaupt, D. (2004). Detection of synovial macrophages in an experimental rabbit model of antigen-induced arthritis: Ultrasmall superparamagnetic iron oxide-enhanced MR imaging. *Radiology* **233,** 149–157.

Mailliard, R. B., Son, Y. I., Redlinger, R., Coates, P. T., Giermasz, A., Morel, P. A., Storkus, W. J., and Kalinski, P. (2003). Dendritic cells mediate NK cell help for Th1 and CTL responses: Two-signal requirement for the induction of NK cell helper function. *J. Immunol.* **171,** 2366–2373.

Martin, R., and McFarland, H. F. (1995). Immunological aspects of experimental allergic encephalomyelitis and multiple sclerosis. *Crit. Rev. Clin. Lab. Sci.* **32,** 121–182.

Martin, R., McFarland, H. F., and McFarlin, D. E. (1992). Immunological aspects of demyelinating diseases. *Ann. Rev. Immunol.* **10,** 153–187.

Mendel, I., Derosbo, N. K., and Bennun, A. (1995). A myelin oligodendrocyte glycoprotein peptide induces typical chronic experimental autoimmune encephalomyelitis in H-2(B) mice – fine specificity and T-cell receptor V-beta expression of encephalitogenic T-cells. *Eur. J. Immunol.* **25,** 1951–1959.

Miller, S. D., Vanderlugt, C. L., Begolka, W. S., Pao, W., Yauch, R. L., Neville, K. L., Katz Levy, Y., Carrizosa, A., and Kim, B. S. (1997). Persistent infection with Theiler's virus leads to CNS autoimmunity via epitope spreading. *Nat. Med.* **3,** 1133–1136.

Miyazaki, A., Hanafusa, T., Yamada, K., Miyagawa, J., Fujinokurihara, H., Nakajima, H., Nonaka, K., and Tarui, S. (1985). Predominance of lymphocytes-T in pancreatic-islets and spleen of pre-diabetic non-obese diabetic (NOD) mice—a longitudinal study. *Clin. Exp. Immunol.* **60,** 622–630.

Moore, A., Grimm, J., Han, B., and Santamaria, P. (2004). Tracking the recruitment of diabetogenic CD8$^+$ T-cells to the pancreas in real time. *Diabetes* **53,** 1459–1466.

Moore, A., Sun, P. Z., Cory, D., Hogemann, D., Weissleder, R., and Lipes, M. A. (2002). MRI of insulitis in autoimmune diabetes. *Magn. Reson. Med.* **47,** 751–758.

Moore, T. L., and Dorner, R. W. (1993). Rheumatoid factors. *Clin. Biochem.* **26,** 75–84.

Morel, P. A., and Feili-Hariri, M. (2001). How do dendritic cells prevent autoimmunity? *Trends Immunol.* **22,** 546–547.

Moreland, L. W., Baumgartner, S. W., Schiff, M. H., Tindall, E. A., Fleischmann, R. M., Weaver, A. L., Ettlinger, R. E., Cohen, S., Koopman, W. J., Mohler, K., Widmer, M. B., and Blosch, C. M. (1997). Treatment of rheumatoid arthritis with a recombinant human tumor necrosis factor receptor (p75)-Fc fusion protein. *New Engl. J. Med.* **337,** 141–147.

Morrissey, S. P., Stodal, H., Zettl, U., Simonis, C., Jung, S., Kiefer, R., Lassmann, H., Hartung, H. P., Haase, A., and Toyka, K. V. (1996). *In vivo* MRI and its histological correlates in acute adoptive transfer experimental allergic encephalomyelitis—quantification of inflammation and oedema. *Brain* **119,** 239–248.

Nagata, M., Santamaria, P., Kawamura, T., Utsugi, T., and Yoon, J.-W. (1994). Evidence for the role of CD8$^+$ cytotoxic T cells in the destruction of pancreatic β-cells in nonobese diabetic mice. *J. Immunol.* **152,** 2042–2050.

Namer, I. J., Steibel, J., Poulet, P., Armspach, J. P., Mauss, Y., and Chambron, J. (1992). *In vivo* dynamic MR imaging of MBP-induced acute experimental allergic encephalomyelitis in Lewis rat. *Magn. Reson. Med.* **24,** 325–334.

Namer, I. J., Steibel, J., Poulet, P., Armspach, J. P., Mohr, M., Mauss, Y., and Chambron, J. (1993). Blood-brain barrier breakdown in MBP-specific T cell induced experimental allergic encephalomyelitis. A quantitative *in vivo* MRI study. *Brain* **116**, 147–159.

Noth, U., Morrissey, S. P., Deichmann, R., Jung, S., Adolf, H., Haase, A., and Lutz, J. (1997). Perfluoro-15-crown-5-ether labeled macrophages in adoptive transfer experimental allergic encephalomyelitis. *Artif. Cell Blood Sub.* **25**, 243–254.

Okon, E., Pouliquen, D., Okon, P., Kovaleva, Z. V., Stepanova, T. P., Lavit, S. G., Kudryavtsev, B. N., and Jallet, P. (1994). Biodegradation of magnetite dextran nanoparticles in the rat—a histologic and biophysical study. *Lab. Invest.* **71**, 895–903.

Owens, T., and Sriram, S. (1995). The immunology of multiple-sclerosis and its animal-model, experimental allergic encephalomyelitis. *Neurol. Clin.* **13**, 51–73.

Pirko, I., Johnson, A., Ciric, B., Gamez, J., Macura, S. I., Pease, L. R., and Rodriguez, M. (2004). *In vivo* magnetic resonance imaging of immune cells in the central nervous system with superparamagnetic antibodies. *FASEB J.* **18**, 179–182.

Raine, C. S. (1984). Analysis of autoimmune demyelination—its impact upon multiple-sclerosis. *Lab. Invest.* **50**, 608–635.

Rausch, M., Hiestand, P., Baumann, D., Cannet, C., and Rudin, M. (2003). MRI-based monitoring of inflammation and tissue damage in acute and chronic relapsing EAE. *Magn. Reson. Med.* **50**, 309–314.

Rolandsson, O., Stigbrand, T., Riklundahlstrom, K., Eary, J., and Greenbaum, C. (2001). Accumulation of 125-iodine labeled interleukin-2 in the pancreas of NOD mice. *J. Autoimmun.* **17**, 281–287.

Salomon, B., Lenschow, D. J., Rhee, L., Ashourian, N., Singh, B., Sharpe, A., and Bluestone, J. A. (2000). B7/CD28 costimulation is essential for the homeostasis of the $CD4^+CD25^+$ immunoregulatory T cells that control autoimmune diabetes. *Immunity* **12**, 431–440.

Santambrogio, P., Levi, S., Cozzi, A., Rovida, E., Albertini, A., and Arosio, P. (1993). Production and characterization of recombinant heteropolymers of human ferritin H-chain and L-chain. *J. Biol. Chem.* **268**, 12744–12748.

Shizuru, J. A., Taylor-Edwards, C., Banks, B. A., Gregory, A. K., and Fathman, C. G. (1988). Immunotherapy of the nonobese diabetic mouse: Treatment with an antibody to T-helper lymphocytes. *Science* **240**, 659–662.

Sipkins, D. A., Gijbels, K., Tropper, F. D., Bednarski, M., Li, K. C. P., and Steinman, L. (2000). ICAM-1 expression in autoimmune encephalitis visualized using magnetic resonance imaging. *J. Neuroimmunol.* **104**, 1–9.

Slavin, A. J., Tarner, I. H., Nakajima, A., Urbanek-Ruiz, I., McBride, J., Contag, C. H., and Fathman, C. G. (2002). Adoptive cellular gene therapy of autoimmune disease. *Autoimmunity Reviews* **1**, 213–219.

Steinman, L. (1996). Multiple sclerosis: A coordinated immunological attack against myelin in the central nervous system. *Cell* **85**, 299–302.

Steward, W. A., Alvord, E. C., Jr., Hruby, S., Hall, L. D., and Paty, D. W. (1985). Early detection of experimental allergic encephalomyelitis by magnetic resonance imaging. *Lancet* **326**, 898.

Swanborg, R. H. (1995). Animal models of human disease—experimental autoimmune encephalomyelitis in rodents as a model for human demyelinating disease. *Clin. Immunol. Immunopathol.* **77**, 4–13.

Tarner, I. H., Nakajima, A., Seroogy, C. M., Ermann, J., Levicnik, A., Contag, C. H., and Fathman, C. G. (2002). Retroviral gene therapy of collagen-induced arthritis by local delivery of IL-4. *Clin. Immunol.* **105**, 304–314.

Theil, E. C. (1987). Ferritin: Structure, gene regulation, and cellular function in animals, plants, and microorganisms. *Annu. Rev. Biochem.* **56**, 289–315.

Todd, J. A., and Wicker, L. S. (2001). Genetic protection from the inflammatory disease type 1 diabetes in humans and animal models. *Immunity* **15**, 387–395.

Tuohy, V. K., Lu, Z. J., Sobel, R. A., Laursen, R. A., and Lees, M. B. (1989). Identification of an encephalitogenic determinant of myelin proteolipid protein for SJL mice. *J. Immunol.* **142,** 1523–1527.

Tuohy, V. K., Sobel, R. A., and Lees, M. B. (1987). EAE induced by myelin proteolipid apoprotein (PLP) in various strains of mice. *J. Neuroimmunol.* **16,** 174.

van Eden, W., and Waksman, B. H. (2003). Immune regulation in adjuvant-induced arthritis: Possible implications for innovative therapeutic strategies in arthritis. *Arthritis Rheum.* **48,** 1788–1796.

van Lent, P. L., van den Bersselaar, L. A., van den Hoek, A. E., van de Loo, A. A., and van den Berg, W. B. (1992). Cationic immune complex arthritis in mice-a new model. Synergistic effect of complement and interleukin-1. *Am. J. Pathol.* **140,** 1451–1461.

Verhoye, M. R., Gravenmade, E. J., Raman, E. R., Van Reempts, J., and Vander Linden, A. (1996). *In vivo* noninvasive determination of abnormal water diffusion in the rat brain studied in an animal model for multiple sclerosis by diffusion-weighted NMR imaging. *Magn. Reson. Imag.* **14,** 521–532.

Vymazal, J., Brooks, R. A., Baumgarner, C., Tran, V., Katz, D., Bulte, J. W., Bauminger, R., and Di Chiro, G. (1996). The relation between brain iron and NMR relaxation times: An *in vitro* study. *Magn. Reson. Med.* **35,** 56–61.

Vymazal, J., Brooks, R. A., Bulte, J. W. M., Gordon, D., and Aisen, P. (1998). Iron uptake by ferritin: NMR relaxometry studies at low iron loads. *J. Inorg. Biochem.* **71,** 153–157.

Wicker, L. S., Miller, B. J., and Mullen, Y. (1986). Transfer of autoimmune diabetes with splenocytes from non-obese diabetic (NOD) mice. *Diabetes* **35,** 855–890.

Winchester, R., Dwyer, E., and Rose, S. (1992). The genetic basis of rheumatoid arthritis: The shared epitope hypothesis. *Rheum. Dis. Clin. North Am.* **18,** 761–783.

Wipke, B. T., Wang, Z., Kim, J., McCarthy, T. J., and Allen, P. M. (2002). Dynamic visualization of a joint-specific autoimmune response through positron emission tomography. *Nat. Immunl.* **3,** 366–372.

Wunder, A., Tung, C. H., Muller-Ladner, U., Weissleder, R., and Mahmood, U. (2004). *In vivo* imaging of protease activity in arthritis: A novel approach for monitoring treatment response. *Arthritis Rheum.* **50,** 2459–2465.

Yu, M., Johnson, J. M., and Tuohy, V. K. (1996). A predictable sequential determinant spreading cascade invariably accompanies progression of experimental autoimmune encephalomyelitis: A basis for peptide-specific therapy after onset of clinical disease. *J. Exp. Med.* **183,** 1777–1788.

Zamvil, S. S., and Steinman, L. (1990). The lymphocyte-T in experimental allergic encephalomyelitis. *Ann. Rev. Immunol.* **8,** 579–621.

Zekzer, D., Wong, F. S., Ayalon, O., Millet, I., Altieri, M., Shintania, S., Solimena, M., and Sherwin, R. S. (1998). GAD-reactive $CD4^{+}$ Th1 cells induce diabetes in NOD/SCID mice. *J. Clin. Invest.* **101,** 68–73.

Index

Contents of Previous Volumes

Volume 48

Volume 49

Volume 51

Volume 52

Volume 53

Volume 54

Volume 55

Volume 56

Volume 57

Volume 58

Volume 62

Volume 63

Volume 65

Volume 66

Volume 67

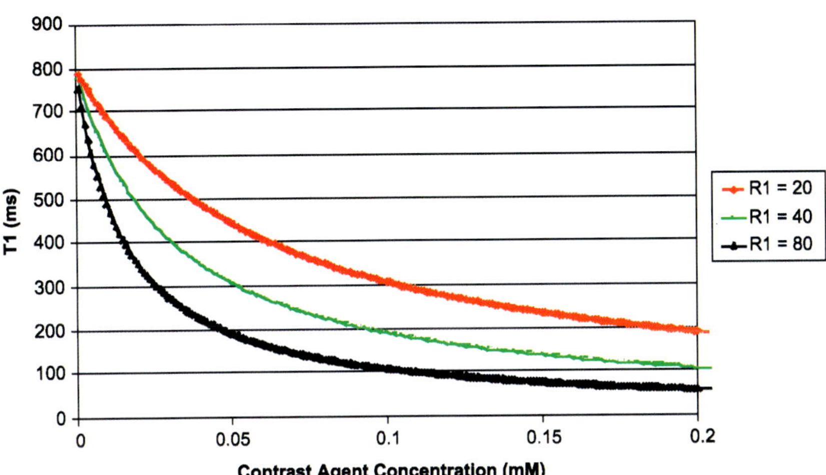

Chapter 1, Figure 1   T1 relaxation time for various R1 values plotted for increasing contrast agent concentrations. $1/T1 = 1/T1_0 + R1 \times$ [contrast agent]; $T1_0$ of native tissue (i.e., myocardium) was assumed to be 800 ms. To achieve a T1 of 200 ms that is usually sufficient for contrast agent detection, $<50~\mu$M Gd is required for R1s ≥ 80 mM$^{-1} \times$ s^{-1}. For an R1 of ≤ 20 mM$^{-1} \times$ s^{-1}, this concentration increases almost 4-fold to $\geq 200~\mu$M Gd.

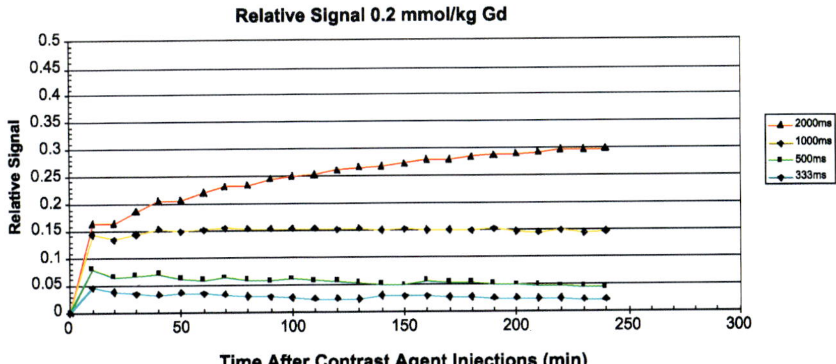

Chapter 1, Figure 4   MR signal strength on IR images for different IR repetition times and with respect to time of contrast administration. Simulation of the Bloch equations based on expected Gd concentrations in blood demonstrating the relative signal versus time after contrast agent injection [minutes] for four different IR repetition times (333–2000 ms). For longer IR repetition times, a higher relative signal level can be expected. Furthermore, the simulation demonstrates that the relative signal level remains constant over a relative long period of time after contrast administration.

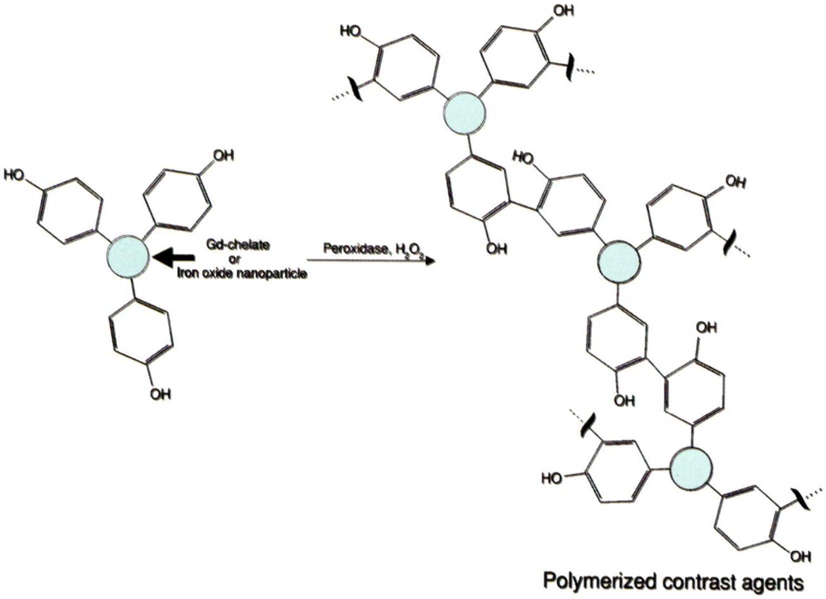

Chapter 1, Figure 7 Principle of enzymatic polymerization of contrast agent. Peroxidase in presence of H_2O_2 results in condensation and polymerization of phenolic contrast agents. Adapted from Perez *et al.*, 2004b.

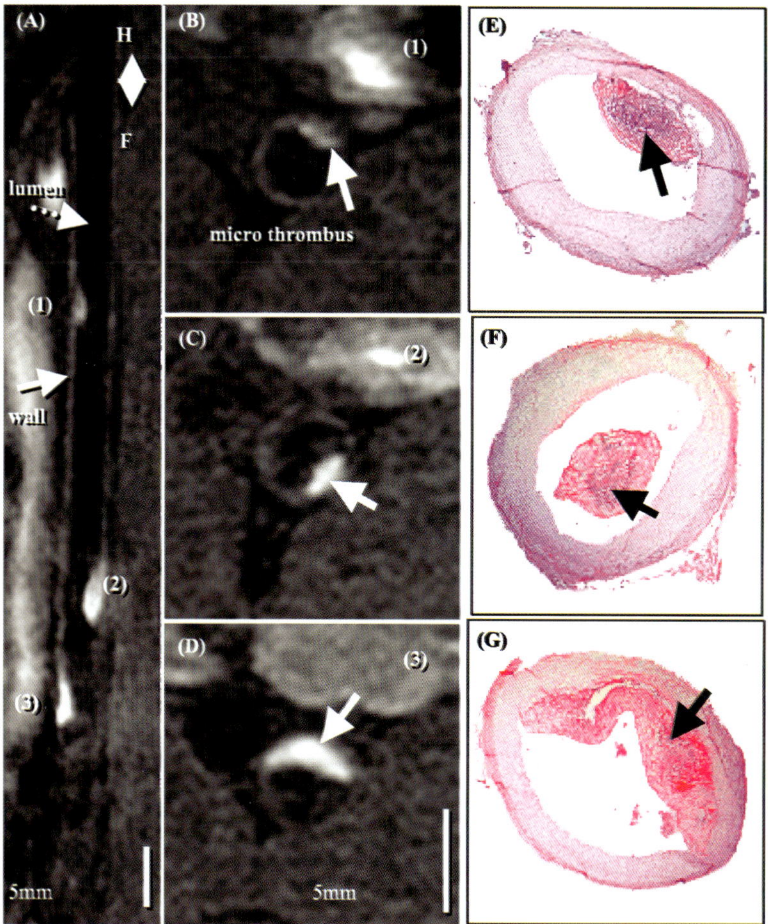

Chapter 1, Figure 8   Fibrin imaging in a rabbit model of plaque rupture. (A) Reformatted view of a coronal 3D data set showing the subrenal aorta approximately 20 hours after EP-1873 administration. Three well-delineated mural thrombi (arrows) can be observed with good contrast between thrombus (numbers), arterial blood (dotted arrow), and the vessel wall (dashed arrow). The in-plane view of the aorta allows simultaneous display of all thrombi showing head, tail, length, and relative location. (B–D) Corresponding cross-sectional views show good agreement with histopathology (E–G). From Botnar *et al.*, 2004b.

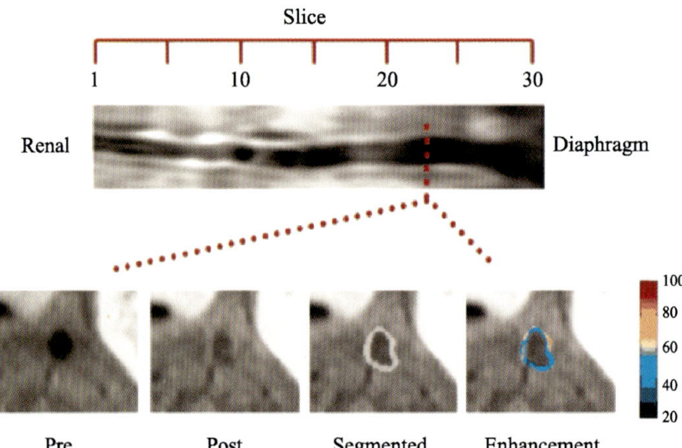

Chapter 1, Figure 9 Imaging of $\alpha v \beta 3$ in early-stage atherosclerosis. *In vivo* spin echo image reformatted to display long axis of aorta from renal arteries to diaphragm of one cholesterol-fed rabbit (top) and at single transverse level (bottom) before (Pre) and after (Post) treatment, after semiautomated segmentation (Segmented, grayish ring; see text for description), and with color-coded signal enhancement (Enhancement) above baseline (in percent). From Winter *et al.*, 2003b.

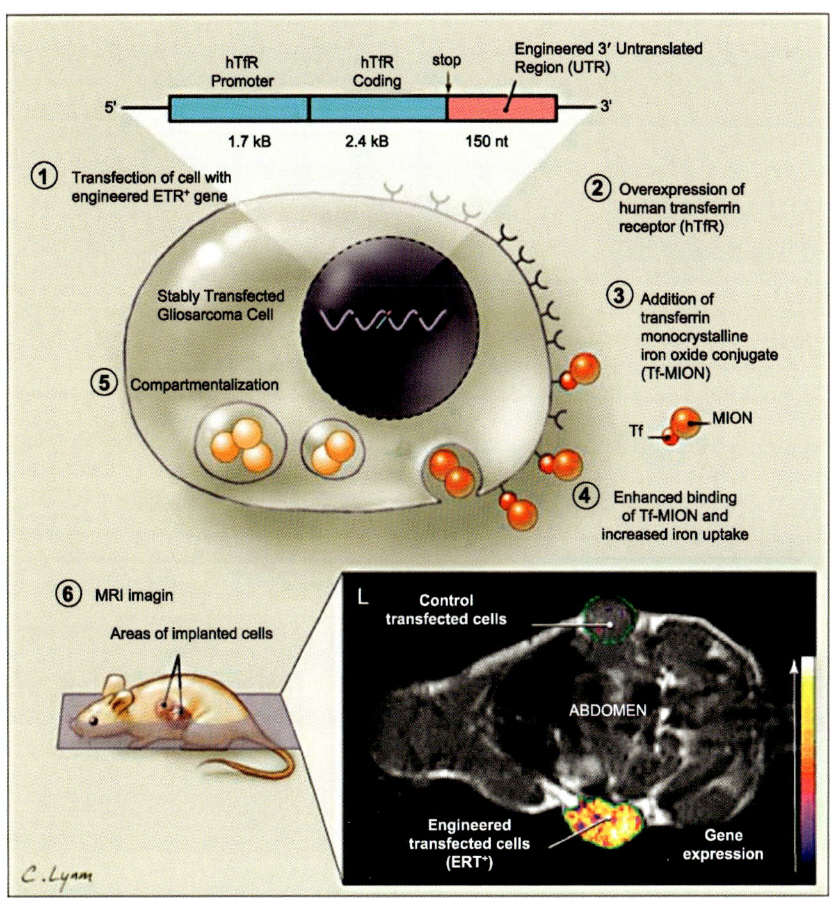

Chapter 1, Figure 10 Imaging ETR expression. Several synergistic steps were used to reveal transgene expression in cells by MR imaging. (A) Overexpression of ETR expression results in an approximately fivefold higher cell uptake. (B) During each ETR-mediated internalization event, several thousand iron atoms enter the cell (MION contains an average of 2064 Fe per 3-nm particle core). (C) Upon cellular internalization and compaction in endosomes, the R2 and R2* relaxivities of superparamagnetic MION further increase approximately fourfold (depicted as a change from red to orange), increasing MR detectability. (D) Cellular internalization of iron does not downregulate the level of ETR expression. The ETR cDNA sequence consists of the hTfR promoter, the coding sequence, and the engineered 3′ UTR-regulatory sequence (top inset). From Tempany and McNeil, 2001.

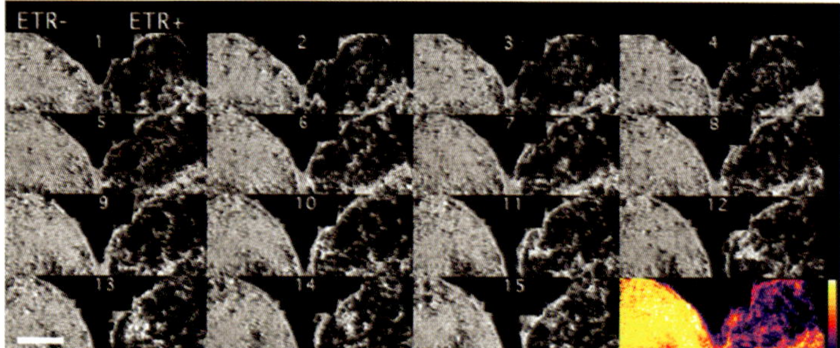

Chapter 1, Figure 12   MR microscopy of excised tumor specimen. ETR− (left) and ETR+ (right) tumor specimen obtained from the same animal depicted in Figure 4 following intravenous injection of Tf–iron oxides. The contiguous MR sections are numbered from 1–15. The ETR− tumor (left half of each image) appears of higher signal intensity (absence of Tf-MION uptake) and is homogenous. The ETR+ tumor (right half of each image) has lower signal intensity (Tf-MION uptake) and appears heterogeneous because of regional differences in Tf–iron oxide uptake. The false color image in the lower right represents a projection image through sections 1–15 to show the marked difference in MR signal intensity between the tumors. The scale bar in the lower left represents 1000 μm. Serial 58-μm thick MR images (gradient echo pulse sequence TR/TE/α: 150/3.7/35 degrees; 7.1 T; 58-μm^3 isotropic voxel resolution). Image and legend from reference Weissleder *et al.*, 2000.

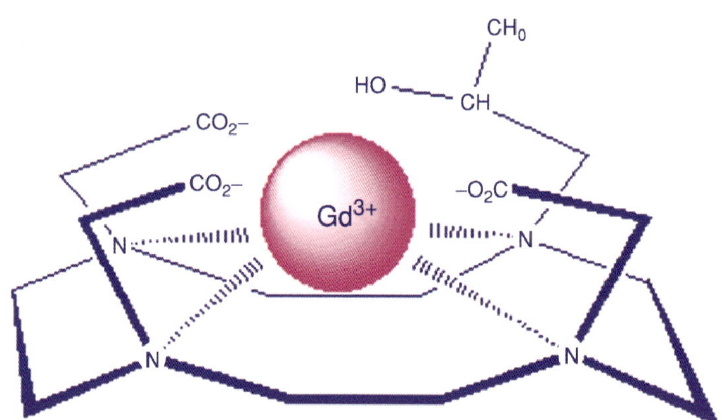

Chapter 2, Figure 1   Structure of the commercially available contrast agent ProHance. A ringlike macrocycle (blue) surrounds and binds a central metal ion (magenta) at several positions (eight, in the example shown).

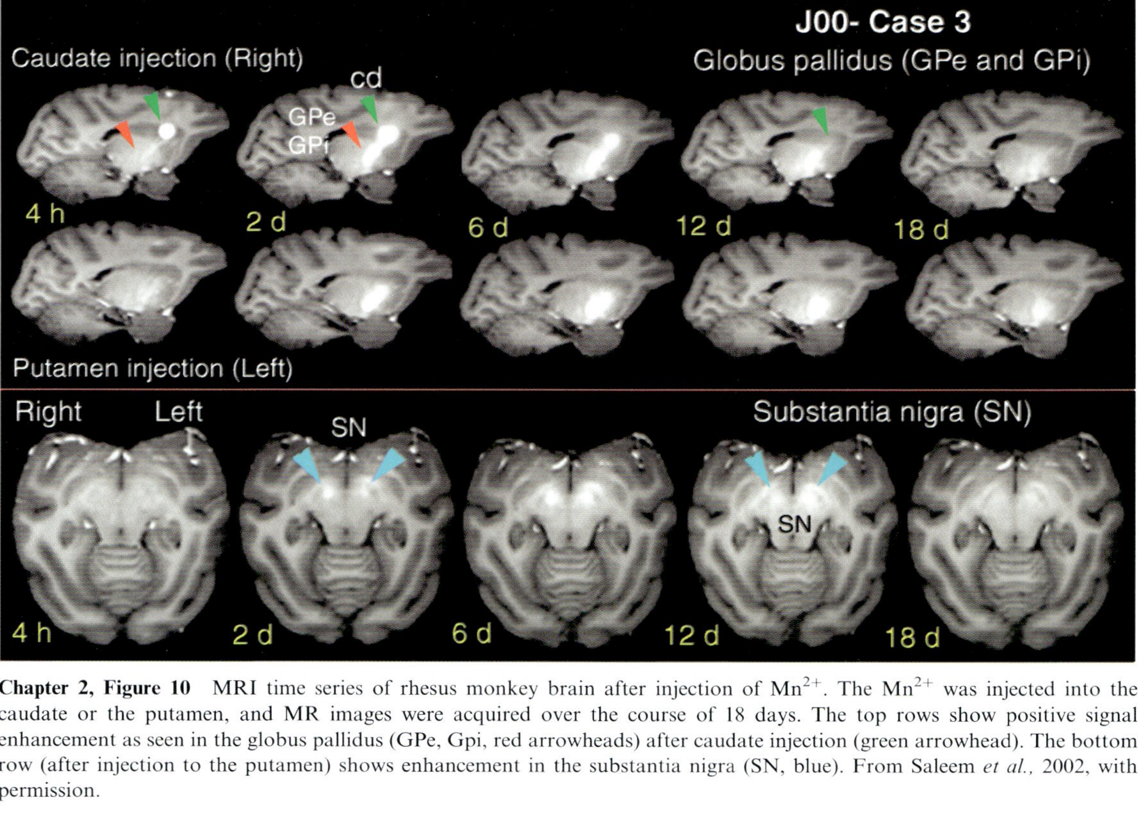

Chapter 2, Figure 10   MRI time series of rhesus monkey brain after injection of Mn^{2+}. The Mn^{2+} was injected into the caudate or the putamen, and MR images were acquired over the course of 18 days. The top rows show positive signal enhancement as seen in the globus pallidus (GPe, Gpi, red arrowheads) after caudate injection (green arrowhead). The bottom row (after injection to the putamen) shows enhancement in the substantia nigra (SN, blue). From Saleem *et al.*, 2002, with permission.

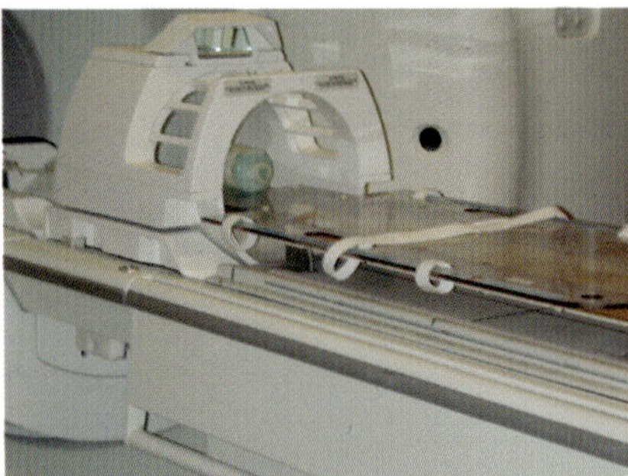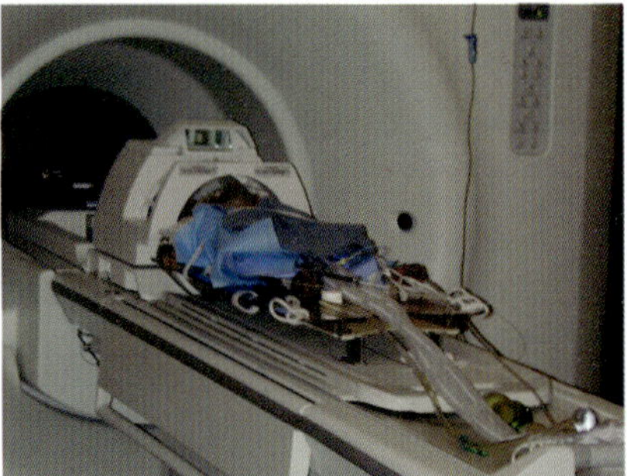

Chapter 4, Figure 3   Monkey cradle and RF coil used for the imaging experiments (left). The cradle was built in-house and consists of two removable (and independent) MR-compatible plates that interface with the patient table and the animal, respectively. As shown on the right, the cradle allows easy positioning of the animal's head at the magnet's isocenter while keeping the animal completely stable on the patient table.

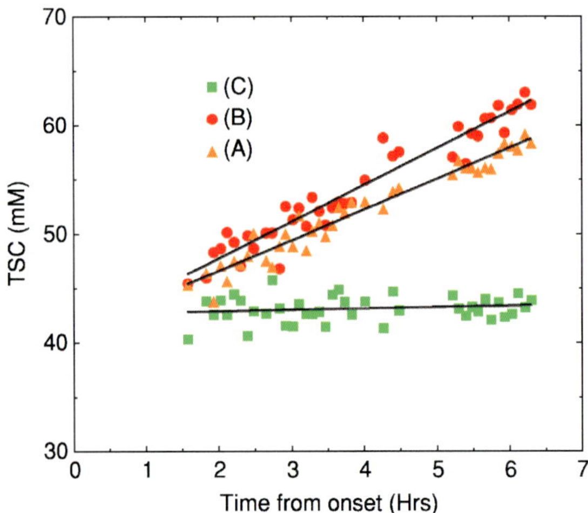

Chapter 4, Figure 7   Time course of TSC for selected ROIs in the monkey brain after endovascular cerebral occlusion of the right MCA territory. Each data point represents a measurement made on a sodium image acquired in 5 minutes. The slopes of the curves demonstrate significant differences in TSC accumulation between the ischemic and nonischemic cortices.

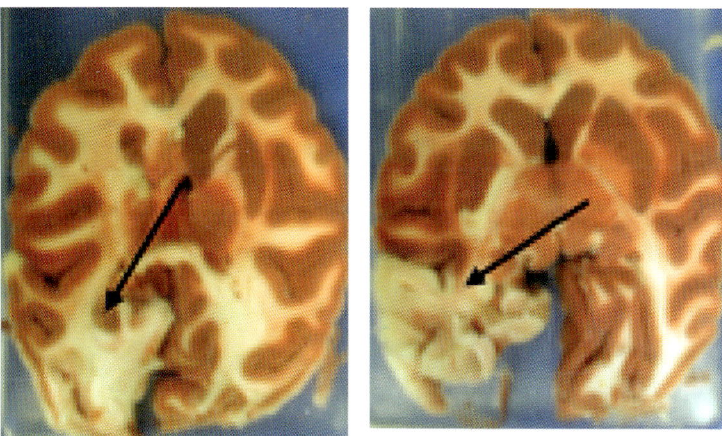

Chapter 4, Figure 14 The TTC-stained sections from the brain of the animal in Fig. 13 (temporary ischemia model). The images demonstrate ischemic damage to the PCA territory which is the only vascular territory that remained occluded upon removal of the balloon catheter.

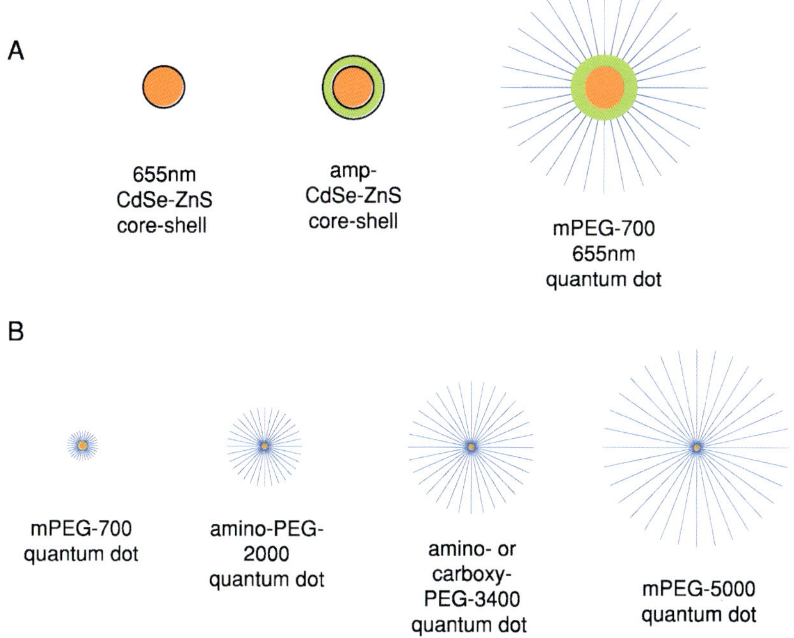

Chapter 5, Figure 1 Quantum dot coatings used for this chapter. (A) 655 nm–emitting cadmium selenide core–zinc sulfide (CdSE-ZnS) shell quantum dot, showing approximate thickness of amphipathic coat (amp) and polyethylene glycol (PEG), average molecular weight 700 (PEG-700), conjugated to an amp-quantum dot. (B) Relative sizes of PEG conjugates used in this chapter. PEG polymer is drawn as if fully extended.

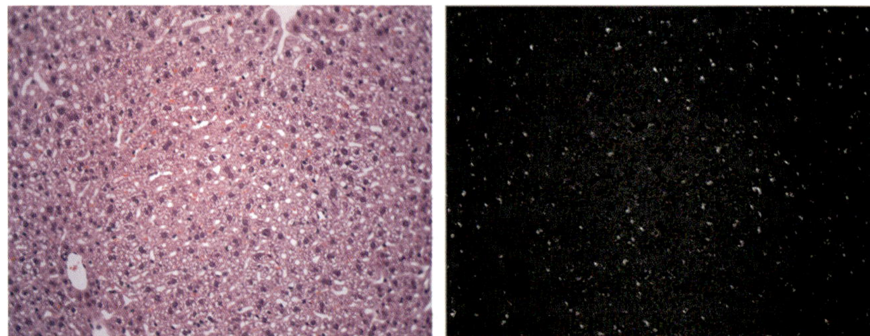

Chapter 5, Figure 4 Liver section from mouse injected using 705 nm–emitting mPEG-5000 quantum dots. The mouse was injected using 300 pmol quantum dots, then necropsied 24 hours later. Tissues were fixed, paraffin-embedded, sectioned, and stained using standard methods. Left, hematoxylin-eosin stain; right, fluorescence. Excitation at 450 nm; emission at 705 nm.

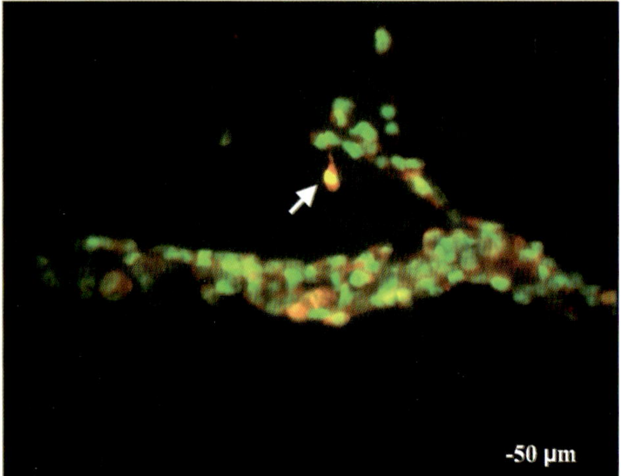

Chapter 6, Figure 1 Extravasation of tumor cells from vessels containing numerous dual-color HT1080 fibrosarcoma cells in a living mouse. HT1080 human fibrosarcoma cells were labeled with histone-H2B-GFP in the nucleus and with retroviral RFP in the cytoplasm. Cells were injected in the heart and visualized in skin vessels by whole-body imaging. From K. Yamauchi and R. M. Hoffman, unpublished data.

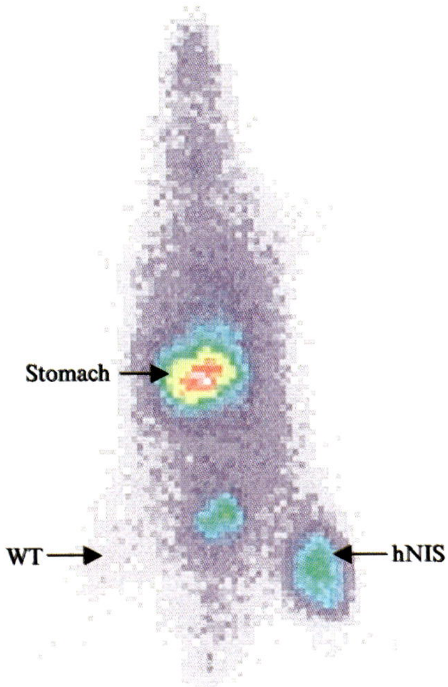

Chapter 7, Figure 3 Scintigraphic image of a tumor-bearing male Copenhagen rat subcutaneously transplanted with hNIS-expressing (hNIS) or wild-type (WT) prostate adenocarcinoma cells (left thigh) at 2 hours after injection of $^{131}I^-$.

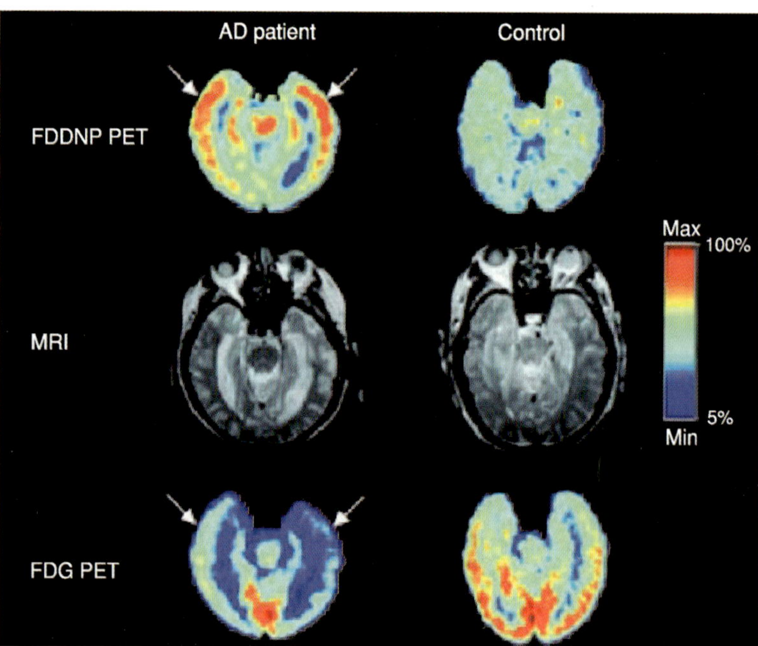

Chapter 8, Figure 1 [^{18}F]FDDNP-PET, MRI, and ^{18}F-labeled deoxyglucose (FDG)-PET images of a patient with AD and a normal subject. The [^{18}F]FDDNP and FDG images of each stage are co-registered to their respective MR images. Areas of FDG hypometabolism are matched with the localization of neurofibrillary tangles and amyloid plaques (APs) resulting from [^{18}F]FDDNP binding (arrows). The [^{18}F]FDDNP images represent activity 25–54 minutes after ^{18}F-FDDNP administration. The FDG images represent activity 20–60 minutes after FDG injection. Reprinted with permission from the American Journal of Geriatric Psychiatry. Copyright 2002, American Psychiatric Association.

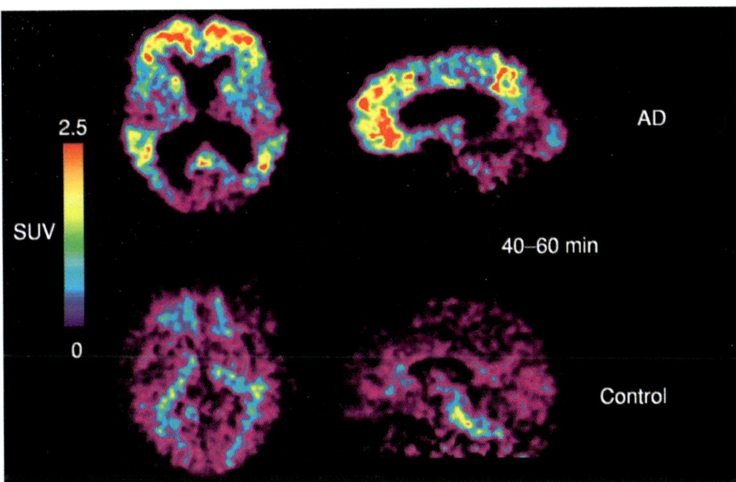

Chapter 8, Figure 2 Transaxial (left column) and sagittal (right column) PIB-PET images expressed as standardized uptake value (SUV) in both a patient with suspected AD (top) and an age-matched control subject (bottom). In the AD image, note the relative intensity of retention in the frontal and temporoparietal cortices and the relative lack of retention in the visual cortex. Reprinted with permission.

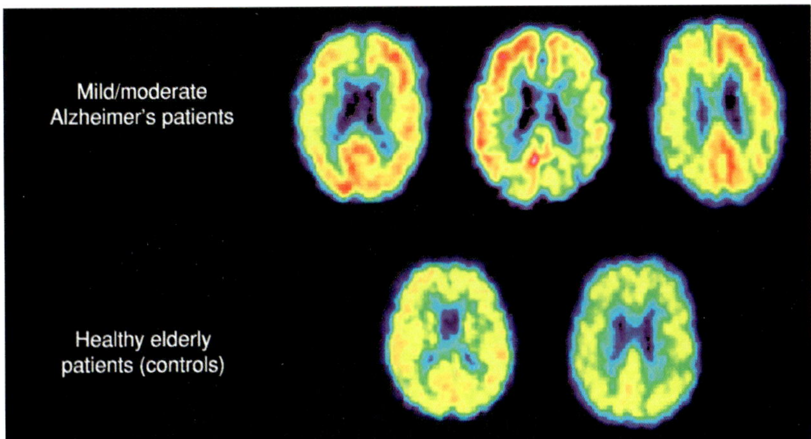

Chapter 8, Figure 3 Parametric images of standardized uptake values obtained by normalizing tissue concentration (nCi/mL) by injected dose per body mass (nCi/g) of PET images summed over 40–120 minutes after injection of 10 mCi of $[^{11}C]$SB-13. Data are shown for representative Alzheimer's disease patients and comparison subjects. (Courtesy N. P. Verhoeff.)

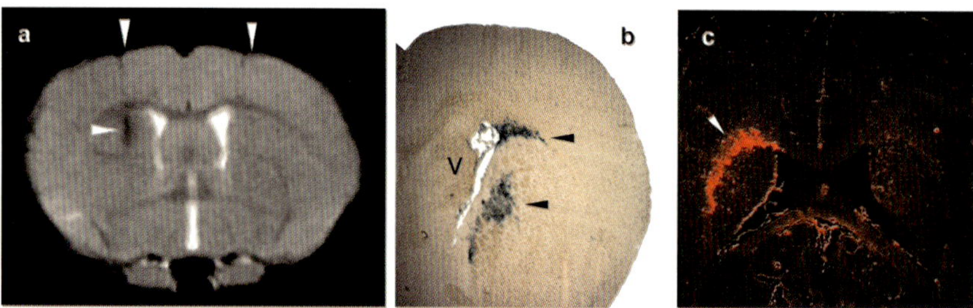

Chapter 9, Figure 3 *In vivo* longitudinal results of MRI reporter expression in the mouse brain. Adenovirus-containing reporter transgenes encoding ferritin subunits were injected into the striatum. Panel a shows the T2-weighted image 5 days after injection showing the injection sites (arrows, MRI reporters left, AdV-lacZ control right). The MRI reporter–transduced cells appear hypointense, while the cells with LacZ show no contrast in MRI. Panel b shows the X-Gal–stained AdV-lacZ–transduced pattern at 5 days after transduction. In b, the staining pattern, similar to the MRI, is predominantly in white matter (top arrow) and striatum (bottom arrow), where v denotes ventricle. Panel c shows immunohistochemistry (IHC) results of ferritin transgene expression in mouse brain slices at 5 days after inoculation. Adenovirus containing the MRI reporters was injected into the left striatum and appears immunopositive in IHC. AdV-lacZ was injected in the contralateral side and shows no staining above background. *In vivo* images (a) were obtained in anesthetized mice at 11.7 T. Coronal slices were acquired at the injection site using a T2-weighted spin-echo sequence with repetition time = 1200 ms, echo time = 35 ms, 0.75-mm slice thickness, and in-plane resolution of 98 μm. For additional details see Genove *et al.*, 2005.